# Disordered Systems and Biological Organization

# NATO ASI Series

## Advanced Science Institutes Series

*A series presenting the results of activities sponsored by the NATO Science Committee, which aims at the dissemination of advanced scientific and technological knowledge, with a view to strengthening links between scientific communities.*

The Series is published by an international board of publishers in conjunction with the NATO Scientific Affairs Division

| | |
|---|---|
| A Life Sciences | Plenum Publishing Corporation |
| B Physics | London and New York |
| C Mathematical and Physical Sciences | D. Reidel Publishing Company Dordrecht, Boston and Lancaster |
| D Behavioural and Social Sciences E Applied Sciences | Martinus Nijhoff Publishers Boston, The Hague, Dordrecht and Lancaster |
| F Computer and Systems Sciences G Ecological Sciences | Springer-Verlag Berlin Heidelberg New York Tokyo |

# Disordered Systems and Biological Organization

Edited by

## E. Bienenstock
Laboratoire de Neurobiologie du Développement
Université de Paris XI, 91405 Orsay/FRANCE

## F. Fogelman Soulié
Laboratoire de Dynamique des Reseaux, L.D.R. CESTA
1, rue Descartes, 75005 Paris/FRANCE

## G. Weisbuch
Groupe de Physique des Solides, École Normale Supérieure
24, rue Lhomond, 75231 Paris/FRANCE

Springer-Verlag Berlin Heidelberg New York Tokyo
Published in cooperation with NATO Scientific Affairs Division

Proceedings of the NATO Advanced Research Workshop on Disordered Systems and Biological Organization held at Les Houches, February 25 – March 8, 1985

ISBN 3-540-16094-9 Springer-Verlag Berlin Heidelberg New York Tokyo
ISBN 0-387-16094-9 Springer-Verlag New York Heidelberg Berlin Tokyo

Library of Congress Cataloging in Publication Data. NATO Advanced Research Workshop on Disordered Systems and Biological Organization (1985 : Les Houches, Haute-Savoie, France) Disordered systems and biological organization. (NATO ASI series. Series F, Computer and systems sciences ; v. 20) "Proceedings of the NATO Advanced Research Workshop on Disordered Systems and Biological Organization held at Les Houches, February 25–March 8, 1985"—Includes index. 1. Biological systems—Congresses. 2. Orderdisorder models—Congresses. 3. Machine theory—Congresses. 4. Combinatorial optimization—Congresses. I. Bienenstock, E. II. Fogelman-Soulié, Francoise, 1948–. III. Weisbuch, G. IV. Series. QH313.N38 1985 574 86-1838 ISBN 0-387-16094-9 (U.S.)

Printing: Beltz Offsetdruck, Hemsbach; Bookbinding: J. Schäffer OHG, Grünstadt
2145/3140-543210

# CONTENTS

# INTRODUCTION

The NATO workshop on Disordered Systems and Biological Organization was attended, in march 1985, by 65 scientists representing a large variety of fields: Mathematics, Computer Science, Physics and Biology. It was the purpose of this interdisciplinary workshop to shed light on the conceptual connections existing between fields of research apparently as different as: automata theory, combinatorial optimization, spin glasses and modeling of biological systems, all of them concerned with the global organization of complex systems, locally interconnected.

Common to many contributions to this volume is the underlying analogy between biological systems and spin glasses: they share the same properties of stability and diversity. This is the case for instance of primary sequences of biopolymers like proteins and nucleic acids considered as the result of mutation-selection processes [P.W. Anderson, 1983] or of evolving biological species [G. Weisbuch, 1984]. Some of the most striking aspects of our cognitive apparatus, involved in learning and recognition [J. Hopfield, 1982], can also be described in terms of stability and diversity in a suitable configuration space. These interpretations and preoccupations merge with those of theoretical biologists like S. Kauffman [1969] (genetic networks) and of mathematicians of automata theory: the dynamics of networks of automata can be interpreted in terms of organization of a system in multiple possible attractors.

The present introduction outlines the relationships between the contributions presented at the workshop and briefly discusses each paper in its particular scientific context.

In the first part of the book, we present some papers about automata theory. The mathematical tools developed there provide a means for analyzing the results further presented in the book. An alternative theoretical approach is introduced in part 2, namely the theory of physical disordered systems, particularly spin glasses; this framework is used throughout the book in many applications to neural networks, pattern recognition, etc. Thus part 3 presents formal models of neural networks. In many instances, a central problem is that of finding a global minimum of a cost or "energy" function , which has many local minima (high degeneracy). A seminal paper [Kirkpatrick et al.,1983] proposed a method inspired from statistical mechanics to solve such problems; various applications are discussed in part 4. More models of biological organization are discussed in part 5.

An index is provided at the end of the volume: the entries are the terms which occur most frequently in more than one paper. This is intended to help the reader in finding further connections between the various contributions. In this introduction, papers within the present volume are printed in boldface; original papers published elsewhere are printed in normal characters; books and proceedings appear in italics.

## I- AUTOMATA THEORY

From its very origins with von Neumann's work, automata theory was intended as a model for both living systems and machines [von Neumann, 1958, 1966]. This line of research focused on the *logical* structure of both artificial and natural machines and emphasized their analogies and differences. One of the most famous artificial machines, proposed by von Neumann, is the self-reproducing automaton (see **Vichniac**).

Later on, work by *Rosenblatt* [1961], *Minsky and Papert* [1969] and *Nilsson*[1965] suggested the use of automata networks as learning machines. The so-called "limitation" theorem for the perceptron (Minsky and Papert) was one of the reason for the decrease of interest in such machines in the following decade.

The recent renewal in this research area is largely due to ideas stemming from physics [Hopfield, 1982; Kirkpatrick, 1983], computer science [*Kohonen*, 1984] and automata theory [*Demongeot et al.*, 1985; *Dobruschin et al.*, 1978; *Robert*, 1986]. New techniques are proposed to solve classical problems, for instance in optimization (e.g., traveling salesman) and in circuit design (placement problems). Closely related is the "connectionist" approach [*Hinton and Anderson*, 1981] to many problems in artificial intelligence and cognitive science: learning, pattern recognition, classification, natural language understanding. Interestingly enough, these new developments have insisted on the relationships between natural computation in the brain and parallel computers or algorithms, in the spirit advocated by von Neumann in the 50's.

Cellular automata were introduced initially by von Neumann [*Burks*,1970; *von Neumann*, 1966] in his self replicating machine; later on, they were used by Conway [*Berlekamp et al.*, 1982] in his game "Life" to provide a simplified model of population dynamics. Since then, cellular automata have been studied from different viewpoints: physics [Wolfram, 1983; Greenberg et al.,1978], theory of computation (cellular automata are shown to exhibit universal computation ability [*Berlekamp et al.* 1982]), dynamical properties [*Farmer et al.*, 1984; *Demongeot et al.*, 1985; Tchuente, 1982]...

This richness of applications is discussed in details in the paper by **G. Vichniac** (pp 3-20) who also presents cellular automata with voting (i.e., majority) rules and irreversible cellular automata. In his paper (pp 21-32), **M. Tchuente** discusses the dynamics of two 1-dimensional cellular automata networks (majority rule; 1-D analog of game Life); such networks may display a complex behavior comparable to the 2-D case; he ends up his presentation by a solution to the classical problem of self stabilization (or firing squad problem) by a cellular network: this generalizes a result by Dijkstra [1974].

The need for a global theory of the dynamical behavior of automata networks had already been stressed by von Neumann («he envisaged a systematic theory... which would contribute in an *essential way* to our understanding of natural systems (natural automata) as well as to our understanding of both analog and digital computers (artificial automata)» [Burks's introduction to *von Neumann*, 1966]. The many contributions to date have not yet completely achieved this goal.

In his paper, **F. Robert** (pp 33-47) presents a substantial building block of such a theory: he introduces the concepts and essential tools provided by iteration theory, especially those which can be derived from the theory of continuous dynamics (see also [*Robert*, 1986]): vectorial boolean "distance", discrete derivative, attractivity. This conceptual framework will be used repeatedly through many of the following papers. **Y. Robert and M. Tchuente** (pp 49-52) focus on the problem of relating the connection graph to the iteration graph, for cellular automata with monotone transition functions and with majority functions.

**H. Hartman and G. Vichniac** (pp 53-57) present an extension of the classical model of cellular automata: all elements may now have different mappings, the network is *inhomogeneous.* Because of the particular role played by non forcing mappings as introduced by Kauffman (see for example his paper in [*Demongeot et al.*, 1985]), they use a network with only AND and XOR mappings. The case of homogeneous networks is well known, especially for networks with only AND mappings (which are linear for the boolean operations) [Goles-Chacc, 1985] or only XOR mappings (linear in $\mathbb{Z}/2$) [Snoussi, 1980]. But introducing a mixture of both mappings produce interesting results: diffraction of patterns if the AND automata are regularly distributed or percolation if they are randomly distributed.

**J.P. Allouche** (pp 59-62) presents some results in number theory obtained by using an automaton which produces a binary sequence formally related to the 1-D cyclic Ising model, with a formal imaginary temperature.

**M. Cosnard and E. Goles** (pp 63-66) show that a formal "neuron" with memory may have a very complex dynamical behavior: they exhibit a discrete approximation of a devil staircase and give conditions which allow the characterization of cycle structure and periods.

**D. Pellegrin** (pp 67-70) presents simulations of boolean networks introduced by S. Kauffman [1969,1970] to model genetic networks (see section 5).

Random automata networks are used by **J. Demongeot and J. Fricot** (pp 71-84) to study the asymptotic behavior of birth and death processes occurring chiefly in epidemiological models.

In his seminal paper on associative memories, Hopfield [1982] laid down the formal analogy between neural networks and spin glasses, based on the notion of an *energy* function of the network's state.

This approach has since been extended and the two papers by F. Fogelman-Soulié and E. Goles-Chacc introduce *Lyapunov functions* in general automata networks. **F. Fogelman-Soulié** (pp 85-100) defines the notion of entropy of a boolean network in analogy to Shannon entropy, and of energy of threshold and majority networks; she proves that these functions are Lyapunov functions for the dynamics. **E. Goles-Chacc** (pp 101-112) generalizes this work and defines the class of *positive* automata, which include as particular cases threshold and majority networks. Some nice examples are presented.

**J. Milnor**'s paper (pp 113-115) is a short review on topological entropy, directional entropy and related notions in cellular automata.

## 2-PHYSICAL DISORDERED SYSTEMS

Physical systems such as liquids or glasses are termed *disordered* because the atoms they are made of are not arranged with strong regularity like in crystals. A geometrical approach to the structural properties of disordered systems is proposed by **R. Mosseri and J.F. Sadoc** (pp 149-152).

On the other hand, the main emphasis in the last years has been on modeling the thermodynamical properties of disordered systems. Theoretical physicists especially favor the spin glass model: particles are represented by *Ising spins* (i.e., +1 or -1 valued), and the details of the structural disorder are not made explicit; only the most important aspects are retained, namely the *random* character of the interactions and the *frustration*. A brief introduction to the spin glass model is given in **M. Mézard**'s paper (pp 119-132), along with a review of some of the latest results in the description of the stable states of the system. Besides their technological applications, disordered systems are of interest to us because they have many different low-energy configurations, which implies that they can exist in a *variety* of *stable* configurations. Further, these ground states are arranged in a particular hierarchical fashion, referred to as the property of *ultrametricity*.

The search for configurations of spins which minimize the free energy of the system can be considered as a problem of combinatorial optimization and methods of operational research have been applied to it [Bièche et al., 1980]. **B. Lacolle**'s paper (pp 133-148) is a description of a symbolic computation approach to the problem, which provides exact results for the computation of the partition function and the free energy.

## 3-FORMAL NEURAL NETWORKS

The idea of modeling neurons by threshold automata can be traced back to Mc Culloch and Pitts [1943]. Recent interest of physicists in neural modeling has been aroused by Hopfield's suggestion [1982] to use networks of such automata for retrieving partially altered or noise-corrupted information. His learning algorithm based on the well known Hebb's learning principle [1949] allows one to store a given set of predefined "reference" states or memories. In physical terms, storing these patterns amounts to shaping the energy landscape.

**P. Peretto and J. J. Niez**'s (pp 171-185) simulation results on short term and long term memory in neural networks is closely related to Hopfield's model; their description of the neural dynamics is set in a probabilistic framework and the simulations make use of a Monte Carlo updating scheme.

Some theoretical results about the performances of Hopfield networks are given by **G. Weisbuch and D. d'Humières** (pp 187-191): scaling laws are derived for the number of memorizable patterns and for the width of the attraction basins. D. Amit [1985] addresses similar problems, within the spin glass formalism.

In their contribution to the present volume, **J. Hopfield and D.W. Tank** (pp 155-170) discuss the advantages of working with continuous rather than discrete variables; they present a number of applications to classical graph theoretic combinatorial problems such as: traveling salesman problem, euclidean match problem, graph coloring. **L. Jackel et al.** (pp 193-196) propose to implement Hopfied's method with "next generation" micro-circuits.

The optimal use of Hopfield's networks (in terms of number of retrievable sequences and width of the attraction basins) is in the case of *uncorrelated* (i.e., nearly orthogonal) reference states. On the other hand, M. Kerszberg and M. Virasoro are interested in non orthogonal attractors, specifically in a hierarchically structured attractor set. **M. Kerszberg**'s (pp 205-208) system is a simple model without learning, which introduces a new notion of distances among classes of input configurations and allows for a test of their ultrametricity. **M. Virasoro** (pp 197-204) modifies Hopfield's learning algorithm in order to obtain an ultrametric structure for attractors, thus minimizing "fatal errors" which would drive the system far from the expected attractors.

Historically, J. Anderson was one of the first contributors to the theory of associative memory in formal neural networks, together with T. Kohonen [1978]. **J. Anderson**'s paper (pp 209-226) in the present volume emphasizes the importance of the non-linearities in these models, and discusses the relative roles of the linear versus nonlinear components in the dynamics. In most models, the linear part is a matrix obtained by the Widrow-Hoff or the pseudo-inverse algorithm [see for instance *R. Duda and P. Hart*, 1973]. In the model presented here (BSB, for "Brain State in a Box"), the non-linearity is a constraint which forces the state to remain within a given hypercube. The state actually converges to a corner of the hypercube. Using three different illustrative data bases, it is shown that the BSB model exhibits simple forms of "cognitive" abilities such as categorization, generalization, and disambiguation.

In the model proposed by **Personnaz and al.** (pp 227-231), the non-linearity is the operation "Sign of " which explicitly constrains the state to the corners of the hypercube. Among all the corners, the favored ones, i.e., the energy minima, correspond to the learned patterns. Thus, the models of Anderson, Hopfield, and Personnaz et al. are closely related to each other; as Anderson puts it, the storage step consists in "learning the corners". In the physicist's language, this amounts to shaping the energy landscape, i.e., creating the appropriate attractor basins.

The "Hierarchical Learning Machine" of **Y. Le Cun** (pp 233-240) again uses the Widrow-Hoff rule, yet in a slightly different situation: the patterns presented to the network during the learning step do not entirely determine the state of the system. The problem created thereby was termed the "Credit Assignment Problem" (CAP) by Hinton et al. [1984]. An original solution to the CAP is proposed here.

The hierarchically structured networks discussed in the paper by **D. d'Humières** (pp 241-245) have their origin in Fukushima's multi layered network [1975]. Starting from a very simple model, the paper sketches a rational explanation of some assumptions that are often made in the context of multilayered networks.

In the neural networks discussed so far, the variables which carry the information, or "meaning", are the activity rates of neurons. This view, termed *rate-coding*, is challenged in the paper by **Ch. von der Malsburg and E. Bienenstock** (pp 247-272) who propose that statistical moments in the firing of neurons -e.g., correlations- may be used to carry and process information. This hypothesis of *statistical coding* is complemented with the postulate of *short-term synaptic plasticity*. In brief, contrasting with rate-coding, the information-carrying variables are connectivity states rather than activity states. The formalism used (simulated annealing) is nevertheless similar to that used in previous papers. This approach, still in a preliminary phase, provides an original solution to the problem of invariant pattern recognition.

The paper by **J. Buhmann and K. Schulten** (pp 273-279) reports on simulations of an associative memory model which is quite classical in its spirit, yet sticks as closely as possible to "real" neurophysiology. The dynamics of membrane potential and spike generation in each cell are described in some detail.

## 4- COMBINATORIAL OPTIMIZATION

Many combinatorial optimization problems are equivalent to the minimization of some real-valued function of a large number of discrete - sometimes boolean- variables. A classical example is the Traveling Salesman Problem (TSP). "Simulated annealing" is a technique inspired from statistical physics, first suggested by Cerny [1982] and Kirkpatrick et al. [1983] to solve such optimization problems. This technique is essentially a Monte-Carlo procedure in which the temperature decreases slowly to zero: a clever probabilistic version of gradient descent. This method is briefly introduced in the paper by **S.A. Solla et al** (pp 283-293). Their paper offers a simple example of a combinatorial optimization problem (1-D placement) which is nicely solved by simulated annealing. The empirical-intuitive discussion of the energy landscape also helps the non-physically trained reader to grasp why the concepts of disorder, frustration and ultrametricity, are relevant in the context of combinatorial optimization.

The concluding "conjecture" of Solla et al., that "if simulated annealing works well in a space, then the low-lying minima are ultrametrically distributed", is somewhat challenged in the paper by **D. Geman and S. Geman** (pp 301-309). Their bayesian approach to image restoration and other applications such as tomography, leads quite naturally to the use of simulated annealing. In this class of problems, the energy landscapes are likely to be very different from TSP, spin glass, euclidean match, and placement landscapes. Indeed, in most image processing applications, although the cost-function to be minimized is far from being convex -which makes the problem non-trivial-, the global minimum that is sought for is essentially unique. Thus, these problems lack the high degeneracy and ultrametricity of the favorite examples of physicists. Interestingly, simulated annealing is still a powerful technique in these cases, and it seems to be spreading rapidly in the domain of image processing. A nice and clear introduction to bayesian decision theory and parameter estimation is given in *Duda and Hart*[1973]

The paper by **B. Gidas** (pp 321-326) is a brief review of some mathematical questions arising in the continuous-value formulation of the annealing algorithm, formally a Langevin equation with time-dependent temperature.

Back to the TSP, **E. Bonomi and J.L. Lutton** (pp 295-299) improve the efficiency of the annealing algorithm by exploiting the natural topology in the problem to cleverly choose the initial state, and restrict the set of attempted moves to reasonable ones, i.e., moves with a decent probability of acceptance.

The paper by **I.G. Rosenberg** (pp 327-331) provides an overview of basic definitions and techniques in boolean optimization. **Y.L. Kergosien** (pp 333-336) discusses the issue of global versus local minima within the framework of catastrophe theory.

## 5-MODELS OF BIOLOGICAL ORGANIZATION

Besides brain modeling, another major application of networks of automata in biology is the theory of evolution -ontogeny and phylogeny. One of the first contributors to this topic was S. Kauffman who devoted much work to the investigation of the "generic" properties of random cellular automata. **S. Kauffman**'s (pp 339-360) contribution to the present volume is chiefly a summary of this approach, and a survey of the most prominent results. The actual issue of mutation and selection in networks of automata is only briefly addressed at the end of the paper.

In connection with Kauffman's random automata networks, **J. Coste and M. Hénon** (pp 361-365) derive the scaling laws concerning the attractors of a random mapping of the set of integers $\{1,...,n\}$ onto itself.

Animal morphogenesis, i.e., the process of evolution of a differentiated organism out of a fertilized egg, is one more potential application of the theory of cellular automata. This topic is addressed by **Y. Bouligand,** (pp 367-384) focusing on the organization of collagen structures in connective tissue. Remarkable patterns of organization are observed in collagen, and some of these patterns can be explained by simple considerations on the topology of defects in periodic structures. Note the comparable role of defects in the paper by **Mosseri and Sadoc.**

The paper by **A.J. Noest** (pp 381-384) provides an interesting application to neurobiology of concepts originating from the physics of disorder. The issue there is that of spike propagation in tissue-culture, i.e., neural networks growing in Petri-dishes. Numerical results suggest that this problem, which is related to percolation, could be representative of a new universality class.

The paper by **M. Schatzman** (pp 385-388) also deals with organization phenomena in the nervous system. The formalism adopted there is that of differential systems, and a unifying scheme is proposed which accounts for a rich variety of behaviors with a relatively simple set of equations. Interestingly, one of these equations is a limiting case of the mean-field theory of a spin system on the circle.

The last two papers, by **H. Axelrad et al**. (pp 389-397) and by **G. Chauvet** (pp 399-402), are concerned with the cerebellum, which is one of the most studied structures in the brain, partly because of its remarkable regular architecture.

*We gratefully acknowledge the financial support of NATO Advanced Research Workshop Program (ARW 588/84), of Société Mathématique de France and of Les Houches Center for Theoretical Physics.*

*Paris December 2, 1985*
**E. Bienenstock, F. Fogelman Soulié, G. Weisbuch**

## 6-REFERENCES

D.H. ACKLEY, G.E. HINTON, T.J. SEJNOWSKI: A learning algorithm for Boltzmann machines. *Cognitive Sci.*, 9, pp147-169, 1985.

J. AMIT, H. GUTFREUND, H. SOMPOLINSKY: Storing infinite number of patterns in a spin glass model of neural networks, *Phys.Rev.Lett.* 55, pp 1530-1533, 1985.

P.W. ANDERSON: Suggested models for prebiotic evolution: the use of chaos. *Proc. Nat. Acad. Sci. USA*, vol. 80, pp 3386-3390, 1983.

E.R. BERLEKAMP,J.H. CONWAY,R.K. GUY: Winning Ways.*Academic Press.* 1982.

I. BIECHE, R. MAYNARD, R. RAMMAL, J.P. UHRY: On the ground states of the frustration models of spin glass by a matching method of graph theory. *J. Phys. A, Math. Gen.*, 13, pp 2553-2576, 1980.

A.W. BURKS: Essays on Cellular Automata. *Univ. Illinois Press.* 1970.

V. CERNY: A thermodynamical approach to the traveling salesman problem: an efficient simulation algorithm. *Preprint Inst. Phys.,Biophys.Bratislava,* 1982.

J. DEMONGEOT, E. GOLES, M. TCHUENTE Eds: Dynamical Systems and Cellular Automata. *Academic Press,* 1985.

E.W. DIJKSTRA: Self stabilizing systems in spite of distributed control. *Comm. ACM,* 17, 11, pp 643-644, 1974.

R.L. DOBRUSHIN, V.I. KRYUKOV, A.L. TOOM Eds: Locally Interacting Systems and their Applications in Biology. Lecture Notes in Math., n° 653, *Springer Verlag.* 1978.

R.O. DUDA, P.E. HART: Pattern classification and scene analysis. *Wiley,* 1973.

D. FARMER, T. TOFFOLI, S. WOLFRAM Eds: Cellular Automata. *North Holland.* 1984.

F. FOGELMAN SOULIE: Contributions à une Théorie du Calcul sur Réseaux. *Thèse, Grenoble.* 1985.

K. FUKUSHIMA: Self organizing multi layered neural networks. *Syst. Comp. Controls,* 6, pp 15-22, 1975.

A.E. GELFAND, C.C. WALKER: Ensemble Modelling. *M. Dekker,* 1984.

A. GILL: Linear Sequential Circuits. *Mc Graw Hill,* New York, 1966.

E. GOLES CHACC: Comportement Dynamique de Réseaux d'Automates. *Thèse, Grenoble.* 1985.

J.M. GREENBERG, B.D. HASSARD, S.P. HASTINGS: Pattern formation and periodic structures in systems modeled by reaction-diffusion equations. *Bull. Amer. Math. Soc.*, vol. 84, n°6, pp 1296-1327, 1978.

D.O. HEBB: The organization of behavior. *Wiley,* 1949.

G.E. HINTON, J.A. ANDERSON Eds. : Parallel Models of associative memory. *Erlbaum Pub.,* 1981.

G.E. HINTON, T.J. SEJNOWSKI, D.H. ACKLEY: Boltzmann machines: constraint satisfaction networks that learn. *Tech. Rep. CMU,* n° CS-84-119, 1984.

T. HOGG, B.A. HUBERMAN: Understanding biological computation: reliable learning and recognition. *Proc. Nat. Acad. Sc; USA*, vol. 81, pp 6871-6875, 1984

J.J. HOPFIELD: Neural Networks and Physical Systems with Emergent Collective Computational Abilities. *Proc. Nat. Acad. Sc; USA*, vol. 79, pp 2554-2558, 1982.

S.A. KAUFFMAN: Metabolic Stability and Epigenesis in Randomly Constructed Genetic Nets. *Theor. Biol.*, 22, pp 437-467, 1969.

S.A. KAUFFMAN: Behaviour of Randomly Constructed Genetic Nets. In "Towards a Theoretical Biology". Ed. C.H. Waddington, *Edinburgh Univ. Press*, vol.3, pp 18-37, 1970.

S. KIRKPATRICK, C.D. GELATT, M.P. VECCHI: Optimization by simulated annealing. *Science*, 220, n° 4598, pp 671-680. 1983.

T. KOHONEN: Associative memory- A system theoretical approach. *Springer Verlag: Communications and Cybernetics*, vol. 17, 1978.

T. KOHONEN: Self Organization and Associative Memory. *Springer Verlag Series in Information Sciences*, vol. 8, 1984.

W.S. MAC CULLOCH, W. PITTS: A logical calculus of the ideas immanent in nervous activity. *Bull. Math. Biophys.* 5, pp 115-133, 1943.

M. MINSKY, S. PAPERT: Perceptrons. *MIT Press*. 1969.

J. von NEUMANN: The computer and the brain. *Yale Univ. P.*, New Haven. 1958.

J. von NEUMANN: Theory of Self Reproducing Automata. A.W. Burks Ed. *Univ. Illinois Press*. 1966.

N.J. NILSSON: Learning machines. *Mac Graw Hill*, 1965.

F. ROBERT: Discrete Iterations. *Springer Verlag*, to appear, 1986.

F. ROSENBLATT: Principles of Neurodynamics: percpetrons and the theory of brain mechanisms. *Spartan Books*. 1961.

T.J. SEJNOWSKI, G.E. HINTON: Separating figure from ground with a Boltzmann machine. in "Vision, brain and cooperative computation", M.A. Arbib, A.R. Hanson Eds., *MIT Press*, 1985.

E.H. SNOUSSI: Structure et comportement itératif de certains modèles discrets. *Thèse Docteur Ing., Grenoble*, 1980.

M. TCHUENTE: Contribution à l'étude des méthodes de calcul pour des systèmes de type coopératif. *Thèse d'Etat, Grenoble*, 1982.

R. THOMAS Ed.: Kinetic logic, a boolean approach to the analysis of complex regulatory systems. Lecture Notes in Biomathematics, vol 29, *Springer Verlag*, 1979.

D.L. WALTZ, J.B. POLLACK: Massively parallel parsing: a strongly interactive model of natural language interpretation. *Cognitive Sci.*, 9, pp 51-74, 1985.

G. WEISBUCH: Un modèle de l'évolution des espèces à trois niveaux, basé sur les propriétés globales des réseaux booléens. *C.R. Acad. Sc. Paris*, 298, pp 375-378, 1984.

S. WOLFRAM: Statistical mechanics of cellular automata. *Rev. of Modern Physics*, 55-3, pp 601-645, 1983.

# CONTRIBUTORS

*For each contribution, the beginning page is given in parentheses.*

**J.P. ALLOUCHE** (59): UER de Mathématiques, Informatique, Univ. Bordeaux I, 351 cours de la Libération, 33 405 Talence Cedex. France.

**J.A. ANDERSON** (209): Center for Neural Science, Brown Univ., Providence RI 02912. USA.

**H. AXELRAD** (389): Laboratoire de Physiologie, CHU Pitié Salpêtrière, 91 Bd de l'Hôpital, 75 634 Paris, Cedex 13. France.

**C. BERNARD** (389): Laboratoire de Physiologie, CHU Pitié Salpêtrière, 91 Bd de l'Hôpital, 75 634 Paris, Cedex 13. France.

**E. BIENENSTOCK** (247): Laboratoire de Neurobiologie du Développement, Bât 440, Univ. de Paris Sud, 91 405 Orsay Cedex. France.

**E. BONOMI** (295): Département de Physique Théorique, Ecole Polytechnique, 91 128 Palaiseau. France.

**Y. BOULIGAND** (367): Laboratoire d'Histologie et de Cytophysique de l'EPHE,67 rue M. Günsbourg, 94200 Ivry sur Seine. France.

**J. BUHMANN** (273): Physik Department, Technische Universitat Munchen. 8046 Garching. West Germany.

**G. CHAUVET** (399): Laboratoire de Biologie Mathématique, Univ. d'Angers, 1 rue Haute de Reculée, 49 045 Angers. France.

**M. COSNARD** (63): TIM3 IMAG, BP 68, 38 402 St Martin d'Hères. France.

**J. COSTE** (361): Laboratoire de Physique de la Matière Condensée, Parc Valrose, 06 034 Nice. France.

**J. DEMONGEOT** (71):TIM3 IMAG,BP68,38402 St Martin d'Hères Cedex. France.

**J. DENKER** (193): AT&T Bell Laboratories, Holmdel, NJ 07 733. USA.

**G. DREYFUS** (227): ESPCI, Laboratoire d'Electronique, 10 rue Vauquelin, 75 005 Paris. France.

**F. FOGELMAN SOULIE** (85): LDR c/o CESTA, 1 rue Descartes, 75 005 Paris. France.

**J. FRICOT** (71):TIM3 IMAG,BP68,38402 St Martin d'Hères Cedex. France.

**D. GEMAN** (301): Department of Mathematics and Statistics, Univ. of Massachussets, Amherst, MA 01 003. USA.

**S. GEMAN** (301): Division of Applied Mathematics, Brown Univ. Providence, RI 02 912. USA.

**B. GIDAS** (321): Division of Applied Mathematics, Brown Univ. Providence, RI 02 912. USA.

**B. GIRAUD** (389): Département de Physique Théorique, CEA, Saclay. France.

**E. GOLES CHACC** (63,101): Dep. Matematicas, Esc. Ingenieria, Univ. de Chile, Casilla 5272, Correo 3, Santiago. Chile.

**H.P. GRAF** (193): AT&T Bell Laboratories, Holmdel, NJ 07 733. USA.

**I. GUYON** (227): ESPCI, Laboratoire d'Electronique, 10 rue Vauquelin, 75 005 Paris. France.

**H. HARTMAN** (53): MIT, Cambridge MA 02 139. USA.

**M. HENON** (361): Observatoire de Nice, 06 003 Nice Cedex. France.

**J.J. HOPFIELD** (155): Divisions of Chemistry and Biology, Caltech, Pasadena, CA 91 125. USA.

**R.E. HOWARD** (193): AT&T Bell Laboratories, Holmdel, NJ 07 733. USA.

**D. d'HUMIERES** (187, 241): Groupe de Physique des Solides, Ecole Normale Supérieure, 24 rue Lhomond, 75 005 Paris. France.

**L.D. JACKEL** (193): AT&T Bell Laboratories, Holmdel, NJ 07 733. USA.

**S.A. KAUFFMAN** (339): Department of Biochemistry and biophysics, Univ. of Pennsylvania, School of Medicine, Philadelphia, Pennsylvania 19 104. USA.

**Y.L. KERGOSIEN** (333): Département de Mathématiques, Bât 425, Univ. de Paris Sud, 91 405 Orsay Cédex. France.

**M. KERSZBERG** (205): Institut für Festkörperforschung des Kernfor schungsanlage, Jülich GmbH, D-5170 Jülich. West Germany.

**B. LACOLLE** (133): TIM3 IMAG,BP68,38402 St Martin d'Hères Cedex. France.

**Y. LE CUN** (233): ESIEE, 89 rue Falguière, 75 015 Paris. France.

**J.L. LUTTON** (295): CNET PAA/ATR/SST, 92 131 Issy les Moulineaux. France.

**C von der MALSBURG** (247): Abteilung für Neurobiologie, Max-Planck-Institut für Biophysikalische Chemie, D-3400 Göttingen. West Germany.

**M.E. MARC** (389): Laboratoire de Physiologie, CHU Pitié Salpêtrière, 91 Bd de l'Hôpital, 75 634 Paris, Cedex 13. France.

**M. MEZARD** (119): Dipartimento di Fisica, Univ. di Roma I, Piazzale A. Moro 2. 00 185 Roma. Italy.

**J. MILNOR** (113): Institute for Advanced Study. Princeton NJ 08 540. USA.

**R. MOSSERI** (149): Laboratoire de Physique des Solides. CNRS. 92 195 Meudon Principal Cedex. France.

**J.J. NIEZ** (171): LETI/MCS CEN Grenoble. BP 85X. 38 401 Grenoble. France.

**A.J. NOEST** (381): Netherlands Institute for Brain Research. Meibergdreef 33. 1105 AZ Amsterdam. The Netherlands.

**D. PELLEGRIN** (67): TIM3 IMAG,BP68,38402 St Martin d'Hères Cedex. France.

**P. PERETTO** (171): DRF/PSC CEN Grenoble. BP 85X. 38 401 Grenoble. France.

**L. PERSONNAZ** (227): ESPCI, Laboratoire d'Electronique, 10 rue Vauquelin, 75 005 Paris. France.

**F. ROBERT** (33): TIM3 IMAG,BP68,38402 St Martin d'Hères Cedex. France.

**Y. ROBERT** (49):TIM3 IMAG,BP68,38402 St Martin d'Hères Cedex. France.

**I.G. ROSENBERG** (327): Mathematics and Statistics. Univ. de Montreal. CP 6128, Succ "A". Montreal Quebec H3C 3J7. Canada.

**J.F. SADOC** (149): Laboratoire de Physique des Solides. Univ. de Paris Sud, 91 405 Orsay Cédex. France.

**M. SCHATZMAN** (385): Département de Mathématiques, Univ. Claude Bernard. 69 622 Villeurbanne Cedex. France.

**K. SCHULTEN** (273): Physik Department, Technische Universitat Munchen. 8046 Garching. West Germany.

**S.A. SOLLA** (283): IBM Thomas J. Watson Research Center. Yorktown Heights. New York 10 598. USA.

**B. SORKIN** (283): IBM Thomas J. Watson Research Center. Yorktown Heights. New York 10 598. USA.

**B. STRAUGHN** (193): AT&T Bell Laboratories, Holmdel, NJ 07 733. USA.

**D.W. TANK** (155): AT&T Bell Laboratories, Murray Hill, NJ 07 974. USA.

**M. TCHUENTE** (21,49): TIM3 IMAG.BP68.38402 St Martin d'Hères Cedex. France.

**G.Y. VICHNIAC** (3,53): Laboratory for Computer Science, MIT, Cambridge MA 02 139. USA.

**M.A. VIRASORO** (197): Dipartimento di Fisica, Univ. di Roma I, Piazzale A. Moro 2. 00 185 Roma. Italy.

**G. WEISBUCH** (187): Groupe de Physique des Solides, Ecole Normale Supérieure, 24 rue Lhomond, 75 005 Paris. France.

**S.R. WHITE** (283): IBM Thomas J. Watson Research Center. Yorktown Heights. New York 10 598. USA.

# 1 AUTOMATA THEORY

# CELLULAR AUTOMATA MODELS
# OF DISORDER AND ORGANIZATION

Gérard Y. Vichniac

*Laboratory for Computer Science*
*Massachusetts Institute of Technology, Cambridge, Massachusetts 02139, USA*

## 1. Introduction

Cellular automata are mathematical objects introduced in 1948 by J. von Neumann and S. Ulam to "abstract the logical structure of life"[1]. Since then, they have established themselves as unique tools to analyze the emergence of global organization, complexity, and pattern formation from the iteration of local operations between simple elements. They have also been extensively used as models of universal computation, and are being increasingly applied to a variety of concepts from physics and chemistry[2]. They are in fact versatile enough to offer analogies with almost all the themes discussed at this meeting (in particular: self-organization, dissipative systems, spatial vs. thermal fluctuations, neural networks, optimization, ergodicity-breaking, and ultrametricity)

Cellular automata are dynamical systems where space and time are discrete, and the dynamical variables are taken from a finite set. Discreteness of space implies here a regular array of sites, or "cells." (This paper concentrates on the two-dimensional square lattice.) Each site is capable of a value (or state) belonging to a finite set. In the important class of *Boolean* automata, this set contains two elements, in other words sites are endowed with one bit only. The two states are generally labelled 0 and 1, on and off, occupied and empty, live and dead, or $\uparrow$ and $\downarrow$ spins.

The system evolves in discrete time according to a local law. The value taken by a cell at time $t + 1$ is determined by the values assumed at time $t$ by the neighboring sites and by the considered site itself:

$$x_{i,j}^{t+1} = f(x_{i,j}^t, x_{i-1,j}^t, x_{i,j-1}^t, x_{i,j+1}^t, x_{i+1,j}^t). \qquad (1)$$

In this equation $x_{i,j}^t$ denotes the state occupied at time $t$ by the site $(i,j)$. The updating is synchronous and the law is uniform: all the cells update their state at the same time and according to the same law. Cellular automata are thus a paradigm for parallel computation[3]. In (1), the transition rule involves the considered site (the "center" cell) and its four adjacent neighbors. These five cells form the so-called von Neumann neighborhood. The rule can also depend on the next-to-nearest neighbors ($x_{i\pm1,j\pm1}$). In that case, the nine-cell template is referred to as the Moore neighborhood.

The purpose of the game is to "play God," i.e., to create and follow the evolution of a variety of synthetic universes, each one characterized by the choice of the transition rule $f$ and of the initial conditions $\{x_{i,j}^{t=0}\}$. Given the ability to simulate cellular automata efficiently, it is always tempting and often instructive to perform "miracles," that is, to suspend at some step the "natural" law of the universe, deliberately modify the configuration at that step (the "hand of God"), and resume the evolution, perhaps with a new

NATO ASI Series, Vol. F20
Disordered Systems and Biological Organization
Edited by E. Bienenstock et al.
© Springer-Verlag Berlin Heidelberg 1986

transition rule. One can, for example, reverse the direction of time if the transition rule is invertible (see below, sections 6 and 7).

We shall concentrate in this paper on the objects just defined, namely deterministic, homogeneous cellular automata. For some applications, however, it is useful to generalize definition (1). One can, for example, add a random number $r_{i,j}^t$ to the list of arguments of $f$, thus defining a *stochastic* rule. Probabilistic cellular automata are strongly reminiscent of the Ising model at finite temperature, a system also characterized by discrete space, discrete variables, and nearest-neighbor interactions. But simultaneous updating, together with full discreteness, make the precise mapping of stochastic cellular automata and the Ising model delicate [4,5,6,7]. Enting[8], and Domany and Kinzel[9] have nevertheless constructed a mapping of a type of cellular automata (without simultaneous updating) in $d$ dimensions and a type of Ising model in $d+1$ dimensions; Kinzel[10] has explored phase transitions in stochastic cellular automata. Toom introduced probabilistic cellular automata with strong nonergodic behavior[11]. Gacs and Reif[12] have recently shown that a 3-dimensional version of Toom's automaton is capable of universal computation that is reliable in the presence of noise. Bennett and Grinstein[13] have measured the phase diagram of Toom's automaton, and have shown why irreversibility is a necessary condition for a noisy cellular automaton to exhibit nonergodic and computationally complex behavior. Another way to depart from (1) is to renounce the uniformity of the rule. In the resulting *inhomogeneous* cellular automata (INCA), different sites can follow different rules to update their state—see [14], this volume. (If one furthermore gives up the geometric regularity of the array, one obtains random Boolean networks discussed in this volume—see e.g., [15]). Finally, Fredkin (see [16,4]) and Margolus[16] have shown how to construct a wealth of invertible rules by modifying the form of (1). Fredkin's method makes $f$ dependent also on $x_{i,j}^{t-1}$, the past value of the considered cell, and thus defines a rule that is second-order in time. Section 7 describes "Q2R," a rule of this type. Margolus's method partitions the array into blocks of cells, and the transition rule maps a block into a block. Propagation of information is achieved by alternating in time two modes of partitioning.

All the mathematical experiments described in this paper were carried out using Toffoli's Cellular Automata Machine (CAM), a dedicated hardware simulator that updates sixty times per second a 256×256 array of cells with up to eight bits per cell[17]. CAM simulates second-order and inhomogeneous Boolean cellular automata by treating them as special cases of first-order homogeneous cellular automata, but with two bits per site. Probabilistic and Margolus automata are simulated by treating a random source and the parity of space-time, respectively, as an additional neighbor.

## 2. Modeling with cellular automata

Cellular automata are fully discrete dynamical systems. One motivation for using them (and, for that matter, of using other discrete systems discussed in this volume—e.g., Boolean networks and spin-glasses) to model natural phenomena is the following: while digital computers are discrete objects, much of science is expressed in the language of continuum mathematics (e.g., the differential equations of physics). In order to overcome this mismatch and let computers do what they do best, i.e., logical manipulations of bits rather than necessarily imprecise floating-point arithmetic, one concentrates on models that are already fully discrete. Such models lend themselves to exact simulation by "finitistic" means (as von Neumann called them[18]); their inner details are under full control and their evolution can be followed step by step. These exactly *computable* models offer a third alternative to the dichotomy between the few (and often very stylized) exactly *solvable* models and the more realistic models that are not solvable at all (only their truncated and rounded-off versions are actually treated numerically). Numerical analysis, an inevitable and often opaque screen between continuous models and their discrete simulations, is eliminated altogether.

For the sake of simplicity and elegance, one usually endows each site with one bit only. (This is in sharp contrast with the computer treatment of partial differential equations, otherwise akin to cellular automata, but where each node of the numerical mesh is given one or several computer words.) One can worry about the poverty of phenomena this Spartan constraint would entail. It turns out, in fact, that Boolean cellular automata display a surprising richness of behaviors.

This richness is so seductive that it has actually been suggested that discrete modeling be taken seriously as a model for basic physical processes. Briefly stated, this literal interpretation of cellular automata views nature itself as computing locally and *digitally* the immediate future. Feynman, Finkelstein, Fredkin, Minsky, Wheeler, Zuse and others (see [19,20,4]) have advocated—or at least seriously considered—this view. Commenting on the game "Life" (described below), Roger Penrose aptly describes the fascination of the apparently unbound modeling power of cellular automata: "It is hard to resist the tempting argument that the game offers a model for our universe—itself presumably governed by rules of the utmost simplicity, yet exhibiting the richness and complexity we observe around us, especially in living things."

This paper takes a less ambitious approach. Following von Neumann (see next section), we shall treat cellular automata just as one generally uses other dynamical systems, i.e., as conceptual analogies, not to be taken literally, but to shed light on the general mechanisms by which man and nature can generate complexity out of simple elements and organization out of disorder.

## 3. Modeling life: self-organization and self-replication

### 3.1 *"Life"*

The best known cellular automaton rule is "Life," invented in 1970 by John H. Conway[21]. It describes multiplication and extinction in a population of "cells." The rule is the following: a "live" cell remains "on" in the next generation if and only if there are presently two or three other "live" cells among its eight outer Moore neighbors. It will die (turn off) by "isolation" if it is surrounded by zero or one live cell, and by "overcrowding" if more than three neighboring sites are occupied. An empty or dead cell will come to life in the next time step if and only if it is surrounded by exactly three live parents. (In Conway's universe, only *ménages à trois* procreate). A remarkable feature of "Life" is the emergence of ordered patterns out of a disordered initial condition (the "primeval soup"), thus escaping the tyranny of the Second Law of thermodynamics (a discussion of this point is given in section 6). Thanks to the loving work of hundreds of "hackers," the zoology of these patterns is abundantly documented[22]. What is more, "Life" has been shown to be capable of universal computation, i.e., given the proper initial configuration, it can simulate the operations of any general-purpose computer.

The standard belief is that billions of steps in a huge array (with, perhaps, a slight amount of noise to simulate cosmic-rays-induced mutations) will give rise to an evolution of increasingly complex structures. One could even imagine that these structures, much like Stanislaw Lem's creatures[23], would become able to communicate with each other, asking questions such as "Does God exist?" or "Do we have free will?"

### 3.2 Self-replicating machines

In his attempt to "abstract the logical stucture of life," von Neumann focused on the function of self-replication, not the way it is achieved in nature. He concluded that an automaton capable of reproducing itself and of constructing another automaton more complex than itself (thus allowing for evolution) must possess the following elements:

1. A *blueprint*, or a whole set of instructions for constructing an offspring, (embodied by a Turing-machine type input tape);

2. A *factory*, to carry out the construction;

3. A *controller*, to make sure the construction follows the plan;

4. A *duplicating machine*, to transmit a copy of the construction program to the offspring (the automaton must provide for the possibility of having grandchildren).

Remarkably enough, von Neumann's scheme anticipated by five years the discovery in 1953 of the basic structure of the reproduction of all organisms larger than viruses. The four elements can be identified respectively with (1.) the genetic material, DNA and RNA; (2.) the ribosomes; (3.) the repressor and derepressor molecules; (4.) the enzymes RNA and DNA polymerases. This mapping was established by F. Dyson[25].

The details of von Neumann's automaton have been completed by A. W. Burks[1,26]. The automaton involves more than 200 000 cells with 29 states each. E. F. Codd[27], a student of Burks's, simplified considerably the original scheme, but neither of these automata have actually been simulated on computers, because of their extreme complexity. The complexity results in part from the requirement of universal *construction*, of which self-replication is only a special case. Universal construction enables the factory to construct almost any configuration specified in the input tape. This capability allows for an overaccelerated evolution: universal construction would not "prevent a cockroach from begetting an elephant"[24] if provided with sufficient information and raw material.

In a recent paper, C. G. Langton (another student of Burks's) lifted the requirement of universal construction to focus on self-replication. He obtained a remarkably simple automaton[28], schematized in Figure 1. In the figure each line is actually three cells thick. A d-shaped pattern at $t = 0$ grows a second loop, completes it at $t = 120$, and divides 8 steps later to create at $t = 151$ a new d-shaped pattern, identical to the original one, but pointing in another direction. The colony grows forever by creating new offspring at its periphery, somewhat like a coral reef. A remarkable aspect of Langton's construction is that the original loop does *not* contain a full replication program

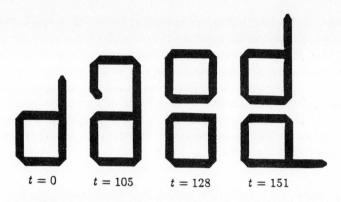

$t = 0$      $t = 105$      $t = 128$      $t = 151$

Figure 1: *Langton's self-duplicating loop.*

(von Neumann's component 1.). Reproduction is achieved, however, by taking advantage of the "physics" of the universe in which the loop lives. The loop, instead of rigidly (i.e., *actively*) implementing a set of instructions, gracefully folds and unfolds itself, *passively* carried by the "momentum" of its parts. This passive mechanism of pattern formation is of course pervasive in the inanimate world. Water molecules, for example, do not contain a blueprint of the complex snow-flakes they can form. J. S. Langer, who is responsible for much of the recent progress in our understanding of snow-flake formation, suggests that living organisms also use the laws of physics to reinvent the missing parts of the blueprint: "We know perfectly well that the genetic material in an animal, for example, can't possibly contain all the information needed to tell where to put every molecule in every cell. There has to be some sort of very simple mechanism that controls when a group of cells stops becoming a neck and starts becoming a head[29]."

To be sure, in order to maintain any meaning of the word "self" in "self-replication," the passive use of the natural evolution law must only be partial, the mechanism must be substantially active. A well-known example of totally passive replication is illustrated in the "XOR" additive rule, where *any* pattern gets indiscriminately duplicated, independently of any "program" (initial condition)—see [14].

### 3.3 Abstract modeling in biology and in physics

Von Neumann's endeavor was unashamedly reductionistic and it was successful . He took the relevant features of life to be the survival, adaptation and reproduction of complex forms of organization. The specific means (viz., organic chemistry) nature has used to achieve this was disregarded as being idiosyncratic and irrelevant. Cellular automata have been used with success to model a major feat of life: the power of self-replication. The study of this feat, as well of that of evolution, is no longer the monopoly of biologists. This actually is the major theme of this meeting. Many of us, mathematicians, physicists, computer scientists and other self-proclaimed experts on "complexity" believe that we have a lot to contribute to the study of life. We are quick to invoke the authority of von Neumann and call biochemistry irrelevant, thus redefining "irrelevant" to mean whatever we are too lazy to learn. Our colleagues the biologists, in urging us to look at the facts they have carefully isolated, remind us that von Neumann himself said that a complex object is the simplest representation of itself. Fair enough. But the point is that physicists, for example, often apply to their own field the same disregard for the facts. A driving force in the progress of theoretical physics for several decades has been the deliberate overlooking of the details of the elementary mechanisms in complex systems. Powerful theories have been constructed just by concentrating on symmetries, and, more recently, on "universality classes." In the rest of this paper, we shall push this approach to an extreme and show how, by deliberately overlooking the apparent continuity of space and time, we can use cellular automata to "abstract the logical structure" of some small regions of the inanimate world.

## 4. Voting rules: metastability/instability transitions, spinodals

Let us consider the simplest "voting rule." Each site has its four nearest neighbors "vote" at time $t$, and responds at time $t+1$ according to peer pressure; and becomes "on" in case of a tie. In the language of the Ising model (where a ↑ or ↓ spin is attached at each site), spins align at time $t+1$ to the local field they feel at time $t$; a bias in the rule acts as an external field in the ↑ direction and assigns the ↑ state if a site is surrounded by two ↑ and two ↓ spins. Since a site becomes "on" if 2, 3, or 4 among its four neighbors are "on," the rule[4] has been called "Q234" or "two-out-of-four (2/4)."

This is the transition rule. We still have to choose the initial condition. One can, for example, independently assign to each site at $t = 0$ an ↑ spin with probability $p$ and a ↓ spin with probability $1 - p$. The subsequent evolution is easy to describe. If one starts with a minority of ↑ spins (to compensate for the external field that induces flips in their favor), many isolated ↑ spins flip during the first steps, as if submerged by the surrounding ↓ sea. But *local density fluctuations* at $t = 0$ trigger the growth of ↑ clusters in the sea of ↓ spins. However, a cluster can grow only by filling concavities; once it reaches a convex shape (here, a rectangle), the growth stops. The rule is *convex-confining*[4]. In other words the growth is bounded by the convex hull of ↑ sites at any time. But two rectangles can overlap and create new concavities, as in Figure 2. Pattern (A) will fill its convex hull (the inner dotted lines) and eventually touch rectangle (B). They will then form a single cluster and fill the new bigger convex hull (the outer dotted lines) which may, in turn, overlap with a third cluster, and so forth.

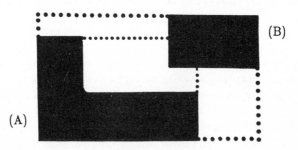

Figure 2: *Overlapping convex hulls*

The outcome of this process depends critically on the ratio of ↑ to ↓ spins at $t = 0$. If the initial density $p$ of ↑ spins is larger than a critical concentration $p_c$ (experimentally found with CAM to be .133 for Q234), then the growth by coalescence of clusters goes on and the whole array rapidly turns to the ↑ state. If, on the other hand, the concentration at $t = 0$ of ↑ spins is less than $p_c$, the coalescence process eventually stops and the system reaches a fixed point (or, because of the parallel updating, a limit cycle of period 2, an illustration of a theorem due to Goles[30]). The system freezes in a pattern made of ↑ rectangular islands in a sea of ↓ sites.

The same phenomenon occurs when the next-to-nearest neighbors vote. The resulting evolution appears more fluid; convex clusters in Moore's neighborhood are truncated octagons rather than rectangles. Figure 3 shows the evolution under M456789 of an initial configuration (Figure 3a) made of a random mixture of ↑ and ↓ spin with a concentration $p = p_c(= .250)$ of ↑. Rule M456789 (also known as[4] "four-out-of-nine (4/9)" and "MGE4") assigns $x_{i,j}^{t+1} = ↑$ if four or more of the nine Moore neighbors are ↑ at time $t$. After 10 steps (Figure 3b), the density of ↑ has dropped to 21% because of the death of isolated ↑. But the surviving ↑ sites are no longer distributed at random. They are correlated, and grow by filling concavities of clusters. After 181 steps, the density is 28%, and the system reaches a fixed point: all the clusters are now convex (Figure 3c).

Since the ↓ phase is not favored by the dynamics (because of the external ↑ field), the sea of the fixed point (Figure 3c) is not stable but only metastable: an infinitesimal noise would eventually erase it, even if all the sites were ↓ at $t = 0$. Figures 3c and 3d illustrate how flipping a single spin at a sensitive location of the metastable ↓ sea can trigger its decay. If some noise flips a ↓ spin next to the large central ↑ cluster, e.g., at the tip of the arrow, it creates new concavities, allowing the large cluster to grow further. Figure 3d shows the configuration at 400 steps after the flip; the 256×256 CAM array will eventually turn all ↑.

Another way to express the metastability is to consider the thermodynamic limit (infinite lattice, infinite time). In an infinite lattice, there will occur with probability one a cluster large enough to grow forever, feeding on and bridging isolated clusters. The system is closely analogous, say, to superheated water—cf. the ↓ sea—in which bubbles of vapor (the thermodynamically stable phase—cf. the ↑ spins) can form. In a finite vessel, homogeneous nucleation (decay of the metastable water phase by formation—resulting from a density fluctuation—of a large enough vapor bubble) is highly improbable. In the thermodynamic limit, on the other hand, there is room for a very large density fluctuation capable of triggering nucleation. The survival of the metastable phase is a transient effect, yet it can last a very long time. J. Cahn[31] has calculated that an infinite volume of superheated water at 110 C would not boil for $10^{10}$ years! (The reason a tea kettle works is *heterogeneous* nucleation that stems from impurities in the water or rough spots on the kettle walls.) At some higher temperature (at the spinodal point) the water ceases to be metastable and boils immediately (spinodal decomposition). Even though metastabilty and instability lead to the same fate (survival of the stable phase only) the two regimes are very different.

Likewise, an infinite number of cellular-automaton steps in an infinite lattice will lead to a homogeneous ↑ for all initial conditions with $p > 0$. But, as in the case of water-vapor metastable equilibrium, the two regimes—metastability ($p < p_c$) and instability ($p > p_c$)—are quite distinct. The critical density $p_c$ is thus a genuine *spinodal point*. No one has so far been able to derive its measured values: .133 for Q234 (Figure 2) and .250 for M456789 (Figure 3). When the number of voting neighbors varies, tantalizingly

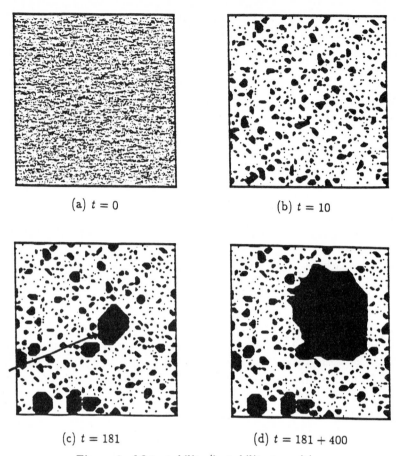

(a) $t = 0$

(b) $t = 10$

(c) $t = 181$

(d) $t = 181 + 400$

Figure 3: *Metastability/instability transition*

Evolution (obtained with CAM) under the "four-out-nine (4/9)" biased majority rule. The ↑ clusters are in black. The configuration (c) is a fixed point. Flipping the ↓ spin shown at the tip of the arrow creates new concavities, and the metastable ↓ sea vanishes at the expenses of the stable ↑ phase (d)—spinodal decomposition at $T = 0$.

simple values are obtained for $p_c[4]$: .0822 for the "two-out-of-five (2/5)" rule, where the center site votes; .266 for (3/6); .191 for (3/7); and .333 for (4/8). It is hard to believe that no simple argument exists to account for these numbers, considering their simplicity and their simple relationships (.266 $= 2 \times$ .133, .191 $\simeq \frac{1}{2}(.133 + .250)$).

Many-body spinodals and metastability/instability transitions are generally believed to pertain to stochastic dynamics only, and thus to be necessarilly tied to temperature. The biased majority cellular automata, being capable of $T = 0$ metastability/instability transitions (characterized by whether $p < p_c$ or $p > p_c$), genuinely extend the concept of spinodal to deterministic dynamics. Randomness is fully inherited from the initial condition; the fluctuations are geometrical, not thermal.

## 5. More voting rules: ordering vs. percolation

Variations on voting rules lead to phenomena in which a disordered initial configuration acquires structure. Let (for example) the center cell vote together with its four neighbors, and define this time a *simple-majority* rule. Our new rule is not biased toward the ↑ or the ↓ state: it is *odd under binary conjugation*[4]. Since the center votes, no tie is possible: the center site becomes "on" at time $t + 1$ if 3, 4, or 5 sites are on at time $t$. The rule has been called[4,5] V345 and "three-out-of-five (3/5)." When one starts with a random initial condition, one observes in the first few steps an ordering process with motion of an interface separating two phases, without conservation of the number of ones—the order parameter (cf. the so-called "model A"[32]). But the ordering stops after a dozen steps, and the system freezes in a percolation pattern[4]. The occurence of freezing can be explained as follows: the center cell's vote counts only if there is a tie among the four adjacent neighbors. When this happens, local stabilization occurs: $x_{i,j}^{t+1} = x_{i,j}^t$, and a global freeze follows.

There are two ways to destabilize the phase boundaries and let the ordering process go on: (i) do not let the center cell vote; and flip a coin in case of a tie. This recovers the Ising spin model at $T = 0$, evolved with the heat-bath Monte-Carlo dynamics or the Glauber dynamics; see the recent study by Grest and Srolovitz[33]. (ii) Another way to destabilize the interface is to let the center cell vote and invert the result of the vote in case of marginal majority, i.e., the center cell will assume the state opposite to that collecting three votes (marginal majority). More precisely, the resulting rule assigns at time $t + 1$ the state that gets 2, 4, or 5 votes among the five von Neumann neighbors at time $t$. (Following [4], the rule can be called V245.) Again, the dynamics is more fluid in Moore's neighborhood, where the counterparts of V345 and V245 are M56789 and M46789, respectively. The modified (or"twisted") rules V245 and M46789 lead to the same macroscopic dynamics as in [33]. One observes an interface motion driven by curvature and effective surface-tension. It is important, however, to note that these rules never invoke random numbers, in contrast to the heat-bath and Glauber dynamics that do so even at $T = 0$. These deterministic cellular automata rely on their own resources for randomness—the randomness at $t=0$, not that of an external heat-bath (cf. [34]).

Since the dynamics steadily decreases the length of the interface, it performs an actual optimization process. This process is somewhat complementary to *simulated annealing*[35]. It is deterministic rather than stochastic, and it starts right away at $T = 0$, rather than reaching this temperature after a slow cooling. V245 and M46789 can be said to perform a *simulated quench*. In fact, any deterministic (i.e., at $T = 0$) manipulation of a configuration that is initially random (i.e., at $T = \infty$) can be seen as a quench process.

## 6. Two kinds of irreversibility

Cellular automata illustrate two kinds of irreversibility: explicit and thermodynamical. Explicit irreversibility simply results from that of the transition rule. Thermodynamical irreversibilty is obtained when a system away from equilibrium relaxes toward equilibrium under a reversible dynamics.

*6.1 Explicit irreversibility: decreasing Lyapunov function*

A fascinating aspect of Conway's game "Life" is the spontaneous emergence of order out of a disordered "primeval soup." The ordering occurs to some extent in all the rules we have considered so far: evolution from a random configuration reaches states that clearly do not look random. A fixed point or a limit cycle of low period is eventually reached. For most *simple* rules, the attractor shows some amount of structure—admittedly defined in a somewhat subjective manner. Cellular automata, much as the Boolean networks discussed in this volume, are capable of self-organization[36,37] akin to the self-ordering behavior of dissipative differential systems.

The creation of order out of chaos appears to violate the Second Law of thermodynamics. But of course cellular automata manage to circumvent (rather than violate) the Second Law in the same way all dissipative systems do: they shrink in time the volume-element of the phase-space. In doing so, they depart from Hamiltonian systems that conserve the volume-element (Liouville's Theorem) and obey the Second Law.

When the phase-space (or state-space) is discrete, this shrinking has dramatic effects: trajectories must *merge*. The dynamics is deterministic in the forward direction of time only. In "Life," for example, most configurations have a multitude of possible antecedents. A region that is empty at time $t$ may have been empty at $t-1$, or inhabited with isolated live cells, or overcrowded with live cells. In a fully discrete system, self-organization entails loss of information. This is not necessarily the case if the dynamical variables $x_{i,j}^t$ take their value in a continuous set. For example, the system $x_{i,j}^{t+1} = \frac{1}{2} x_{i,j}^t$ (with $x_{i,j}$ real) admits the zero configuration as a fixed point, but does not suffer any loss of information: the past can be recovered by $x_{i,j}^{t-1} = 2\, x_{i,j}^t$. (Mergers will of course occur if the system is iterated in finite-precision arithmetic, say, on a digital computer[38].)

In many simple irreversible cellular automata, the emergence of structure is enhanced by an increase in the *symmetry* of patterns. In "Life," for exemple, symmetric "traffic lights" and "pinwheels" appear spontaneously. In M46789, clusters become more and more circular (a fact that accounts in part for the effective surface tension of the dynamics). The symmetry increase is again a direct consequence of irreversibility. A symmetric object can have symmetric and asymmetric antecedents, but will remain symmetric if the rule is deterministic and intrinsically symmetric (as in the above *counting* rules, that depend on the number of 1 cells in the template, but not on the position of these cells). In a dynamics which is deterministic in the forward direction of time only, symmetry is a conserved attribute, but lack of symmetry is not.

For a *generic* cellular automaton rule, constructed by filling at random a look-up table, the attractors should be scattered evenly in the state-space—most of them should lie in the huge grey region of Figure 4. For example in CAM, with $N$=65536 cells, less than one configuration out of a million has a proportion of ones out of the 49%–51% range. This argument, which relies on the narrowness of the binomial distribution for large $N$, applies of course also to observables other than density. As a consequence, no conspicuous structure should emerge from a typical rule. This is what one actually observes with random rules in the Moore neighborhood. In the von Neumann neighborhood, in contrast, some order occurs more often than not[39]. This follows from the fact that *accidental regularities* or an *imbalance* in the number of 0's and 1's are more likely in a table with $2^5 = 32$ entries (for the von Neumann neighborhood) than in a table with $2^9 = 512$

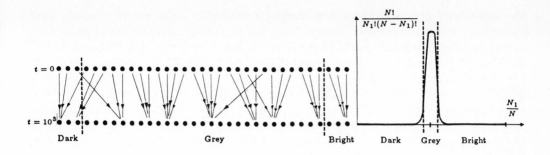

Figure 4: *Mergers of trajectories in irreversible cellular automata.*
Some configurations at $t = 0$ and $t = 1000$ are represented with dots. They are ordered from
left to right according to their number $N_1$ of 1 cells. For $N$ large, an overwhelming majority of
configurations have about $N/2$ "on" cells, and look uniformly grey. Dark (with few 1's) and bright
(with many 1's) configurations occupy the two tails of a binomial distribution, shown on the right.
Configurations at $t = 1000$ with numerous antecedents are representative of an attractor.

entries (for the Moore neighborhood). In this respect, "Life" is highly atypical (it was
certainly not constructed randomly!). The rule is invariant under permutation of the
outer neighbors; furthermore it returns a 1 in 140 cases and a 0 in 372 cases (an imbalance
corresponding to ten standard deviations). The regularities in the "Life" table and its
large on/off imbalance are responsible for the emergence of sparse 1 inhomogeneities.
This argument accounts for the vanishing of the emergence of order in one-dimensional
cellular automata[36] and Boolean nets[40] when the number of inputs increases. Since
there are $2^{k+1}$ counting rules (out of $2^{2^k}$ possible rules) with $k$ inputs, the ratio of these
rules (that are invariant under permutation of the inputs) drops from $1/2$ to $1/16$ when
$k$ varies from two to three. In this respect, when $k = 2$, most randomly generated rules
are atypical, and emergence of order should be expected. A more elaborate reasoning,
based on the ratio of *forcing* functions, accounts for the short periods and the spatial
structure of the ordered attractors for $k$ small[15,40,41,14].

In analogy with continuous dynamical systems where a (negative) Lyapunov exponent measures the rate of convergence of different initial conditions toward a common
attractor, a Lyapunov function can be introduced[40], the variations of which measure
the amount of merging in Boolean systems:

$$S^{(N)}(t) = - \sum_{X=0}^{2^N-1} P^{(N)}(X,t) \log P^{(N)}(X,t). \tag{2}$$

In this expression, $P^{(N)}(X,t)$ is the probability that a configuration $X = (x_1, x_2, \ldots, x_N)$
is reached at time $t$, as determined by the exact bookkeeping of all the mergers. (For
an actual Lyapunov *exponent* for cellular automata, see [42]). The superscripts $(N)$
indicate that all the $N$ degrees of freedom (here, the sites) are taken into account in
the definition of $P^{(N)}(X,t)$. Fogelman[40] has shown that synchronous updating (as in
cellular automata) makes $S^{(N)}$ a nonincreasing function of time:

$$S^{(N)}(t+1) \leq S^{(N)}(t). \tag{3}$$

The capacity of noninvertible cellular automata to display order results precisely from their ability to forget their own history. Since the microscopic interactions of physics are believed to admit no merging of trajectories (an assumption weaker than time-reversal invariance), self-organizing cellular automata like "Life" actually "cheat"; they avail themselves of an "invisible garbage truck" (according to an expression of Tom Toffoli's) which at each time step takes away obsolete information, namely, any record pointing to the actual antecedent among all the possible ones. In nature, it might be that organization is cheap; the really hard task is getting rid of disorder, which, among other things, keeps track of the past. Living organisms (e.g., individuals, termite colonies, the Earth) maintain their organized state by exporting their disorder.

The actual *emergence* of structure is a consequence of a *shift of level* of description. Although random-looking, the initial system fully encodes (and thus "knows") the structure that a deterministic computation unveils; it takes an external observer to appreciate any newness in the result of this computation[43]. The emergence of structure decreases the number of available *microstates* but may increase the number of observed *macrostates*. Under self-ordering, fewer bits are needed to specify the former—this is the meaning of (3), but more information is required to describe the latter. For example, in the ordering rules of the last section, a multitude of random looking initial configurations converge to two possible fully aligned states. On the other hand, all these "grey" initial configurations are equivalent to an observer concerned with average quantities only. He will view the process as an evolution from *the* uniform macrostate to one of two aligned states. Similarly, our observer will *gain* information by quenching a magnet from *the* paramagnetic phase (with magnetization $< \sigma >= 0$) to one of the two ferromagnetic phases ($< \sigma >= +1$ or $< \sigma >= -1$). The ordered states revealed by an irreversible evolution are few, but *recognizable*. As H. Atlan emphasized to me, the central unsolved problem is the complete mathematical characterization of this recognizability.

### 6.2 Thermodynamical irreversibility: increasing reduced entropy

Equation (2) has the familiar form of the definition of entropy. (For various kinds of entropy in cellular automata, see J. Milnor's article in this volume and references therein). Inequality (3), however, is in apparent contradiction with the Second Law which requires that entropy be a *nondecreasing* function of time. Again, the difficulty is related to the noninvertibility of the dynamics. If the transition rule is invertible, the evolution admits no mergers and $S^{(N)}$ is constant:

$$S^{(N)}(t+1) = S^{(N)}(t). \tag{4}$$

This equation is evident: an invertible evolution step corresponds to a permutation of the configurations (labeled by the dummy index $X$), and thus will not affect the value of the sum in (2).

In the last section, we saw that a discrete evolution law (e.g., the $f$ of equation (1)) is necessarily noninvertible if it leads to ordering. Conversely, absence of mergers preserves the density of points, because of the conservation of information and its incompressibility in a discrete state-space. In other words, reversibility alone buys Liouville's theorem for free in fully discrete dynamics[44]. Reversible cellular automata are thus a discrete counterpart of Hamiltonian systems. We shall show soon that they indeed obey the Second Law.

The physical world actually satisfies Equation (4) (or a continuous equivalent). The fully correlated, $N$-body entropy of a fully specified, isolated system is a constant of the motion. The entropy which is subjected to the thermodynamic *increase* is a coarse-grained one. Nature has several subtle ways to coarse-grain the entropy [45,46,47,48]. A theorist's favorite way to alter $S^{(N)}$ is to cut out many-body correlations, thus defining a mean-field approximation.

Cellular automata are well suited to illustrate this reduction of the entropy. It is achieved by introducing the marginal distribution $p_i^{(1)}(x, t)$, a one-body (or one-site) probability defined by summing $P^{(N)}$ over $N - 1$ degrees of freedom

$$p_i^{(1)}(x,t) = \sum_{x_1=0}^{1} \cdots \sum_{x_{i-1}=0}^{1} \sum_{x_{i+1}=0}^{1} \cdots \sum_{x_N=0}^{1} P^{(N)}(x_1, \ldots, x_N, t), \tag{5}$$

where there is no summation over the variable $x_i$. The marginal, or *reduced* distribution $p_i^{(1)}(x, t)$ measures the probability that site $i$ is in state $x$ at time $t$. (We use a single index $i$ to label the sites of the array, irrespective of its dimension). The quantity $p_i^{(1)}(x = 1)$ has a simple physical meaning. It is the local density of the value 1, or the local expected value—formally given by $\sum_{x=0}^{1} x \, p_i^{(1)}(x)$. A reduced site entropy is readily defined as

$$S_i^{(1)}(t) = -\sum_{x=0}^{1} p_i^{(1)}(x,t) \log p_i^{(1)}(x,t), \tag{6}$$

and the total reduced entropy reads

$$S^{(1)}(t) = \sum_i S_i^{(1)}(t). \tag{7}$$

It corresponds (up to the sign) to Boltzmann's H-function (see, e.g., [50]), and obeys (near equilibrium)

$$S^{(1)}(t+1) \geq S^{(1)}(t), \tag{8}$$

in accordance with the Second Law. Remember that the inequality holds only for a reversible rule (the assumption of microreversibilty is crucial in the proof of Boltzmann's H-theorem[50]). Again, inequalities (8) and (3) are in opposite directions, the latter pertains to irreversible systems only.

There is a cellular-automaton experiment that illustrates how (4) and (8) are both obeyed. (To interpret this experiment, we assume *self-averaging*, i.e., we shall consider a specific system, perform spatial averages on it and equate such averages to the ensemble averages used to define $S^{(N)}$ and $S^{(1)}$.) Start with an initial configuration made out of a few 1 cells, perhaps forming some message, in a background of 0 cells. Apply many steps of an appropriate reversible rule. Almost any reversible rule will do, as long it is nonlinear and nonconfining, to allow the growth of our initial pattern. For the purpose of the experiment, asymmetric laws are better. With periodic boundary conditions (as in CAM), the growing pattern will eventually interact with itself, and any trace of the initial message will appear lost. In fact, the subsequent configurations will look random (in particular, the density of 1 cells will remain very close to 1/2.) Indeed, for an ergodic enough transition rule, the system will sample evenly a nonnegligible fraction of the $2^N$ possible configurations, where uniform-looking configurations overwhelmingly dominate. Most of its trajectory will be within the "grey" region of Figure 4.

A cryptologist might view the initial configuration as a *plaintext* that becomes encrypted by evolution, resulting in a *ciphertext*. A physicist might think of Gibbs's example of stirring of a drop of ink in a glass of water. After stirring, the mixture becomes uniformly grey. The original drop cannot be recovered; the process is *for all practical purposes* irreversible. (See, however [46,48,49] for a discussion of spin-echo experiments and the associated apparent paradox.) But in principle one should be able to "unstir" the mixture by exactly reversing the momenta of all the molecules involved—this is Loschmidt's paradox[51]. This would require (i) access to all the degrees of freedom, (ii) total isolation of the system, and (iii) infinite-precision specification (also called "excavation") of the configuration. This feat is out of reach of the physicist, who is equipped with macroscopic instruments, and thus can control few-body correlations only.

The synthetic universes represented by cellular automata are mathematical objects, and thus readily fulfill the above three requirements. Point (iii) emphasizes the importance of discreteness. The problem of the excavation of all the digits that specify a configuration is simply absent: a cell is in state 0 or 1, not in state .999.... A finite amount of information suffices to characterize a finite configuration with infinite precision. As a consequence, Loschmidt's paradox is easily realized: an equal number of steps of the reverse rule recovers the initial configuration. The simplicity of the initial configuration has not vanished, but only diffused into many-body correlations. The very possibility of recovering exactly the initial condition shows that these many-body correlations are not lost; on the contrary, they are readily available. Hence equation (4) is satisfied. On the other hand, the observer's eye cannot keep track of these complex correlations. It automatically averages over them, and follows chiefly the evolution of the density $p_i^{(1)}(1)$ of the 1 cells (the brightness of the monitor screen). The lattice-average density, in our experiment, increases steadily until it reaches its equilibrium value. This relaxation toward equilibrium clearly shows the "arrow of time," in accordance with (8). Notice, however, that the process will not appear irreversible forever. The finiteness of the system forbids an infinite, aperiodic evolution. The absence of trajectory mergers implies that a reversible evolution follow a cycle that contains the initial configuration. At the end of the huge, but finite period (Poincaré's recurrence time), equation (8) will be violated—this is Zermelo's paradox[51].

In the Margolus neighborhood[16], one can readily define rules that conserve the number of ones, which thus can be identified with "particles." Reduction (5) is then an instance of standard reduction to *hydrodynamical* variables—defined in general as densities of conserved quantities. Actually, Margolus and Toffoli have constructed rules which, in the limit of vanishing mesh size, reproduce hydrodynamic equations, very much like the lattice-gas model of Hardy, de Pazzis and Pomeau[52]. A remarkable aspect of these models is that continuously changing quantities are represented by the variations in the density $p_i^{(1)}$ of individual bits[53], rather than by multibit computer words used in finite precision arithmetic.

## 7. Ergodicity-breaking and ultrametricity in Q2R

Q2R is a reversible cellular automaton constructed with Fredkin's prescription[16,4]. The rule involves the four adjacent neighbors and also the *past* value of the considered cell. The center cell will take a time $t + 1$ the value it assumed at time $t - 1$ unless two out its four nearest neighbors are "on" at time $t$, in which case the center cell will take the binary opposite of its past value. In the language of the Ising model (with the spin variables $\sigma_i \in \{+1, -1\}$), we have $\sigma_i^{t+1} = \sigma_i^{t-1}$ unless the local field due to the four Ising neighbors vanishes, in which case $\sigma_i^{t+1} = -\sigma_i^{t-1}$ .

Pomeau[54] has shown that the quantity

$$E^t = \sum_{<i,j>} \sigma_i^t \sigma_j^{t+1} \tag{9}$$

is a constant of the motion, i.e., $E^{t+1} = E^t$. The sum is taken over neighboring pairs only, the invariant has the form of an Ising energy, except that the values of the spins are taken at two successive times. Though similar in form, Pomeau's conserved quantity is distinct from Goles's algebraic invariants[30].

For initial conditions made out of rectangles of ones at $t=0$ and $t=1$, Q2R moves on a surface in state-space characterized by a large number of valleys separated by high potential barriers[4], akin to the deep minima of free energy in a spin-glass. Since the energy is conserved in the reversible Q2R dynamics, the system wanders through the microcanonical ensemble rather than through the canonical ensemble, and one should view the barriers as entropy bottlenecks rather than as genuine energy barriers. The height of the barriers (or the narrowness of the bottlenecks) increases with the number of sites. In the thermodynamic limit, therefore, the system is trapped in the first visited valley, which carries a value of an *order parameter*, just as an Ising system close to $T = 0$ is trapped in one of the two minima of the free energy, labelled by a value $\pm|M|$ of the magnetization, which is conserved exactly in the thermodynamic limit. But unlike the Ising case, the minima in Q2R cannot be related by a symmetry operation, there is *ergodicity-breaking*, a property that characterizes spin-glasses. Ultrametricity[55] is another property relevant to spin-glasses; it has been invoked to interpret Parisi's mean-field theory of these materials. In general, ultrametricity measures the amount of hidden hierarchy. According to recent measurements by G. Sorkin[56], Q2R exhibits an exceptionally large amount of ultrametricity. Prior to these measurements, combinatorial optimization problems (in particular the one-dimensional placement problem and the travelling salesman problem[35]) were believed to be the most ultrametric systems that are not explicitly hierarchical[57].

## Acknowledgements

I am grateful to H. Atlan, C. Bennett, M. Droz, R. Giles, S. Kauffman, W. Klein, N. Margolus, G. Marx, H. Orland, N. Packard, Y. Pomeau, T. Toffoli, and S. Wolfram for fruitful discussions. I wish to thank E. Bienenstock, F. Fogelman, and G. Weisbuch for their hospitality at this stimulating meeting; and C. De Dominicis and G. Toulouse, the organizers of the March '85 CECAM workshop at Orsay, who helped make my trip possible. I also would like to thank J.-P. Meyer and the CMI, Genève, for their logistical support, N. Margolus, T. Toffoli, and D. Zaig for fine-tuning the CAM simulator, and P. Tamayo and D. Zaig for their assistance in the preparation of the figures. This work was supported by grants from DARPA (N0014-83-K-0125), NSF (8214312-IST), and DOE (DE-AC02-83ER13082).

# References

[1] A. W. Burks, *Essays on Cellular Automata* (University of Illinois Press, 1970).

[2] D. Farmer, T. Toffoli, and S. Wolfram (eds), *Cellular Automata* (North-Holland, 1984).

[3] T. Toffoli, *Cellular Automata Mechanics*, Tech. Rep. no. 208, Comp. Comm. Sci. Dept., The University of Michigan (1977).

[4] G. Y. Vichniac, *Physica* 10D (1984) 96–115; reprinted in [2].

[5] B. Hayes, *Scientific American,* 250:3 (1984), 12–21.

[6] M. Creutz, to appear in *Ann. Phys. (N. Y.).*

[7] R. C. Brower, R. Giles, and G. Vichniac, to appear.

[8] G. Enting, *J. Phys. C* 10 (1977) 1379–1388.

[9] E. Domany and W. Kinzel, *Phys. Rev. Lett.* 53 (1984) 311–314.

[10] W. Kinzel, *Z. Phys. B* 58 (1985) 229–244.

[11] A. L. Toom, in *Locally Interacting Systems and their Applications in Biology,* R. L. Dobrushin, V. I. Kryukov, and A. L. Toom (eds), Lecture Notes in Mathematics, no 653 (Spinger-Verlag, 1978); and in *Multicomponent Random Systems,* R. L. Dobrushin (ed), in *Adv. in Probabilities*, Vol. 6 (Dekker, 1980), pp. 549–575.

[12] P. Gacs and J. Reif, in *Proceedings of the Seventeenth ACM Symposium on the Theory of Computing, Providence, R. I., 1985* (ACM, 1985), pp. 388–395.

[13] C. H. Bennett and G. Grinstein, *Phys. Rev. Lett.* 55 (1985) 657–660.

[14] H. Hartman and G. Y. Vichniac, this volume.

[15] S. A. Kauffman, this volume; *Physica* 10D (1984) 145–156; reprinted in [2].

[16] N. H. Margolus, *Physica* 10D (1984) 85–95; reprinted in [2].

[17] T. Toffoli, *Physica* 10D (1984) 195–204; reprinted in [2].

[18] H. M. Goldstine and J. von Neumann, *unpublished* (1946), reprinted in *J. von Neumann's Complete Works*, Vol. 5 (Pergamon, 1961), pp. 1–32.

[19] Proceedings of the Conference on *Physics of Computation, Part II: Computational models of physics*, R. Landauer and T. Toffoli (eds), *Int. J. Theor. Phys.* 21 (1982) no 6–7.

[20] R. P. Feynman, *The Character of Physical Law* (MIT Press, 1967), pp. 57–58.

[21] M. Gardner, *Scientific American* 223:4 (1970) 120–123.

[22] E. R. Berlekamp, J. H. Conway, and R. K. Guy, *Winning Ways*, vol 2. (Academic Press, 1982) Chap. 25; M. Gardner, *Wheels, Life and other Mathematical Amusements* (Freeman, 1982); W. Poundstone, *The Recursive Universe*, (W. Morrow, 1985).

[23] S. Lem, "Non Serviam," in *A Perfect Vacuum: Perfect Reviews of Nonexistent Books* (Harcourt Brace Jovanovich, 1978).

[24] R. Penrose, "Lessons From the Game of Life," Review of W. Poundstone's book[22], *The New York Times*, March 17, 1985, VII, 34:2.

[25] F. Dyson, *Disturbing the Universe* (Harper and Row, 1979).

[26] J. von Neumann, *Theory of Self-Reproducing Automata* (edited and completed by A. W. Burks), (University of Illinois Press, 1966).

[27] E. F. Codd, *Cellular Automata* (Academic Press, 1968).

[28] C. G. Langton, *Physica* 10D (1984) 135–144; reprinted in [2].

[29] J. S. Langer, in *Discover*, Jan. 1984, p. 76.

[30] E. Goles, this volume; and in *Comportement dynamique de réseaux d'automates*, Thèse, Grenoble 1985.

[31] J. Cahn, in *Critical Phenomena in Alloys, Magnets, and Superconductors*, R. E. Mills, E. Ascher and R. J. Jaffee (eds) (McGraw-Hill, 1971).

[32] B. Halperin and P. C. Hohenberg, *Rev. Mod. Phys.* 49 (1977) 435–479.

[33] G. Grest and P. Srolovitz, *Phys. Rev.* B30 (1984) 5150–5155.

[34] P. Grassberger, *Physica* 10D (1984) 52–58; reprinted in [2].

[35] S. Kirkpatrick, C. D. Gelatt Jr. and M. P. Vecchi, *Science* 220 (1983) 671–680; S. Kirkpatrick and G. Toulouse, *J. Physique*, in press.

[36] S. Wolfram, *Physica* 10D (1984) 1–35; reprinted in [2].

[37] N. H. Packard and S. Wolfram, *J. Stat. Phys.* 38 (1985) 901–946.

[38] G. Y. Vichniac, *SIAM J. Alg. Disc. Meth.* 5 (1984) 596–602.

[39] N. H. Margolus, private communication.

[40] F. Fogelman-Soulié, this volume; and *Contribution à une théorie du calcul sur réseaux*, Thèse, Grenoble (1985).

[41] A. E. Gelfand and C.C. Walker, *Ensemble Modeling* (Marcel Dekker, 1984).

[42] N. H. Packard, in *Dynamical Systems and Cellular Automata*, J. Demongeot, E. Goles, and M. Tchuente (eds) (Academic Press, 1985), pp. 123–137.

[43] H. Atlan, F. Fogelman-Soulié, J. Salomon, and G. Weisbuch, *Cybernet. Systems*, 12 (1981) 103–121.

[44] T. Toffoli and N. Margolus, in preparation.

[45] I. Oppenheim and R. D. Levine, *Physica* 99A (1979) 383–402.

[46] J. E. Mayer and M. G. Mayer, *Statistical Mechanics* (2nd ed.) (Wiley, 1977).

[47] R. C. Tolman, *The Principles of Statistical Mechanics* (Oxford University Press, 1938; Dover, 1979).

[48] R. Balian, Y. Alhassid, and H. Reinhardt, to appear in *Physics Reports*.

[49] R. G. Brewer and E. L. Hahn, *Scientific American* 251:6 (1984) 50–57.

[50] K. Huang, *Statistical Mechanics* (Wiley, 1963).

[51] P. C. W. Davies, *The Physics of Time Asymmetry* (University of California Press, 1977).

[52] J. Hardy, O. de Pazzis, and Y. Pomeau, *Phys. Rev. A* 13 (1976) 1949–1961.

[53] T. Toffoli, *Physica* 10D (1984) 117–127; reprinted in [2].

[54] Y. Pomeau, *J. Phys. A* 17 (1984) L415–L418.

[55] See the papers of Mézard, Solla *et al.* and Virasoro, this volume; and M. Mézard, G. Parisi, N. Sourlas, G. Toulouse, and M. Virasoro, *Phys. Rev. Lett.* 52 (1984) 1156–1159.

[56] G. Sorkin, private communication.

[57] S. A. Solla, G. B. Sorkin, and S. R. White, this volume.

# DYNAMICS AND SELF-ORGANIZATION IN ONE-DIMENSIONAL ARRAYS

Maurice TCHUENTE

CNRS-IMAG Laboratoire TIM3
BP 68 38402 Saint Martin d'Hères cédex, France

## 1 INTRODUCTION

A *one-dimensional (1-D for short)* array is a collection of identical finite state machines indexed by integers x of $\mathbb{Z}$, and where any cell x can directly receive informations from its neighbours $x + i$, $i = -n$, ... , n, where n is a positive integer called the scope of the array. Each machine can synchronously change its state at discrete time steps as a function of its state and the states of its neighboring machines.

Such a cell-structure was first introduced by Von Neumann (1966) in order to prove the existence of a machine capable of *self-reproduction*. Since then, this model has beeen applied in a great variety of areas (see Tchuente (1982)) for a detailed bibliography. In this paper, we are interested in two questions.

First of all, we consider a 1-D array as a *discrete dynamical system*, i.e., given a neighborhood structure and a local transition function, we study the global properties of the array by considering the evolution from the set of all possible finite initial configurations.

Secondly, we are interested in a *synthesis* problem. More precisely, given a global behaviour corresponding to a *synchronization* property with *self-repair mechanism*, we show how the cell structure of the individual cells of the array must be designed in order to realize this global property.

## 2 DYNAMICS OF MAJORITY NETWORKS

In this section, we study the dynamical behaviour of uniform one-dimensional *majority* networks, $N = (\mathbb{Z}, V, \{0,1\}, f)$, i.e. binary networks where

$$f(q_{-n}, \ldots, q_{-1}, q_0, q_1, \ldots, q_n) = \begin{cases} 1 & \text{if } \Sigma q_i \geq n+1 \\ 0 & \text{otherwise} \end{cases}$$

In such a network, a cell x assumes a state q at time $t+1$, if and only if, at

NATO ASI Series, Vol. F20
Disordered Systems and Biological Organization
Edited by E. Bienenstock et al.
© Springer-Verlag Berlin Heidelberg 1986

time t, more than half of its neighbours $x+i$, $i = -n, ..., n$ are in state q. As pointed out by Unger (1959), such transformations are used for instance as a smoothing process in *pattern recognition* algorithms: a cell in state 0 whose neighborhood contains a majority of cells in state 1, corresponds to a hole in an otherwise black area, and must be filled; similarly, isolated ones, i.e. cells in state 1 surrounded by cells in state 0 must be eliminated. Such automata may also be used in *numerical simulations* of *physical* phenomena related for instance to *percolation* (see the paper of Vichniac in these proceedings).

Our analysis is based on a block decomposition method which leads to a *macroscopic approach*. More precisely, any finite configuration c is identified with its support (i.e. the cells such that $c(x) = 1$) and is denoted

$$c = U_{1 <= i <= p} [ a_i, b_i ], \text{ where } a_i <= b_i \text{ and } b_i + 1 < a_{i+1} \text{ for any } i.$$

Hereafter, the intervals $[a_i, b_i]$ are called the *component*s of c.

LEMMA 1  If $c = U_{1 <= i <= p} [a_i, b_i]$ and $F(c) = U_{1 <= j <= q} [a'_j, b'_j]$, then

(i) $c(a'_j + n) = 1$ and $c(b'_j + n + 1) = 0$ for any j

(ii) $q <= p$

Proof: We denote $S(x) = \Sigma_{-n <= i <= n} c(x+i)$. Let us first note that, for any x,

$$S(x) - S(x-1) = c(x+n) - c(x-n-1)$$

(i) Since $s(a'_j) >= n+1$ and $S(a'_j - 1) <= n$, it follows that

$$S(a'_j) - S(a'_j - 1) = c(a'_j + n) - c(a'_j - n - 1) >= 1$$

hence

$$c(a'_j + n) = 1 \text{ and } c(a'_j - n - 1) = 0$$

by considering $S(b'_j)$ and $S(b'_j + 1)$ it can be shown in a similar way that

$$c(b'_j + n + 1) = 0.$$

(ii) From property (i) above, it follows that the $a'_j + n$, $j = 1, ..., q$ belong to distinct components of c, hence $q <= p$ □

LEMMA 2  If $c = U_{1 <= i <= p} [a_i, b_i]$, and $F(c) = U_{1 <= i <= q} [a'_i, b'_i]$, then the following conditions are equivalent:

(i) $b_i - a_i >= n$ for $i = 1, ..., p$ and $a_{i+1} - b_i >= n+2$ for $i = 1, ..., p-1$

(ii) $F(c) = c$

(iii) $p = q$

Proof: It is easily verified that (i) implies (ii) and (ii) implies (iii). Let us now show that (iii) implies (i). From lemma 1, $a'_j + n$ belongs to $[a_j, b_j]$ for any j. Taking j = 1 , it follows that

$$a_1 <= a'_1 < a'_1 + n <= b_1 ,$$

hence

$$b_1 - a_1 >= n \text{ and } [a_1 , b_1] \subseteq [a'_1 , b'_1].$$

On the other hand, from the proof of lemma 1,

$$b'_1 + n < a_2$$

hence

$$b_1 <= b'_1 < b'_1 + n < a_2 \text{ and } a_2 - b_1 >= n+2 .$$

The rest of the proof can easily be carried out by induction on i ☐


**PROPOSITION 3** If $c = U_{1 <= i <= p} [a_i , b_i]$ , then $F^p(c) = F^{p+1}(c)$, i.e. any finite configuration with p components evolves in p steps towards a stable configuration.

Proof: The proof is trivial from the preceding lemma, since the number $p_r$ of components of $F^r(c)$ is a decreasing sequence starting from p ☐

**COMMENTS**: This result is quite powerful since it gives a very nice bound for the transient length of the iteration. In Tchuente (1977), this technique has been applied to an automaton of dimension 2 in order to prove that any finite configuration evolves either towards a stable configuration or a cycle of length two. For the case n = 1, it is easily verified that

$$F(c) = c' \text{ and } F(c') = c,$$

where $c(x) = 1$ iff $x = 0$ (modulo 2) and $c'(x) = 1$ iff $x = 1$ (modulo 2.) As a consequence, proposition 3 cannot be extended to infinite configurations.

A theorem of Goles and Tchuente (1984) shows that this result cannot be extended to finite structures where any cell x evolves according to a function $f_x$ defined by $f_x(q_{-n} , ... , q_n) = 1$ iff $b_x <= \Sigma q_i$ , where $b_x$ varies from cell to cell, because, in such a situation, the structure can admit cycles of length two.

So far, we have considered only structures with nearest-neighbour interactions. The next result deals with the case where any cell x receives inputs from x-q, x-p, x, x+p, x+q, with p < q.


**PROPOSITION 4** Let N be a 1-D majority network with neighborhood index V = (-q,-p,0,p,q). If q = 2p then any finite configuration is convergent, otherwise the finite configurations can be partitionned into two classes $C_1 U C_2$ , such that any c of $C_i$, i = 1,2, evolves towards a cycle of length i.

Proof: If q = 2p, then, as illustrated in fig. 1, the structure can be partitionned into two sub-structures corresponding to a majority network with nearest-neighbour interaction, and the result follows from proposition 3.

Fig. 1   p = 2  and  q = 4

For q ≠ 2p, a theorem of Goles and Olivos (1980) ensures that the structure cannot admit cycles of length greater than two. Since the null configuration c (i.e. c(x) = 0 for any x) is trivially stable,  we just need to exhibit a cycle of length two; the general construction is quite lengthy and may be found in Tchuente (1982). In the lines below, we just give a sketch of the construction.

*Step 1 :* consider the infinite configuration c = c'Uc" where

   c'(x) = 1 iff x = -p (modulo q)  and c"(x) = 1 iff x = 2p (modulo q)

It is easily verified that

   F(c') = c' , F(c") = c" , and F(c) = c

*Step 2 :* consider the infinite configurations

   $c_1$ = c U {p,q}   and $c_2$ = c U {0,p+q}

It is easily seen that

   $c_1 \neq c_2$ , $F(c_1) = c_2$ , and $F(c_2) = c_1$ .

*Step 3 :* In this step, we construct from $c_1$ and $c_2$, two finite configurations $c'_1$ and $c'_2$ such that

   $c'_1 \neq c'_2$ , $F(c'_1) = c'_2$ , and $F(c'_2) = c'_1$

This construction is performed by replacing in c' and c", "the infinite legs" as follows (for u = -p , 2p, and for appropriate values of n) :

   {u+vq ; v >= n} is replaced by {u+nq, u+(n+1)q, u+nq+p, u+(n+1)q+p }

   {u-vq ; v >= n} is replaced by {u-nq, u-(n+1)q, u-nq-p, u-(n+1)q-p }

(see the examples below for an illustration).□

**COMMENT** The difference between this result and the theorem of Goles and Olivos (1980) about general symmetric threshold networks, is that we can assert the existence of cycles of length two.

   In the following examples, the states of oscillating cells are darkened.

**EXAMPLE 1** ( p = 1 and q = 3)

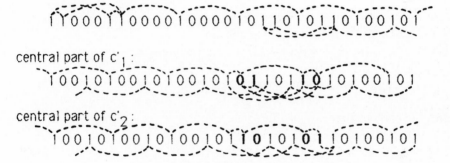

$c'_1$ : 1 1 0 1 1 0 0 1 0 0 1 0 0 1 0 0 1 0 1 1 1 0 1 0 0 1 0 0 1 0 0 1 1 0 1 1

$c'_2$ : 1 1 0 1 1 0 0 1 0 0 1 0 0 1 0 0 1 1 0 1 0 1 0 0 1 0 0 1 0 0 1 1 0 1 1

**EXAMPLE 2** ( p = 2 and q = 3 )

$c'_1$ : 1 0 1 1 0 1 0 0 1 0 0 1 0 0 1 0 0 1 1 1 1 0 0 1 0 0 1 0 0 1 0 1 1 0 1

$c'_2$ : 1 0 1 1 0 1 0 0 1 0 0 1 0 0 1 0 1 1 0 0 1 1 0 1 0 0 1 0 0 1 0 1 1 0 1

**EXAMPLE 3** ( p = 1 and q = 5 )

left end of $c'_1$ and $c'_2$ :

1 1 0 0 0 1 1 0 0 0 0 1 0 0 0 0 1 0 1 1 0 1 0 1 1 0 1 0 0 1 0 1

central part of $c'_1$ :

1 0 0 1 0 1 0 0 1 0 1 0 0 1 0 1 0 1 1 0 1 1 0 1 0 1 0 0 1 0 1

central part of $c'_2$ :

1 0 0 1 0 1 0 0 1 0 1 0 0 1 0 1 1 0 1 0 1 0 1 0 1 1 0 1 0 0 1 0 1

the right end of $c'_1$ and $c'_2$ is similar to the left end.

## 2 THE GAME OF LIFE FOR BOUNDED 1-D STRUCTURES

We are now going to study a class of 1-D uniform bounded stuctures where the local transition function is similar to the game of *LIFE* introduced by Conway (1970) i.e.

$f(q_{-n}, \dots, q_{-1}, q_0, q_1, \dots, q_n) = 1$ if and only if

$(q_0 = 0$ and $n+1 <= \Sigma q_i <= n + k_0)$ or $( q_0 = 1$ and $n+1 <= \Sigma q_i <= n + k_1)$

These rules have a very nice interpretation: state 1 corresponds to a living cell and state 0 corresponds to a dead cell. Transition from state 0 to state 1 is a *birth*; the rule says that a dead cell x becomes alive at time t+1 if and only if at time t, it is not *isolated* i.e.,

$n+1 <= S(x) = c(x-n) + \dots + c(x+n),$

and its neighborhood is not *overpopulated* , i.e.,

$$S(x) = c(x-n) + \ldots + c(x+n) <= n + k_0 .$$

Similar conditions hold for *survival*, i.e. for a cell which is alive at time t and remains alive at time t+1.

With these rules , a finite configuration does not expand during iteration i.e. if

$$c(x) = 0 \text{ for } x < a \text{ or } x > b$$

then the same property holds for F(c). As a consequence, we cannot expect to observe complex objects such as the gliders, exhibited in the game of *LIFE* , Conway (1970). We just intend to show on this simple class of automata, how birth and survival conditions must be chosen in order to yield non trivial evolutions.

**PROPOSITION 5** If $k_0 >= k_1$ , then any finite configuration evolves in a finite number of steps towards a stable configuration $c = U_{1<=i<=p} [a_i , b_i]$ such that $n <= b_i - a_i <= n + k_1 - 1$ and $n + 2 <= a_{i+1} - b_i$ for any i.

Proof: Let us consider a finite configuration

$$c = U_{1<=i<=p} [a_i , b_i] \text{ such that } F^r(c) = c \text{ for some } r >= 1.$$

We denote

$$c = c^0 \text{ and } F^r(c) = c^r = U_{i>=1}[a_i^r , b_i^r] .$$

It is easily verified that

$$a_1^r = a_1 \text{ for any } r >= 0,$$

thus

$$c^r(x) = 1 \text{ for any } r >= 0 , \text{ and } a_1 <= x <= a_1 + n.$$

On the other hand, for any x of $[a_1 , b_1^r]$,

$$S^r(a_1+n)-S(x) = c^r(a_1)+c^r(a_1+1)+ \ldots +c^r(x-1-n) - ( c^r(a_1+1+n)+ \ldots +c^r(x+n))$$

$$= 1 + 1 + \ldots + 1 - ( c^r(a_1+1+n) + \ldots + c^r(x+n))$$

$$>= 0$$

thus

$$n+1 <= S^r(b_1) <= S^r(a_1+n) <= n + k_1 \text{ for any } r >= 0 \text{ and } a_1 <= x <= b_1^r.$$

hence

$$b_1^{r+1} >= b_1^r \text{ for any } r$$

and, consequently,

$$b_1^r = b_1 \text{ for any } r.$$

Since, for any r

$$S^r(b_1) - S^r(b_1+1) = c^r(b_1-1-n) - c^r(b_1+1+n) = 1 - c^r(b_1+1+n) >= 0$$

the situation can be summarized as follows:

$$n <= S^r(b_1+1) <= S^r(b_1) <= n + k_1 <= n + k_0 \text{ and } c^r(b_1) = c^{r+1}(b_1+1) = 0$$

Thus

$$S^r(b_1+1) = n \text{ and } a_2{}^r - b_1 >= n+2 \text{ for any } r$$

It can easily be shown by induction that

(*)   $a_i{}^r = a_i$, $b_i{}^r = b_i$, $n <= b_i - a_i <= n+k_1-1$ and

$$a_{i+1} - b_i >= n+2 \text{ for } r>=0 \text{ and } i>=1$$

Conversely, it is a trivial matter to verify that any finite configuration c which verifies (*), is stable, i.e. F(c) = c.
This ends the proof of the proposition □

**PROPOSITION 6** If $n - k_1 + 1 < k_0 < 2k_1 - n$ then the automaton admits cycles of length two.

Proof  It is sufficient to consider the following configurations:

```
...0 1 1 ... 1 1 0 0 ... 0 0 1 1 ... 1 1 0 ... 0 1 1 ... 1 1 0 0 ... 0 0 1 1 ... 1 1 0 ...
   <------>  <------->  <------>  <---->  <------>  <------->  <------>
    n+1      n-k₁+1      k₁      n-k₀     k₁       n-k₁+1      n+1
```

```
...0 1 1 ... 1 1 0 0 ... 0 0 1 1 ... 1 1 1 ... 1 1 1 ... 1 1 0 0 ... 0 0 1 1 ... 1 1 0 ...
   <------>  <------->  <------>  <---->  <------>  <------->  <------>
    n+1      n-k₁+1      k₁      n-k₀     k₁       n-k₁+1      n+1
```

**COMMENTS**    These results show that, among the class of automata studied here, those which exhibit the more complex behaviour are those where *conditions for birth are more restrictive than those for survival.* This observation is in accordance with the rules of the game of *LIFE,* and are probably a sufficient condition for the generation of complex behaviours in structures of arbitrary dimension.

For $k_0 >= k_1$, as shown in proposition 5, the stable configurations have a very simple structure. On the contrary, it is shown in Tchuente (1982), that the set of stable configurations is much more complex when $k_0 < k_1$ .

This observation illustrates the fact that, as in the classical studies on continuous iterations, the structure of fixed points can have a great influence on the behaviour of iterative sequences. An approach based on boolean distance has been developped by Robert (1985) in order to exploit systematically the analogy between continous and discrete iterations.

# 4 SELF-STABILIZING NETWORKS

Since the pioneering work of Von Neumann (1966) on *SELF-REPRODUCING* automata, it is well known that networks of finite-state machines, with simple local interactions can exhibit very complex global behaviours. The most famous synchronization problem in cellular structures is the so-called *FIRING SQUAD SYNCHRONIZATION PROBLEM*. This problem first arose in connection with causing all parts of a self-reproducing machine to be turned on simultaneously, and can be stated as follows.

Consider a linearly connected set of identical finite-state machines. One of the end machines is called the *General* and the other machines of the array are called *soldiers*. The system is synchronous, i.e. at time t+1, every machine of the array numbered i enters a new state $q(i,t+1)$ which is a function $f(q(i-1,t), q(i,t), q(i+1,t))$ of its state and the states of its two immediate neighbours at time t. The problem is to define the set of states and the local transition function of the system in such a way that the General can cause all the soldiers of the array to go into one particular terminal state (i.e. they fire their guns) simultaneously. At time t = 0, all the soldiers are assumed to be in the same quiescent state. When the General undergoes the transition into the state labeled *"fire when ready"* he does not take any initiative afterwards. The signal can propagate down the line no faster than one soldier per unit of time, and the problem is how to get all coordinated in rythm. The difficulty of the problem follows from the fact that the same kind of soldier with a fixed number of states is required to be able to do this, regardless of the length n of the firing squad. In particular, the soldier with m states must work correctly even when the size n of the network is much bigger than m. Two soldiers, the General and the farthest from the General, are allowed to be slightly different from the other soldiers in being able to act without having soldiers on both sides of them, but their structure must also be independant of n.

This synchronization problem was solved by Waksman (1966) and Balzer (1967) for the case of linearly connected arrays. Subsequently, Rosenstiehl (1966), and Romani (1976) have studied the extension to arbitrary networks.

In this section we are interested in *SELF-STABILIZATION*, a dual problem introduced by Dijkstra (1974) and which can be stated as follows -For each machine of the array, one or more *privileges* are defined i.e. boolean functions of its own state and the states of its neighbours
- the machines are activated in discrete time steps t = 1, 2, ... . An arbitrary number of machines may be activated at time t, and as t goes to infinity, any machine is activated an infinite number of times (hereafter we say that such a sequence of activations is regular).

- when a machine with a privilege is activated, it performs a *move* in order to transmit the privilege to one of its neighbors.
- In each *legitimate state*, exactly one privilege is present, and each possible move will bring the system again in a legitimate state.
- for any initial state, and for any regular sequence of activations, the system ultimately reaches a legitimate state.

These conditions can be summarized by saying that the system exhibits a *self-repair* behaviour: indeed, when once in an illegitimate state, it cannot remain so forever, since it ultimately evolves towards a legitimate state. Dijkstra has proposed three very nice solutions to this problem, but his algorithms need quite complicated proofs.

We are now going to show that a recurrent approach can lead very easily to a variety of interesting solutions.(see Tchuente (1979) for more details).

In the following examples, the solutions are given by exhibiting the complete transition graph of the network. For sake of clarity, the state of the privileged automaton is darkened.

**EXAMPLE 1 : n = 2**      $P_1$—$P_2$

$$
\begin{array}{ccc}
0\,0 & \to & 1\ 0 \\
\uparrow & & \downarrow \\
0\ 1 & \leftarrow & 1\ 1
\end{array}
$$

**EXAMPLE 2 : n = 3**      (linearly connected network)

$P_1$—$P_2$—$P_3$

$$
\begin{array}{l}
0\,0\,0 \to 1\,0\,0 \to 1\ 1\ 0 \to 1\ 1\ 1 \\
\ \uparrow \qquad\qquad\qquad\qquad\qquad \downarrow \\
0\ 1\ 0 \leftarrow 0\ 1\ 1 \leftarrow 0\,0\ 1 \leftarrow 1\ 0\ 1
\end{array}
$$

**EXAMPLE 3 : n = 4**      (ring-connected network)

$$
\begin{array}{cc}
P_1 - P_2 \\
|\qquad\ | \\
P_4 - P_3
\end{array}
$$

$$
\begin{array}{l}
0\,0\,0\,0 \to 1\ 0\ 0\ 0 \to 1\ 1\ 0\ 0 \to 1\ 1\ 1\ 0 \quad 0\ 1\ 0\ 1 \to 1\ 1\ 0\ 1 \to 1\ 0\ 0\ 1 \to 1\ 0\ 1\ 1 \\
\ \uparrow \qquad\qquad\qquad\qquad\quad \downarrow \qquad\quad \uparrow \qquad\qquad\qquad\qquad\quad \downarrow \\
0\,0\,0\ 1 \leftarrow 0\,0\ 1\ 1 \leftarrow 0\ 1\ 1\ 1 \leftarrow 1\ 1\ 1\ 1 \quad 0\ 1\ 0\ 0 \leftarrow 0\ 1\ 1\ 0 \leftarrow 0\,0\ 1\ 0 \leftarrow 1\ 0\ 1\ 0
\end{array}
$$

The recurrent algorithm is based on the following result:

**THEOREM 7** Let us consider two self-stabilizing networks with a unique common automaton A (hereafter called the central cell). We assume that A is boolean in both algorithms. It is possible to design for the union of these networks, the following self-stabilizing algorithm (the original networks are called left and right branches) :
- the state of A is a couple (q,q') of {0,1}x{L,R}, where L and R stand respectively for left and right. A is privileged in the whole network if one of the following conditions hold:
  - it is privileged in the left branch with respect to q, and q' = L; the corresponding move is q' := R.
  - it is privileged in the right branch with respect to q and q' = R; the correspondirg move is q' := L.
  - it is privileged in both branches with respect to q. the corresponding move changes q and q' :
    if q = 0 then q := 1 else q := 0 ; if q' = L then q' := R else q' := L;
- the condition for the privilege of a cell of the left branch which is a neighbour of A, is strengthened by adding the extra condition [ q' = L]; an analogous property holds for the cells of the right branch.
- for all other automata, the condition for privilege is unchanged.

<u>Proof</u> If q' = L, then the left branch evolves independently of the right branch; as a consequence, the self-stabilization of the left branch guarantees that A will later be privileged in the left branch; since the corresponding move is q' := R, we can now apply the same argument to the right branch. This shows that, in the algorithm associated with the whole network, the central automaton alternatively sends the privilege to the left and to the right branch. Since, from the hypothesis, the two branches are self-stabilizing, the system will evolve towards a configuration where there is at most one privilege in each branch.

If this privilege is unique, then it will go alternatively to the left and to the right branch, and the algorithm is valid; if there is exactly one privilege in each branch, then these privileges will ultimately "collide" in the central automaton and, from that moment A will send a single privilege alternatively to the left and to the right.

This ends the proof of the theorem. For more details as well as for some extensions the interested reader may refer to Tchuente (1979). □

## COROLLARY 8

(i)   For any positive integer $N = 2n + 3$, $n >= 0$, there exists a linearly-connected self-stabilizing network based on the recurrent approach, and where n automata are binary and the others have four possible states.

(ii) For any positive integer $N = 3n + 1$, $n >= 1$, there exists a recurrent self-stabilizing algorithm associated with a network of size N, and where there are exactly $2n + 2$ binary automata, and $n - 1$ automata with four possible states.

<u>Proof</u>   These algorithms are obtained from the solutions of examples 2 and 3 above by applying the construction of theorem 7 (see the fig. 4).

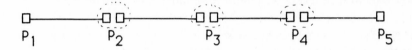

Fig. 2   a linear network of size 5

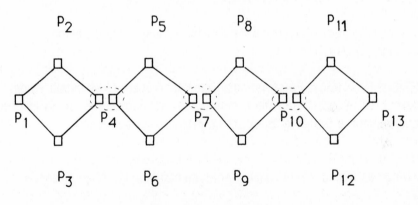

Fig. 3   n = 4,   N = 13

**COMMENT**   The recurrent approach appears to be a very powerful tool; indeed the solution presented here for linear networks needs less states than that of Dijkstra (1974), and on the other hand, as shown in Tchuente (1979) it stabilizes two times faster.

## 5 REFERENCES

BALZER R. (1967): An 8-state minimal time solution to the Firing Squad Synchronization Problem. Information and Control, **10**, pp 22-42

CONWAY J. (1970): Mathematical games, Scientific American, pp 120-123 (paper written by M. Gardner)

DIJKSTRA E.W. (1974): Self-stabilizing systems in spite of distributed control, Communications of the ACM, **(17), 11,** pp 643-644

FOGELMAN SOULIE F. (1985) Contribution à une théorie du calcul sur réseaux, Thèse d'état, USMG-INPG, Grenoble

GOLES E. and OLIVOS J. (1980): Comportement itératif des fonctions à multiseuil, Information and Control, **(45), (3),** pp 300-313

GOLES E. and TCHUENTE M. (1984): Iterative behaviour of one-dimensional threshold automata, Discrete Applied Mathematics, **8**, pp 319-322

GOLES E. (1985): Comportement dynamique de réseaux d'automates, Thèse d'état, USMG-INPG, Grenoble.

ROBERT F. (1985): Discrete Iterations, Academic Press, to appear.

ROMANI F. (1976): Cellular automata synchronization, Information Sciences, **10**, pp 299-318

ROSENSTIEHL P. (1966): Existence d'automates d'états finis capables de s'accorder bien qu'arbitrairement connectés et nombreux, International Computation Centre Bulletin, **5**, pp 215-244

TCHUENTE M. (1977): Evolution de certains automates cellulaires uniformes binaires à seuil, Séminaire d'Analyse numérique, **265** , Grenoble

TCHUENTE M. (1981): Sur l'auto-stabilisation dans un réseau d'ordinateurs, RAIRO Theoretical Computer Science, **(15), 1,** pp 47-66

TCHUENTE M. (1982): Contribution à l'étude des méthodes de calcul pour des systèmes de type coopératif, Thèse d'état, USMG-INPG, Grenoble

UNGER S.H. (1959): Pattern detection and recognition, Proc. IRE, pp 1737-1752

VON NEUMANN J. (1966): Theory of Self-Reproducing Automata, A.W. Burks ed., University of Illinois Press.

WAKSMAN A. (1966): An optimum solution to the firing squad synchronization problem, Information and Control, **9**, pp 66-78

# BASIC RESULTS FOR THE BEHAVIOUR OF DISCRETE ITERATIONS.

François ROBERT

Université de Grenoble
IMAG / INPG
BP 68   F-38402 St Martin d'Hères

## 1.Introduction

There has been an improved effort for about 20 years in studying the **dynamical behaviour of discrete, iterative systems**. The reason for this is probably that different (but conceptually similar) **discrete models** are presently of interest in various domains of science, such as : **Physics** (spin glass problems, see (10),(12),(14), for example), **Chemistry** (diffusion reactions (9) ), **Biomathematics** (neural networks, genetic nets, (10), (11), (14), (18), (20)), **Computer Science** (pattern recognition, associative memories (7), (10), cellular automata (1), (2), (9), (19), (21), (22), (23), cellular arrays for systolic computation in V.L.S.I. systems (13), (15), (17), (19) and so on : see especially (4), (5), (6)).

In these discrete, iterative models, **there is a need for new mathematical tools**, allowing some coherent analysis of the behaviour. **Simulations** have been used for a long time ( (7), (9), (10), (11), (12), (22), (23)...), providing interesting informations. **New theorems**, also, are needed, and seeked for.

Recent results ( (3), (7), (8), (19), (22), (23) ) show interesting progress in some cases, for example when a notion of **energy** can be attached to the system, by analogy with classical physics. This is the case, indeed, for "threshold automata" (see also the papers by F.FOGELMAN and E.GOLES in this volume).

In what follows, I wish to show, at an elementary level, how well-known behavioural results from the domain of **continuous iterations** can be adapted, up to a certain extent, to the context of **discrete iterations**.

We will make use mainly of a **discrete metric tool** (a boolean vectorial distance) coupled with **a notion of discrete derivative** (analogous to a Jacobian matrix ).

These tools allow basic results to be established for discrete iterations (see (16) for an extended analysis ). Here we will examine only some of them.

These results, in turn, have been applied in different contexts : In **boolean networks**, where the discrete derivative can be used for studying **forcing notions** (see (7) ) and in **threshold automata**, where **attractive**

**fixed points** are seeked for in designing **associative memories** (7).

## 2.Basic notions

In what follows, we always consider the n-cube $C_n = \{0,1\}^n$ as our basic space. In examples, $C_n$ is the set of all possible **configurations** of zeros and ones (or $\cdot$ and $*$) at the **nodes** (often called **cells**) of a (k , k) grid (with $n=k^2$) :

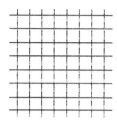

$$n = k^2$$

Of course, card ( $C_n$ ) = $2^n = 2^{k*k}$, which is huge, even for k small ( k=5 , $2^{25}$= 33 554 432 ; k=10 , $2^{100} * 10^{30}$ ).

An element x of $C_n$ is then an n-tuple x = $(x_1, \ldots, x_n)$ of zeros and ones. Given such an x, the **immediate neighbourhood** (or **first neighbourhood** ) of x , denoted by $V_1(x)$, is defined as

$$V_1(x) = (x, \tilde{x}^1, \ldots, \tilde{x}^n)$$

where $\tilde{x}^i = (x_1, \ldots, \overline{x}_i, \ldots, x_n)$ is a **first neighbour** of x (i= 1, 2, .., n), with the convention that $\overline{0} = 1$, and $\overline{1} = 0$.

Notice that, given a configuration x at the nodes of the grid, a first neighbour of x is nothing else than x itself where **at most one cell** has been changed ( $0 \rightarrow 1$ or $1 \rightarrow 0$ )

**Example** ( k = 3 )

```
  *   ·   ·              *   ·   ·
  ·   *   *              ·   *   ·
  *   *   ·              *   *   ·
```
Configuration x           A first neighbour of x

Of course it is possible to define wider neighbourhoods, such as, for example, the second, the third, or ...the i[th] neighbourhood of an element x of $C_n$.

Now, we will consider a given, fixed mapping, say $F$, from $C_n$ into itself.

The problem is (remember that Card( $C_n$ ) is huge ) to say something about the behaviour of the following discrete iteration in $C_n$:

$$x^0 \text{ given in } C^n.$$
$$x^{r+1} = F(x^r) \quad (r = 0, 1, \ldots)$$

that we detail in

$$x^0 = (x_1^0, x_2^0, \ldots, x_n^0) \text{ given in } C^n.$$
(P) $\quad x_i^{r+1} = f_i(x_1^r, x_2^r, \ldots, x_n^r) \quad (i = 1, 2, \ldots, n).$

For obvious reasons, the functions $f_i$ ( from $C_n$ into $C_1 = \{0, 1\}$ ) are sometimes referred to as the **local transition functions**, whereas $F$ is known as the **global transition function**.

**(P)**, in fact, is an elementary model for the **parallel processing** of an automata network defined by the given operator $F$. A **serial mode of operation** for this automaton would be

$$y^0 = (y_1^0, y_2^0, \ldots, y_n^0) \text{ given in } C_n.$$
(S) $\quad y_i^{r+1} = f_i(y_1^{r+1}, \ldots, y_{i-1}^{r+1}, y_i^r, \ldots, y_n^r) \quad (i = 1, 2, \ldots, n).$

This **serial** mode of operation on $F$ corresponds exactly to a **parallel** mode of operation on an operator $G = (g_i)$ defined by

$$g_1(y) = f_1(y)$$
$$g_i(y) = f_i(g_1(y), \ldots, g_{i-1}(y), y_i, \ldots, y_n)$$
$$(i = 1, 2, \ldots, n).$$

Here, $G$ stands for **Gauss-Seidel** (operator associated to $F$).

Of course, such a serial mode of operation is defined up to a permutation of the indexes $i$ (there are $(n-1)!$ different serial modes for one parallel mode of operation). More general kinds of serial, or serial-parallel, or even chaotic (random) modes of operation can be also considered ( see (7), (16) ).

As $C_n$, though large, is finite, all these iterative processes finally end up into a **cycle**, or even a **fixed-point** ( that is, a **stable configuration** ) after generally some **transitory phase**. The problem is now to say something interesting about these cycles, fixed-points, or about the **transitory length**.

Notice that the fixed-points remain unchanged when passing from any mode of operation to another: they are simply the fixed-points $(v = F (v))$ of the given operator F on $C_n$. However, **the cycles generally differ** from one mode of operation to another (see Example 1 below).

In current examples, the local transition functions we consider are actually **local**, that is **each $f_i$ depends only on some variables $x_j$**. Typical examples are the well-known Von Neumann and Moore neighbourhoods on the ( k , k ) grid :

```
          *                    *   *   *
      *   o   *                *   o   *
          *                    *   *   *
```

| **Von Neumann** | **Moore** |
|---|---|
| $f_i$ depends only on | $f_i$ depends only on $x_{i-k-1}, x_{i-k}, x_{i-k+1}$ |
| $x_{i-k}, x_{i-1}, x_i, x_{i+1}, x_{i+k}$ | $x_{i-1}, x_i, x_{i+1}, x_{i+k-1}, x_{i+k}, x_{i+k+1}$ |

Indeed, the dependance of the $f_i$ on the $x_j$ can be visualized equivalently

- either by a **connectivity graph** (of n vertices $P_i$ ) where an arc joins $P_j$ to $P_i$ if $f_i$ really depends on $x_j$ .

- or by an **incidence matrix** : that is an (n x n) boolean matrix $B(F) = (b_{ij})$ with

$$b_{ij} = 1 \quad \text{if there is an arc from } P_j \text{ to } P_i$$
$$b_{ij} = 0 \quad \text{otherwise.}$$

Notice that the **connectivity graph** for F has n vertices whereas the **iteration graph** for F (that is the graph drawn on $C_n$ by joining any x to F(x) ) has $2^n$.

**Example 1**  n = 3 , with

$$F ( x ) = \begin{array}{l} f_1( x_1 , x_2 , x_3 ) = x_1.x_2 + \overline{x_3} \\ f_2(x_1 , \quad , x_3 ) = x_1.x_3 \\ f_3( \quad , x_2, \quad ) = \overline{x_2} \end{array} \qquad \text{(boolean notations)}$$

The incidence matrix for F and its connectivity graph are the following :

$$B(F) = \begin{matrix} 1 & 1 & 1 \\ 1 & 0 & 1 \\ 0 & 1 & 0 \end{matrix}$$

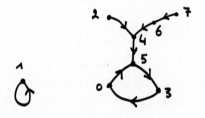

The tabulation for F and its iteration graph are

| $x$ | | $F(x)$ | |
|---|---|---|---|
| 0 | 0 0 0 | 1 0 1 | 5 |
| 1 | 0 0 1 | 0 0 1 | 1 |
| 2 | 0 1 0 | 1 0 0 | 4 |
| 3 | 0 1 1 | 0 0 0 | 0 |
| 4 | 1 0 0 | 1 0 1 | 5 |
| 5 | 1 0 1 | 0 1 1 | 3 |
| 6 | 1 1 0 | 1 0 0 | 4 |
| 7 | 1 1 1 | 1 1 0 | 6 |

This iteration graph can be redrawn as

There is one fixed point and one cycle.

Of course, this corresponds to the **parallel** mode of operation on F.

The Gauss–Seidel operator (associated to the basic **serial** mode of operation on F) is then defined by

$$G(x) = \begin{array}{l} g_1(x_1, x_2, x_3) = f_1(x_1, x_2, x_3) = x_1.x_2 + \overline{x_3} \\ g_2(x_1, x_2, x_3) = (x_1.x_2 + \overline{x_3}).x_3 = x_1.x_2.x_3 \\ g_3(x_1, x_2, x_3) = \overline{x_1.x_2.x_3} = \overline{x_1} + \overline{x_2} + \overline{x_3} \end{array}$$

The incidence matrix for G and its connectivity graph are

$$B(G) = \begin{matrix} 1 & 1 & 1 \\ 1 & 1 & 1 \\ 1 & 1 & 1 \end{matrix}$$

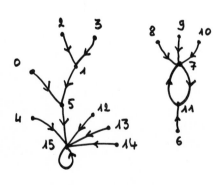

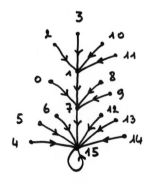

The tabulation for G and its iteration graph are

| $x$ | | | $G(x)$ | | | |
|---|---|---|---|---|---|---|
| 0 | 0 0 0 | | 1 0 1 | | | 5 |
| 1 | 0 0 1 | | 0 0 1 | | | 1 |
| 2 | 0 1 0 | | 1 0 1 | | | 5 |
| 3 | 0 1 1 | | 0 0 1 | | | 1 |
| 4 | 1 0 0 | | 1 0 1 | | | 5 |
| 5 | 1 0 1 | | 0 0 1 | | | 1 |
| 6 | 1 1 0 | | 1 0 1 | | | 5 |
| 7 | 1 1 1 | | 1 1 0 | | | 6 |

One verifies in passing that G does have the same fixed point as F. However G has **no cycle**.

In the general case, the problem of comparing the shapes of the iteration graphs for F and G is an interesting , but difficult question (see(16)).

Here is just an example

**Example 2** n = 4

$$f_1( x ) = x_2 \qquad\qquad g_1( x ) = x_2$$
$$f_2( x ) = x_1 + \overline{x_3} \quad \text{from which} \quad g_2( x ) = x_2 + \overline{x_3}$$
$$f_3( x ) = x_1 + x_2 \quad \text{we get :} \quad g_3( x ) = x_2 + \overline{x_3} \quad \text{(boolean notations)}$$
$$f_4( x ) = 1 \qquad\qquad g_4( x ) = 1$$

The iteration graphs for F and for G are now shown below

iteration graph for F

iteration graph for G

These two graphs actually differ, but they do have however some common features. 15 is a common fixed point for F and G. There is one cycle in the iteration graph for F, and no cycle in the iteration graph for G.

## 3. A Boolean vectorial distance on $C_n$

We will make use of the following boolean vector distance d on $C_n$

$$x = (x_1, \ldots, x_n)$$

$$y = (y_1, \ldots, y_n)$$

$$\longrightarrow \quad d(x, y) = \begin{vmatrix} \vdots \\ d_i(x_i, y_i) \\ \vdots \end{vmatrix}$$

where $d_i(x_i, y_i) = 0$ if $x_i = y_i$, 1 otherwise.

Notice that $d(x,y)$ is to be considered as a boolean vector and that we get the usual axioms for the distance d

$$d(x, y) = 0 \text{ iff } x = y$$
$$d(x, y) = d(y, x)$$
$$d(x, y) + d(y, z) \quad \geqslant \quad d(x,z)$$

      ↑            ↑

    Boolean sum     componentwise in $(0,1)^n = C_n$

    ( 1 + 1 = 1 )      ( $0 \leqslant 0 \leqslant 1 \leqslant 1$ )

Equiped with this topological tool, $C_n$ is complete and the topology is discrete; the converging sequences are the (endly) stationnary sequences. Notice that

$$\forall x \in C_n, \quad d(x, \widetilde{x}^i) = e_i = (0, \ldots, 1, \ldots, 0)^t$$

## 4. The discrete derivative

The **discrete derivative** of F at a point x in $C_n$ is the boolean matrix $\quad F'(x) = (f_{ij}(x))$ defined by

$$f_{ij}(x) = 1 \text{ if } f_i(x) \neq f_i(\widetilde{x}^j)$$
[ that is if $f_i(x_1, \ldots, x_n) \neq f_i(x_1, \ldots, \overline{x}_j, \ldots, x_n)$ ]
$$f_{ij}(x) = 0 \text{ otherwise.}$$

This derivative will play the role of the **Jacobian** of F. Computing F'(x) needs $n^2 + n$ evaluations of functions $f_j$. In Example 1 above we get, for instance

$$F'(0) = \begin{matrix} 0 & 0 & 1 \\ 0 & 0 & 0 \\ 0 & 1 & 0 \end{matrix} \quad \text{and} \quad F'(7) = \begin{matrix} 1 & 1 & 0 \\ 1 & 0 & 1 \\ 0 & 1 & 0 \end{matrix}$$

There are some basic and interesting relations between the discrete derivative F'(x), the boolean vector distance d and the incidence matrix B(F) of F

**Proposition 1** (cf (16))

1. $\forall\ x \in C_n$,  $\quad F'(x) \ \leqslant\ B(F)$

2. $\mathbf{Sup}\ \{F'(x)\} = B(F)$
   $\quad x \in C_n$

3. $\forall\ x \in C_n$, $\forall\ j$, $\quad d(F(x), F(\tilde{x}^j))) = F'(x) \cdot \underbrace{d(x, \tilde{x}^j)}_{\substack{e_j}}$

$$j^{th}\text{ column of } F'(x)$$

Point 3 shows that for any x and one of its first neighbours, the distance $d(F(x), F(\tilde{x}^j))$ can be evaluated exactly. When y is no longer a first neighbour of x, $d(F(x), F(y))$ can only be bounded:

**Proposition 2** (cf (16))

$\forall\ x, y \in C_n$, $d(F(x), F(y)) \ \leqslant\ \mathbf{Sup}\{F'(z)\} \cdot d(x, y)$
$\qquad\qquad\qquad\qquad\qquad z \in [x, y]$

for an (arbitrary) **chain** $[x, y]$ **of minimal length** on the n-cube $C_n$.

Moreower we get a basic inequality

**Proposition 3** (cf (16))

$\forall\ x, y \in C_n$, $d(F(x), F(y)) \ \leqslant\ B(F) \cdot d(x, y)$

which can be obtained from Proposition 1 (point 2) and Proposition 2 – or either proved directly.

## 5. Application : local convergence in the first neighbourhood of a fixed point (cf(16)).

Let $\S = F(\S)$ be a **fixed point** of F in $C_n$ (that is a **stable configuration** for F ). $\S$ will be said **attractive** ( in its first neighbourhood $V_1(\S)$ ) if

1) $F(V_1(\S)) \subset V_1(\S)$

2) for any $x_0$ taken from $V_1(\S)$, the iteration
$$x^{r+1} = F(x^r)$$
(which remains in $V_1(\S)$, according to 1 )

reaches $\S$ in at most n steps. Then $x^n = F^n(x^0) = \S$

This notion of attraction can be characterized by using the tools we introduced above:

## Proposition 4
In order that $\S$ will be attractive in its first neighbourhood, it is **necessary and sufficient** that
1) $F'(\S)$ has at most one 1 per column.
2) The boolean spectral radius of $F'(\S)$ is zero; that is, there exists a permutation matrix P such that the matrix $P^t F'(\S) P$ is **strictly** lower triangular.

## Example 3
n=3 ; $\S = (1\ 1\ 0)$ and

$$F'(\S) = \begin{array}{ccc} 0 & 0 & 0 \\ 0 & 0 & 1 \\ 1 & 0 & 0 \end{array}$$

$F'(\S)$ satisfies the required conditions. One verifies that we necessarily have

Indeed $\S$ is attractive in its first neighbourhood $V_1(\S)$.

This notion of attractive fixed point is meaningful in the context of automata networks. In fact, it means that **a stable configuration would attract its first neigbours in at most n steps, by passing only through 1$^{rst}$ neighbours of S in the meantime.**

## 6. Massive neighbourhoods

It is possible to define **attraction** in wider subsets than the 1$^{rst}$ neigbourhood of a fixed point  S  of F. **Characterizations** of these notions become quickly  cumbersome. However one can give simple **sufficient** conditions of attraction in a **massive** neighbourhood of a fixed point.

**Definition** A  **massive**  neighbourhood  V  of an element x of $C_n$ is by definition a  subset  of  $C_n$  containing  x , and such that:

If  u ∈ V , then every  y ∈ $C_n$   such that d(x , y) ≤ d(x , u) is also in V.

Then we have :

**Proposition 5**
Let V be a massive neighbourhood of a fixed point S of F and let
$$W = \{ d(S , x) ; x ∈ V \}$$
the associated (massive) neighbourhood of 0 in $C_n$.
Furthermore, let      M = **Sup** F'(z)
                                  z ∈ V
If we have                M . W ⊂ W
                                  $\rho$ ( M ) = 0    (boolean spectral radius)
then  S is attractive in  V .

**Example 4**   n=3 ;  S= (0,0,0) = 0
V = W = {0 , 1 , 2 , 4 , 5}   in a massive neighbourhood of 0 :

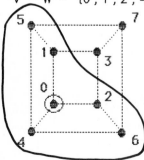

with
$$f_1(x) = x_2 . (\bar{x}_1 + x_1 . x_3)$$

$$f_2(x) = 0$$

$$f_3(x) = x_1 + x_2$$

we get
  1) $F(0) = 0$ : $0$ is a fixed point for F.

$$2) M = \textbf{Sup} \{F'(z)\} = \begin{matrix} 0 & 1 & 0 \\ 0 & 0 & 0 \\ 1 & 1 & 0 \end{matrix}$$
$$z \in W$$

We now have
$$\begin{cases} M \cdot W = \{0, 1, 5\} \subset W \\ \rho\, (M) = 0 \end{cases}$$

We may easily verify the local convergence towards $0$ in the massive neighbourhood $W = \{0, 1, 2, 4, 5\}$ of $0$ since the iteration graph of F is the following :

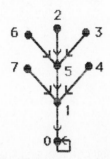

( black dots denote the elements of W ).

## 7. Attractive cycles cf (16)
Let $(\mathbb{S}_1, \mathbb{S}_2, \mathbb{S}_p)$ be **a cycle of length p** for F , that is
$F(\mathbb{S}_1) = \mathbb{S}_2$, $F(\mathbb{S}_2) = \mathbb{S}_3, \ldots F(\mathbb{S}_p) = \mathbb{S}_1$ with $\mathbb{S}_i$ all different.

The notion of attractive fixed point can be extended to a cycle ; we only quote a basic result.

**Proposition 6** In order that the cycle $(\mathbb{S}_1,\ldots, \mathbb{S}_p)$ of F will be attractive in its first neighbourhood, it is necessary and sufficient that:

  1) $F'(\mathbb{S}_i)$ has at most one 1 per column $(i=1,2,\ldots,p)$
  2) the boolean spectral radius of the matrix
  $\qquad F'(\mathbb{S}_p)\ldots F'(\mathbb{S}_1)$ is equal to zero

Here is an example

**Example 5**   n=3 ; F is defined by its table

| $x$ | | | $F(x)$ | | | |
|---|---|---|---|---|---|---|
| 0 | 0 0 0 | | 0 1 0 | | 2 |
| 1 | 0 0 1 | | 1 1 0 | | 6 |
| 2 | 0 1 0 | | 0 1 1 | | 3 |
| 3 | 0 1 1 | | 1 0 0 | | 4 |
| 4 | 1 0 0 | | 0 0 1 | | 1 |
| 5 | 1 0 1 | | 1 1 1 | | 7 |
| 6 | 1 1 0 | | 0 0 1 | | 1 |
| 7 | 1 1 1 | | 0 0 0 | | 0 |

$V_1(1)$          $V_1(6)$

F admits an attractive cycle of length 2 : $\{1 , 6\}$

Indeed one verifies that

$$F'(1) = \begin{matrix} 0 & 0 & 1 \\ 0 & 1 & 0 \\ 1 & 0 & 0 \end{matrix} \quad \text{and} \quad F'(6) = \begin{matrix} 0 & 0 & 0 \\ 1 & 0 & 0 \\ 0 & 0 & 1 \end{matrix}$$

both have at most one 1 per column, and that

$$F'(1).F'(6) = \begin{matrix} 0 & 0 & 1 \\ 1 & 0 & 0 \\ 0 & 0 & 0 \end{matrix} \quad \text{and} \quad F'(6).F'(1) = \begin{matrix} 0 & 0 & 0 \\ 0 & 0 & 1 \\ 1 & 0 & 0 \end{matrix}$$

both have a boolean spectral radius zero.

## 8. Conclusions

We tried to show in this paper how it is possible to study **elementary but basic behaviours of discrete iterations**, with tools such as **a vectorial boolean distance** and **the use of a discrete derivative** ( more can be said : see (16) for a contraction theory, for comparisons between different iteration graphs, for a Newton method in $C_n$ and so on ).

Of course, it is not claimed here that these tools are the only ones to be used : Other tools, such as **invariants**,     or

**energy/entropy** notions, for example, revealed to be powerful (see(7),(8) and also, more generally , (3),(4),(5),(6),(19),(22),(23), and the bibliography of (16) ).

As a conclusion, we would simply focus again on the fact that iterative properties of discrete dynamical systems and automata nets constitute an interesting research area, where **new tools** and **new ways of understanding** need actually to be elaborated.

# 9. References

(1) A.W.BURKS :    Essays on cellular Automata., University of Illinois Press (1970).

(2) E.F. CODD :    Cellular Automata., Academic Press (1968).

(3) M. COSNARD, E. GOLES : Dynamique d'un automate à mémoire modélisant le fonctionnement d'un neurone., C.R.A.S. , t. 299, I N$^0$ 10, (1984).    (see also the papers by M. COSNARD  and  E. GOLES in the present volume )

(4) M. COSNARD, J. DEMONGEOT, A. LEBRETON, Editors : Rythms in biology and other fields of Applications. Springer Verlag, Lecture notes in Mathematics n°49 (1983).

(5) J. DELLA DORA , J. DEMONGEOT, B. LACOLLE , Editors : Numerical methods in study of critical phenomena., Springer Verlag (1981).

(6) J. DEMONGEOT, E. GOLES,  M.TCHUENTE, Editors : Dynamic behaviour of automata networks, Academic Press (1984).

(7) F. FOGELMAN :    Contributions à une théorie du calcul sur réseaux. Thesis, Grenoble 1985.( see also the paper by F. FOGELMAN in the present volume. )

(8)  E. GOLES : Comportement dynamique de réseaux d'automates. Thesis, Grenoble 1985. ( see also the paper by E. GOLES in the present volume. )

(9) J.M. GREENBERG, B.D. HASSARD, S.P. HASTINGS: Pattern formation and periodic structures in systems modelled by reaction diffusion equations. Bull . Am. Math. Soc. 84. 6, p. 1296-1327 (1978).

(10) J. J. HOPFIELD : Neural networks and physical systems with emergent collective computational abilities. Proc. Nat. Acad. Sc. U.S.A. (79), p. 2554-2558 (1982).

(11) S. KAUFFMAN : Behaviour of randomly constructed genetic nets.
in <u>Towards a theoretical biology.</u> Vol3, Edinburgh University Press, p. 18-46 (1970).

(12) S. KIRKPATRICK : Models of disordered materials. In <u>Ill condensed matter</u>, Les Houches, North Holland (1979).

(13) C.A. MEAD, M.A. CONWAY : <u>Introduction to V.L.S.I systems</u> Addison Wesley , (1980).

(14) P. PERETTO : Collective properties of neural networks; A statistical physics approach. (to appear in Biological Cybernetics.)

(15) P. QUINTON : The systematic design of systolic arrays. (to appear).

(16) F. ROBERT : <u>Discrete iterations</u> Springer Verlag (to appear).

(17) Y. ROBERT : Thesis, Grenoble (to appear).

(18) R. SHINGAI : Maximum period of 2-dimensional uniform neural networks Inf and Control (11) , 324-341, (1979).

(19) M. TCHUENTE : <u>Contribution à l'étude des méthodes de calcul pour des systèmes de type coopératif.</u> Thesis, Grenoble (1982).
( see also the paper by M. TCHUENTE in the present volume .)

(20) R. THOMAS : <u>Kinetic Logic.</u>
Lecture Notes in biomathematics, Vol 29, Springer Verlag (1979).

(21) J. VON NEUMANN : <u>Theory of self reproducing automata.</u>
A.W. Burks Editor; University of Illinois Press (1966).

(22) S. WOLFRAM (Editor) : <u>Cellular automata.</u> Los Alamos Science (1984).

(23) S. WOLFRAM : Statistical mechanics of cellular automata.
Rev. Mod. phys. 55, n° 3, 601-642 (1983).

# ON SOME DYNAMICAL PROPERTIES OF MONOTONE NETWORKS

Yves ROBERT and Maurice TCHUENTE

CNRS - Laboratoire TIM3/IMAG
BP 68 - 38402 St Martin d'Hères Cedex - FRANCE

## 1 INTRODUCTION

Cellular automata were first introduced in the 1950's by John Von Neumann. Informally, a cellular automaton can be viewed as a discrete dynamical system whose global behaviour is generated by the (simple) local interactions of its elementary cells, hereafter called the sites of the automaton. As pointed out by Wolfram (1984), cellular automata have arisen in several disciplines, because they provide examples in which the generation of complex behaviour by the cooperative effects of simple components may be studied. No unified mathematical framework has been yet developped to modelize the iterative behaviour of general networks of automata, but some tools such as algebraic operators and Lyapounov functions (Goles (1985), Fogelman (1985)), modular arithmetic and polynomial algebra (Martin (1983)), arithmetic in finite fields (Gill (1966), Tchuente (1982)) have been introduced to analyze special classes of cellular automata. In this paper, we characterize the dynamics of some monotone cellular automata. First we use a morphism technique to derive the iterative behaviour of automata whose transition functions are generalized majority functions. Then, we study some relationships between the connection-graph and the iteration-graph of boolean automata with memory, assuming that the transition function of each site is monotone with regard to its inputs.

## 2 NOTATIONS AND DEFINITIONS

Throughout the paper, the set of sites of the cellular automaton $\mathscr{C}$ that we study is denoted $\mathscr{S} = \{S1, S2, ..., Sn\}$. The value $Si(t)$ of $Si$ at step $t$ belongs to $Q = \{0,1,2,...,p\}$. For $t \geq 0$, we let $\mathscr{S}(t) = (S1(t),...Sn(t)) \in Q^n$ be the configuration of $\mathscr{C}$ at step $t$; $\mathscr{S}(0)$ is called the initial configuration of $\mathscr{C}$. The transition function of $\mathscr{C}$ is a function
$$F = (f_1,...,f_n) : x = (x_1,...,x_n) \in Q^n \dashrightarrow F(x) \in Q^n$$
such that each $f_i$ depends only on a restricted number of variables $x_j$. We let $FN(i)$ be the set of indexes $j$ such that $f_i$ depends on $x_j$: $FN(i)$ stands for the Functional Neighbourhood of the site $i$.

NATO ASI Series, Vol. F20
Disordered Systems and Biological Organization
Edited by E. Bienenstock et al.
© Springer-Verlag Berlin Heidelberg 1986

Finally, the connection-graph $G_c(F)$ is defined by $G_c(F) = (\mathscr{S}, \Gamma)$, where $(j,i) \in \Gamma \Leftrightarrow j \in FN(i)$, and the iteration-graph of F is the functional digraph $G_i(F) = (Q^n, F)$ where any vertex $x \in Q$ has a unique out-going arc $(x, F(x))$.

## 3 USING A MORPHISM TECHNIQUE

In this section, we consider a cellular automaton $\mathscr{C}$ whose transition function F is defined as follows: for $t \geq 0$ and $1 \leq i \leq n$, let $k(i)$ be an arbitrary nonnegative integer $\leq |FN(i)|$ and let $\{Sj(t); j \in FN(i)\}$ be ordered as $x_1 \leq x_2 \leq ... \leq x_{|FN(i)|}$. Then we define $Si(t+1) = x_{k(i)}$. That is to say, the next value of site i is the $k(i)$-th value of the current states of its functional neighbourhood when totally ordered. Assuming that $G_c(F)$ is symmetric, that is $i \in FN(j) \Leftrightarrow j \in FN(i)$, we can state the following:

**proposition 1** : if $G_c(F)$ is symmetric, then $G_i(F)$ is a subgraph of a tree.

proof : we prove that any elementary circuit of $G_i(F)$ is of length two; we proceed by contradiction. So, assume there exists an elementary circuit $[x_1, x_2, .., x_T] = [\mathscr{S}(t+1), \mathscr{S}(t+2), ..., \mathscr{S}(t+T)]$ of $G_i(F)$ with $T \geq 3$; for $q \in Q$ we introduce the function $\varphi_q : Q \dashrightarrow \{0,1\}$, $\varphi_q(x) = 1$ if $x \geq q$, $\varphi_q(x) = 0$ if $x < q$. The proof of the following two lemmas is straigthforward :

lemma 1 : there exists $q \in Q$ such that $\{ \varphi_q(x_i) ; 1 \leq i \leq T \}$ is a periodic sequence of $\{0,1\}^n$ of length greater than 2.

Let $A = (a_{ij})$ be the n by n matrix defined by $a_{ij} = 1$ if $j \in FN(i)$ and $a_{ij} = 0$ otherwise. We introduce the threshold function $H = (h_1, ..., h_n) : \{0,1\}^n \dashrightarrow \{0,1\}^n$ defined by $h_i(y_1, ..., y_n) = \quad$ 1 if $\Sigma_{1 \leq i \leq n} a_{ij} \cdot y_j \geq |FN(i)| - k(i)$.

$$0 \text{ otherwise}$$

lemma 2 : for any $q \in Q$, $\varphi_q \circ F = H \circ \varphi_q$

Thus $\varphi_q \circ F^t = H^t \circ \varphi_q$ for all $t \geq 1$, and the contradiction follows from a theorem of Goles (1985) which states that any elementary circuit of $G_i(H)$ is of length less than or equal to two.

## 4 MONOTONE BOOLEAN AUTOMATA

In this section we concentrate on boolean networks whose transition functions are monotone. Thus $Q = \{0,1\}$ and we assume that the transition function $F : (f_1, ..., f_n) : \{0,1\}^n \dashrightarrow \{0,1\}^n$ satisfies to:

$(x_i \leq y_i , 1 \leq i \leq n) \Rightarrow (f_i(x) \leq f_i(y) , 1 \leq i \leq n)$

We consider automata with memory, i.e. we assume that the value $Si(t+1)$ of a site i can depend on the value of $Sj(t)$, $j \in FN(i)$, and on $Si(t)$, $Si(t-1)$, $Si(t-2)$, ..., $Si(t-k)$. The number k (independent of i) is the length of the memory. Any network $\mathcal{C}$ with n sites $S1,...,Sn$ and a k-length memory can be viewed as an automaton with $(k+1)*n$ sites and no memory, by replacing each site Si by $(k+1)$ sites $Si_0, Si_1,...,Si_k$ arranged along a circuit of length k+1. Each site $Si_j$, $j \geq 1$, is only a transfer cell which copy the previous value of the site $Si_{j-1}$, i.e. $Si_j(t) = Si_{j-1}(t-1)$ for $t \geq 1$.

**proposition 2** : let $\mathcal{C}$ be a cellular automaton with a memory of length 1. If $G_c(F)$ is contained in a chain, then $G_i(F)$ is contained in a tree.

proof : owing to the above interpretation, we can consider $\mathcal{C}$ as a cellular automaton with no memory; the associated connection-graph is then contained in a caterpillar, and the result follows from a theorem of Robert (1985a).

In the following two paragraphs, we illustrate the relation between the maximal degree of the connection-graph $G_c(F)$ and the maximal length of an elementary circuit of the iteration graph $G_i(F)$. We restrict ourselves to the case of a <u>single site</u> monotone automaton with a memory of length k.

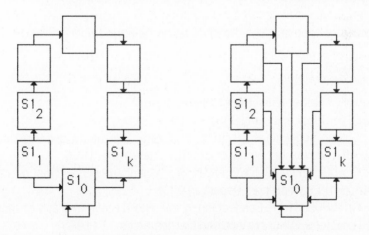

Single site with k-length memory (k=7)

$S1_0$ depends on itself, $S1_1$ and $S1_k$     $S1_0$ depends on itself, $S1_1$ and $S1_k$

**Figure 1**                              **Figure 2**

CASE 1

We consider the automaton whose connection-graph is depicted in figure 1. $S1_1$, $S1_2$, ..., $S1_k$ are transfer cells, and for any $t \geqslant 1$, the value $S1_0(t+1)$ is a monotone function of $S1_0(t)$, $S1_1(t) = S1_0(t-1)$ and $S1_k(t) = S1_0(t-k)$. Thus $FN(1_0) \subset \{ 1_0, 1_1, 1_k \}$.

**proposition 3** : let T be the length of an elementary circuit of $G_i(F)$. Then either $T = 2$ or T divides $k+1$.

CASE 2

Consider now the automaton of figure 2: its connection-graph is of maximal degree $k+1$. Such an automaton has been extensively studied in the literature when the transition function is a linear function over the finite field $Z/qZ$, q prime (Gill (1966)) or when the transition function is a threshold function (see the paper of Cosnard in these proceedings). We deal here with a general <u>monotone</u> transition function. For small values of k ($k \leqslant 5$), all elementary circuits of $G_i(F)$ are of length smaller than or equal to $k+1$. However, we show that there exists elementary circuits of $G_i(F)$ whose length grows exponentially with k :

**proposition 4** : there exist single monotone boolean automata with k-length memory whose iteration-graph contain elementary circuits of length growing exponentially with k (as k goes to infinity)

The proof of propositions 3 and 4 can be found in Robert (1985b).

5. REFERENCES

FOGELMAN F, Thesis, Grenoble (1985)
GOLES E., Thesis, Grenoble (1985)
GILL A., Linear sequential circuits, Analysis, Synthesis and Applications, Mc Graw-Hill Series in Computer Science, New York, 1966
MARTIN O., ODLYZKO A. , WOLFRAM S., Algebraic properties of cellular automata, Bell Laboratories Report (1983)
ROBERT Y., TCHUENTE M., Connection-graph and iteration-graph of monotone boolean functions, Discrete Applied Mathematics 11 (1985)
ROBERT Y., TCHUENTE M.,Dynamical Behaviour of monotone networks, RR 521, IMAG Grenoble, March 1985
TCHUENTE M., Thesis, Grenoble (1982)
WOLFRAM S., Preface to Physica 10D (1984), Elsevier Science Publishers

# INHOMOGENEOUS CELLULAR AUTOMATA (INCA)

H. Hartman and G. Y. Vichniac

*Massachusetts Institute of Technology, Cambridge, MA 02139, USA*

## 1. Introduction

A cellular automaton in 2 dimensions, as considered in this paper, consists of a regular array of sites (or cells), with a value, or state (0 or 1) at each site. With an evolution law in discrete time, a cellular automaton is a fully discrete dynamical system. The law is local: the value assumed by any one site at time $t+1$ is determined by the values taken at time $t$ by the neighboring sites. (For reviews, see Vichniac, this volume, and references therein.) These systems are homogeneous in the sense that all sites evolve according to the same transition rule.

S. Kauffman analyzed Boolean networks containing instances of all sixteen Boolean function of two inputs at the nodes of a random directed graph. The major result was that those systems are capable of self-organization. A network of $N$ Boolean elements does not wander through its $2^N$ possible states, but after a short transient, enters a limit cycle of very short period ($\sqrt{N}$ on the average). Moreover the number of possible asymptotic behaviours (limit cycles) is also remarkably small (see Kauffman, this volume and [1]). Recently Atlan *et al.*[2] have put Kauffman's system on a directed 2-dimensional regular array and exhibited the emergence of organization into spatially well defined subnets; and the asymptotic behaviour of the network was robust under noise, surgery, and immersion (see the articles by Fogelman, Goles, and Weisbuch in this volume).

In this contribution, we are proposing to merge random Boolean networks with cellular automata. In order to do this, we have simplified the Boolean networks by using only two Boolean functions and nondirected links between sites of the lattice. We use the two Boolean functions Exclusive-OR (XOR) and AND with five and four inputs, and distribute them at random on the nodes of a regular 2-dimensional array.

We take advantage of the observation by Kauffman that the behaviour of Boolean nets is essentially controlled by the ratio of "forcing," (or "canalizing") to nonforcing functions. By removing directionality of the links and reducing the number of Boolean function from sixteen to two, we are left with the simplest possible model. What is more, the model lends itself to analysis in term of standard percolation theory. In our system the XOR sites are nonforcing as the knowledge of the output is not "forced" by one of the inputs whereas the next value at an AND site is forced by a 0 at any of the four neighbours.

## 2. The XOR rule

Next to Conway's game "Life," probably the best known cellular automaton rule is the "Exclusive-OR," or "XOR" rule, invented by Fredkin. It is easy to define: sites

NATO ASI Series, Vol. F20
Disordered Systems and Biological Organization
Edited by E. Bienenstock et al.
© Springer-Verlag Berlin Heidelberg 1986

will assume in the next time step the Exclusive-OR of the states presently occupied by their neighbors. A site will be 1 at time $t+1$ if there is, in its neighborhood at time $t$, an odd number of sites in the 1 state; it will take the value 0 otherwise. In fact, the rule comes in different flavors, depending on the definition of the neighborhood of a site. In the square lattice, for example, one can take the 4-input XOR of the states occupied by the four adjacent neighbouring sites, or, alternatively, take the 5-input XOR of those neighboring states and of the value of the considered site itself. The rules are also known as the "sum mod 2 with four (or five) neighbors," and the "Q13" (or "V135") rules[3].

Though very simple, Fredkin's rules give rise to complex phenomena. In particular they are capable of duplicating any initial pattern. In all its variants, the XOR rule is linear, i.e., the evolution of the sum (modulo 2) of two patterns is equal to the sum (modulo 2) of their evolutions. The linearity of the rule makes it share properties of linear partial differential equations. It is actually a discrete form of the heat equation over the integers modulo 2. Discreteness, however, renders its behaviour akin to that of the wave equation[2]. We shall now see how the introduction of inhomogeneities gives rise to reflection and diffraction phenomena.

## 3. XOR with inhomogeneities of AND

The homogeneity of the system can be broken by stipulating that some sites, in calculating their next state, will not XOR the values of their neighbors but AND them instead. (The other sites will continue to use the XOR rule to determine their future state). Note that those sites following the 4-input AND function will become 1 only if all four of its adjacent neighbors are in the 1 state. The AND function is thus in most cases equivalent to a constant 0, and acts as a reflective boundary condition to the 1's travelling along the XOR sites (though leading to almost identical geometric patterns, we use AND instead of the simpler constant 0 function because of the universal computation capabilty of the XOR-AND elements). We carried out our studies using CAM, an high-performance simulator where inhomogeneous rules can be realized using a parameter bit array in addition to the value array, as described by Toffoli[4].

Figure 1 shows a cellular automaton configuration. Each site is characterized by two attributes: first its state (0 or 1), a dynamical variable that may change from generation to generation; and second the quenched transition rule (XOR or AND) it follows to update its state. In the figure, sites in the shaded region use a 4-input AND rule, while sites at the right use a 5-input XOR function. Values 0 and 1 are represented by blanks and dots, respectively. Figure 1 shows the evolution after 32 steps of 3x3 block of 1's (on the right) and of a L-shaped pattern. The block gave birth to four offsprings, and the leftmost descendent of the letter L has been reflected against the AND wall. Note that the convex hulls (the "territories") associated with the blocks the L's do overlap, but the two evolutions follow their own courses, as in the superposition of two traveling waves. Instead of a wall of AND-sites we can form four AND walls enclosing a square of XOR-sites. For some special sizes of the square, the evolution enters a limit cycle of very short period: patterns bounce back and forth inside of the regular cavity.

In Figure 2, we introduced at $t = 0$ an extra AND inhomogeneity. It is now, after 32 steps, the origin of a diamond-shaped diffraction pattern. If again we close the square with three more AND walls, the presence of the additional AND site, by destroying the square symmetry, multiplies the period of the motion by several orders of magnitudes, i.e., it effectively changes a regular motion into a chaotic one.

## 4. Random mixtures of XOR and AND rules, percolation

When all sites follow the XOR rule, the evolution originating from a "seed" made out of one site with value 1 in a background of 0's grows without bounds as a diamond shape. If we now distribute at random XOR and AND rules over the lattice the matters obviously differ because the 1's must find their way around the AND-sites (that most often act like 0 boundary conditions.) We observed that the evolution depends sharply on the relative concentration (fixed in time) of XOR and AND rules. Above a critical concentration $p_c(\text{XOR})$ of sites following the XOR rule, growth is similar to that in an AND-free space. To be sure, the duplicating effect is lost, together with the perfect symmetry, but the growth is still fast, and the diamond shape is not destroyed, but merely rounded: the AND inhomogeneitites scatter and multiply the impinging 1's, in a discrete analogue of the Huyghens Principle. Below the critical concentration, on the other hand, the growth out of a seed is always bounded and irregular.

At the exact concentration $p_c(\text{XOR})$ of sites that update their states with the XOR function, growth are very dendritic, and one may obtain a cluster that span the whole array. There is percolation of 1's on XOR sites. The clusters obtained with our XOR-AND mixture are always subsets of standard percolation clusters of statistical mechanics. The latter are obtained by replacing the XOR rule with a 5-input OR function. Our critical concentration $p_c(\text{XOR})$ must then be larger or equal to the threshold $(p_c(\text{OR})=.59)$ for site percolation in the 2-dimensional square lattice. The equality should hold because of the infinite rate of growth above $p_c(\text{OR})$ of all quantities that measure cluster connectedness. The equality is supported by our measurements. Whereas the standard percolation problem involves no dynamics (it is purely geometric), the XOR rule imports some dynamics, as the clusters of 1's are always active, and involve long periods, readily explained by the mechanism illustrated in figure 2. The 1's are simply bounced around the wall of a box of a very complicated shape.

## 5. Conclusion

In our merger of cellular automata and Boolean networks we take advantage, in the most economical way, of the observation by Kauffman that the behaviour of Boolean nets is dominated by the ratio of forcing over nonforcing functions. Our model lends itself to analysis in terms of percolation clusters: the properties of organization into spatially well defined subnets, the existence of a stable core common to all attractors, and robustness of the attractors under noise, surgery, and immersion are in our case but a simple consequence of the non-connectedness of percolation clusters: If you tamper with one cluster, other clusters will not be affected.

INCA are also related to other complex systems. They offer a multidimensional extension of shift registers, made up with delays, multiplication (mod 2) and addition (mod 2). In this system, capable of parallel universal computation, one can view the distribution of rules as the wiring and program, and the initial distribution of values as the input data.

The partition of the lattice into active and stable regions suggests that our XOR-AND model shares properties with the $\pm J$ Ising spin-glass at $0$ $K$. The latter is formally equivalent to a more complicated INCA, involving 16 Boolean functions (to incode the 16 possible combinations of the ferro/antiferromagnetic couplings on the 4 links at each bond.)

## Acknowledgements

We are grateful to F. Fogelman, S. Kauffman, N. Margolus, T. Toffoli, and G. Weisbuch for helpful discussions, and to Pablo Tamayo and David Zaig for their assistance in the preparation of the figures. This work was supported (H. H.) by grants from NASA (NAGW-114); and (G. Y. V.) from DARPA (N0014-83-K-0125), NSF (8214312-IST) , and DOE (DE-AC02-83ER13082).

## References

[1]  S. A. Kauffman, *J. Theoret. Biol.*, 22, 437–467 (1969).

[2]  H. Atlan, F. Fogelman-Soulié, J. Salomon, and G. Weisbuch, *Cybernet. Systems*, 12, 103–121 (1981).

[3]  G. Y. Vichniac, *Physica* 10D, 96–116 (1984).

[4]  T. Toffoli, *Physica* 10D, 195–204 (1984).

**Figure 1.** Configuration after 32 steps in which the replication of a 3x3 block and an L pattern have taken place. Shaded region follows the 4-input AND rule. Non-shaded region follows the 5-input XOR rule. Notice that the leftmost L has been reflected by the AND wall.

**Figure 2.** The setup is the same as in Figure 1; however an additional AND site has been introduced at $t = 0$ in the XOR region. This leads, after 32 steps, to a diamond-shaped pattern.

**Figure 3.** A large cluster of 1's which was seeded by a single 1. The 5-input XOR and 4-input AND rules are distributed at random, the frequency of XOR being .59, the frequency of AND being .41.

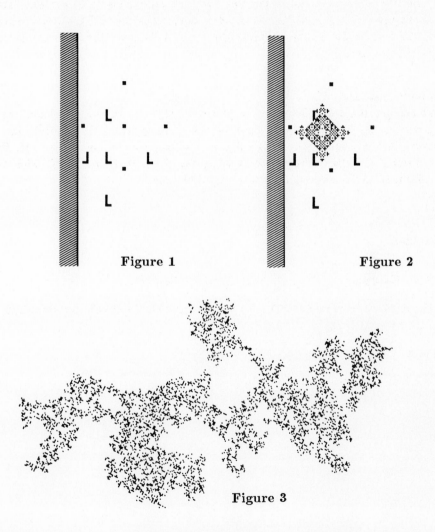

Figure 1

Figure 2

Figure 3

# THE ISING MODEL AND THE RUDIN-SHAPIRO SEQUENCE

Jean-Paul ALLOUCHE
UA 226 - U.E.R. de Math. de Bordeaux I
351, cours de la Libération
33405 TALENCE CEDEX (France)

(from a joint paper with Michel MENDÈS FRANCE)

Abstract. -   Provided purely imaginary temperatures are allowed, the
1-dimensional cyclic Ising model is identified with the Rudin-
Shapiro sequence.

Introduction. -  In this paper, which is a short form of [1], we start with
the following "2-automaton" (see [2]) :

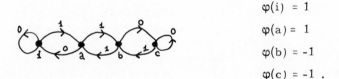

$$\varphi(i) = 1$$
$$\varphi(a) = 1$$
$$\varphi(b) = -1$$
$$\varphi(c) = -1 \ .$$

Our aim is to prove that the binary sequence generated by this machine is
closely related to the 1-D cyclic Ising model. Let us recall briefly the
formalism of this Ising model and the working way of the above machine :

## I. -  A few words about the cyclic 1-dimensional Ising model

Consider  N  points or sites, $0, 1, 2 \ldots$ N-1, on a circle. At each
site  q  the spin $\mu_q$  is equal to $\pm 1$ . For a given configuration, i.e. a
given choice of $\mu_q$  for each  q , the interaction energy $E(\mu)$  is defined
by

$$E(\mu) = -J \sum_{q=0}^{N-1} \mu_q \mu_{q+1} - H \sum_{q=0}^{N-1} \mu_q \ ,$$

NATO ASI Series, Vol. F20
Disordered Systems and Biological Organization
Edited by E. Bienenstock et al.
© Springer-Verlag Berlin Heidelberg 1986

where J is the coupling constant and H the external field. The partition function is $Z(N, \beta J, \beta H)$ :

$$Z(N, \beta J, \beta H) = \sum_{\mu} \exp(-\beta E(\mu)) ,$$

where $\beta = \dfrac{1}{kT}$ , k is the Boltzman constant, $T > 0$ is the absolute temperature.

## II. - The Rudin-Shapiro sequence

### 1) The Rudin-Shapiro sequence

We consider the 2-automaton given above (see [2]). It generates the Rudin - Shapiro binary sequence as follows : let n be an integer represented by its binary expansion. Feed the machine, starting from i , with the digits of n , one after the other ; this finally gives a letter i, a, b or c and, by $\varphi$ , a number equal to $\pm 1$ , which is by definition $\varepsilon_n$ . For instance : if $n = 19 = \overline{1 0 0 1 1}_2$ , one has

$$n.i = \overline{1 0 0 1 1} . i = \overline{1 0 0 1} . a = \overline{1 0 0} . b = \overline{1 0} . c = \overline{1} . c = b$$

hence $\varepsilon_{19} = \varphi(b) = -1$ .

It is not hard to check that $\varepsilon_n = (-1)^{u(n)}$ , where $u(n)$ is the number of 11's in the binary expansion of n .

### 2) The modified digits of n and the cyclic Rudin-Shapiro sequence

$N \geq 1$ is a given integer. For every n such that $0 \leq n < 2^N$ , let $n = \sum_q e_q(n) 2^q$ be the binary expansion of n ( $e_q$ is equal to 0 or 1 ) ; define the "modified digits" of n by :

$$e'_q(n) = \begin{cases} e_q(n) & \text{if } q \leq N-1 \\ e_o(n) & \text{if } q = N \end{cases} .$$

The cyclic Rudin-Shapiro sequence is the number $u(n, N)$ of 11's in the string $e'_N e'_{N-1} \cdots e'_o$ . In other words, if necessary put extra 0's on the left of n to get a binary word of length N , cycle it and count the number of 11's in this cyclic word.

## III. - Imaginary temperatures

In number theory we are interested in the sum $\sum\limits_{0 \le n \le N} \exp(2i\pi x u(n))$.
This sum is closely related to the sums $S_N$ , where

$$S_N = \sum_{0 \le n < 2^N} \exp(2i\pi\, x u(n, N)) \quad .$$

To evaluate this last sum, define for $0 \le n < 2^N$ the numbers $\mu_q(n)$ by

$$\mu_q(n) = 1 - 2\, e'_q(n)$$

( $e'_q$ are the "modified digits" of $n$ , hence $\mu_q = \pm 1$ ).

A straightforward computation yields :

$$u(n, N) = \frac{N}{4} - \frac{1}{2} \sum_{q=0}^{N-1} \mu_q(n) + \frac{1}{4} \sum_{q=0}^{N-1} \mu_q(n)\, \mu_{q+1}(n)$$

hence

$$S_N = \exp(i\pi\frac{Nx}{2}) \sum_{0 \le n < 2^N} \exp(\, i\pi x(\frac{1}{2} \sum_{q=0}^{N-1} \mu_q(n)\, \mu_{q+1}(n) - \sum_{q=0}^{N-1} \mu_q(n))) \quad .$$

The bijection between the integers in $[\, 0\, , 2^N[$ and the $2^N$ choices of
$\mu_0 \cdots \mu_{N-1}$ in $\pm 1$ gives :

$$S_N = \exp(i\pi \frac{Nx}{2})\; Z(N, \frac{i\pi x}{2}\, , -i\pi x) \quad ,$$

where $Z$ is the partition function of the $1$-D cyclic Ising model with
$\beta J = \frac{1}{2} i\pi x$ and $\beta H = -i\pi x$ , i. e. with <u>an imaginary temperature</u>.

## IV. - Application to a number-theoretical result

$S_N$ can be calculated by Kramers and Wannier's method [3] ; the
transfer matrix is :

$$L = \begin{pmatrix} \exp(-\dfrac{i\pi x}{2}) & \exp(-\dfrac{i\pi x}{2}) \\[2mm] \exp(-\dfrac{i\pi x}{2}) & \exp(\dfrac{3i\pi x}{2}) \end{pmatrix}$$

with eigenvalues $\lambda_1, \lambda_2 = (\cos \pi x \pm (- \sin^2 \pi x + e^{-2i\pi x})^{1/2})$ ; hence :

$$Z(N, \frac{i\pi x}{2}, -i\pi x) = \mathrm{Tr}(L^N) = \lambda_1^N + \lambda_2^N \quad .$$

We can deduce the following result :

THEOREM. - If x is real not integer, then :

$$\left| \sum_{0 \le n \le N} \exp\left(2i\pi\, x\, u(n)\right) \right| \le C(x)\; N^{\alpha(x)}$$

where α (x) is given explicitly by :

$$\alpha(x) = \frac{1}{\text{Log } 2} \cdot \text{Log max}_{\pm} \left| \cos \pi x \pm \left(- \sin^2 \pi x + e^{-2i\pi x}\right)^{1/2} \right|$$

hence $\frac{1}{2} \le \alpha(x) < 1$ .

COROLLARY. - For every irrational x the sequence (x u(n)) is equi-distributed modulo 1 .

-:-:-:-

## REFERENCES

[1] J.-P. ALLOUCHE et M. MENDÈS FRANCE, Suite de Rudin-Shapiro et modèle d'Ising, Bull. Soc. Math. France 1985 (à paraître).

[2] G. CHRISTOL, T. KAMAE, M. MENDÈS FRANCE et G. RAUZY, Suites algébriques, automates et substitutions, Bull. Soc. Math. France 108 (1980), 401-419.

[3] H. A. KRAMERS, G. H. WANNIER, Statistics on the two-dimensional ferromagnet, I et II, Physic. Rev. 60 (1941), 252-262, 263.

# DYNAMICAL PROPERTIES OF AN AUTOMATON WITH MEMORY

Michel COSNARD and Eric GOLES CHACC

CNRS-Laboratoire TIM3/IMAG
BP68 - 38402 St Martin d'Hères-FRANCE

## 1 INTRODUCTION

Automata networks have been often used to model elementary dynamical properties of neuronal networks (Mc Culloch and Pitts (1943)). Caianiello et al. (1967) generalized the above model in order to introduce the refractory character of the neural response. They proposed to use a memory associated to the system and obtained the following discrete iteration scheme, in order to simulate the response of a single neuron:

$$x_{n+1} = \mathbf{1} \left( \sum_{i=0}^{k-1} a_i x_{n-i} - b \right) \qquad (1)$$

where $x_n$ belongs to $\{0,1\}$, $a_i$ and $b$ are real parameters, $k$, the size of the memory, a given integer and $\mathbf{1}(u)$ is 1 if $u$ is non negative and 0 otherwise.

Fisrt we present a complete study of the case $a_i = a^i$, with $0 \leq a \leq 1/2$. For each value of $k$, we show that there exists a unique globally attracting cycle whose length is less than $k+1$. The graph of the rotation number of this cycle versus $b$ is a discrete approximation of a devil staircase. On the other hand, we study some particular cases: palindromic memory structure and reversibility versus irreversibility of the automaton. Finally, we propose various conjectures concerning the general case.÷

## 2 GEOMETRIC MEMORY

Equation (1) is completely defined by the pair $(a,b)$ where $a=(a_0,...a_{k-1})$ and $b$ is the threshold. It can be transformed into the k-system:

$$T : \{0,1\}^k \mapsto \{0,1\}^k \quad x=(x_0,...,x_{k-1}) \mapsto T(x)=(f(x), x_0,...,x_{k-2})$$

where $f(x) = \mathbf{1} \left( \sum_{i=0}^{k-1} a_i x_i - b \right)$. T is a special shift acting on k-vectors.

We consider first the case $a_i=-a^i$ and $b=-\theta$ with $0<a\leq 1/2$ and $\theta \geq 0$.

$$x_{n+1} = 1 \left( \sum_{i=0}^{k-1} \theta - a^i x_{n-i} - b \right) \quad (2)$$

It is clear that the dynamical properties of (2) depends heavily on the position of $\theta$ with respect to the different partial sums of $a^i$. We shall not discuss in details the bifurcation structure of (2). We only present some results whose proofs can be found in Cosnard and Goles (1984).

**proposition 1** : For all $\theta$ and $k$, (2) admits a unique globally attracting cycle. The period of this cycle is less than or equal to $k+1$. If $p \leq k+1$, then there exists $\theta$ such that (2) admits a p-cycle.

In order to understand the sequence of periods, we define the rotation number of a cycle. Let C be a cycle of length q and let p be the number of 1 in C. The rotation number of C is defined to be $\rho(C)=p/q$. We consider $p/q$ as a fraction but do not allow for reduction. For brevity, we shall use $\rho(\theta)$ instead of $\rho(C)$ if, for some value $\theta$, (2) admits a cycle C.

**proposition 2** : 1) If C is a q-cycle, then $\rho(C)=p/q$ is irreducible.
2) Let $p/q$ be an irreducible fraction with $q \leq k+1$. Then there exists a unique q-cycle C such that $\rho(C)=p/q$.
3) Considered as a function of $\theta$, $\rho$ is monotone increasing from **R** onto the set of irreducible fractions with denominators less than $k+2$.

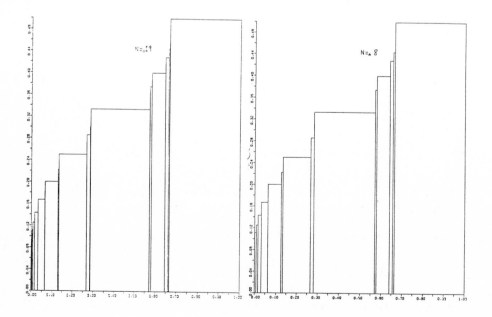

Figure 1 shows the graphs of $\rho$ for k=8 and k=19. The preceding results give us a way to obtain the bifurcation structure of (2). Consider the set of irreducible fractions with denominators not greater than k+1 and list them in increasing order. We obtain the Farey sequence of order k+1. The list of their denominators is the ordered list of the periods of the attracting cycles of (2). Using their numerators, it is possible to obtain the corresponding values of $\theta$.

From a more general point of view, we have shown that a simple iteration formula can lead to a very rich structure of bifurcations. Remark, however, that the length of the cycles is linear in k compared to the $2^k$ possible states.

## 2 REVERSIBLE SYSTEMS

We shall say that the system (a,b) is reversible if for all y in $\{0,1\}^k$, there exists a unique vector such that Tx=y, i.e. T is a bijection.

T is a shift if for any $(x_0,...,x_{k-1})$ we have $T(x_0,...,x_{k-1})=(x_1,...,x_{k-1},x_0)$.

T is an antishift if for any $(x_0,...,x_{k-1})$ we have $T(x_0,...,x_{k-1})=(x_1,...,x_{k-1},x'_0)$ where $x'_0=1-x_0$.

**proposition 3** : T is reversible if and only if T is a shift or an antishift.

The next result gives us a characterization of shifts and antishifts and allows us to know from the shape of the memory in which case we are.

**proposition 4** :

T is a shift if and only if $a_{k-1} > \sum_{i=0}^{k-2} |a_i|$ and $\sum_{a_i>0}^{k-2} a_i < b \le a_{k-1} + \sum_{a_i<0}^{k-2} a_i$ .

T is a antishift if and only if $a_{k-1} < \sum_{i=0}^{k-2} |a_i|$ and $a_{k-1} + \sum_{a_i<0}^{k-2} a_i < b \le \sum_{a_i>0}^{k-2} a_i$ .

**proposition 5** : 1) A reversible system has only cycles of length p such that p divides k if $a_{k-1}$ is positive and 2k if $a_{k-1}$ is negative.

2) Equation (1) cannot have cycles of length $2^k$.

## 3 PALINDROMIC MEMORY

Let us consider now the case where (a,b) are such that $a_i=a_{k-i-1}$, i.e. the memory coefficients form a palindromic word. Without loss of generality, we may assume that for any x we have $\sum a_i x_i - b \neq 0$. In this case we introduce the following operator :

$$E(y_0,...,y_k) = -\sum_{j=0}^{k-1} y_j \sum_{s=t+1}^{k} a_{s-j-1} y_s + b \sum_{j=0}^{k} y_j$$

Let us apply the operator to a trajectory $x_n$ of the system

$$E(x_n,...,x_{n-k}) = -\sum_{j=0}^{k-1} x_{n-j} \sum_{s=t+1}^{k} a_{s-j-1} x_{n-s} + b \sum_{j=0}^{k} x_{n-j}$$

Since the memory coefficients have a palindromic structure, we have:

$$\Delta_n E = E(x_n,...,x_{n-k}) - E(x_{n-1},...,x_{n-k-1})$$

$$= -(x_n - x_{n-k-1})(\sum_{s=1}^{k} a_{s-1} x_{n-s} - b)$$

and if $x_n \neq x_{n-k-1}$, we have $\Delta_n E < 0$. Hence E is a Lyapunov function associated to the system (see Goles(1985) for a general use of Lyapunov functions in automata theory) from which we deduce the following result.

**proposition 6**: If T is the length of a cycle of a palindromic system, then T is a divisor of k+1.

## 4 CONJECTURES

From the preceding results and some experiments, we can formulate the two general conjectures:

1 if the memory coefficients are non negative, then the lengths of the cycles of (1) are less than or equal to k

2 in the general case, the lengths of the cycles of (1) are less than or equal to 2k.

## 5 REFERENCES

CAIANIELLO E.R. et al., Kybernetik,4 (1967).
COSNARD M., GOLES E., C. R. Acad. Sc.,299, 10 (1984) pp 459-461.
GOLES E., Thesis, Grenoble (1985).
Mc CULLOCH W., PITTS W., Bull. Math. Biophys., 5 (1943) pp 115-133.
NAGAMI H. et al., Math. Biosc., 18 (1973)
NAGUMO J., SATO S., Kybernetik, 3 (1972) pp 155-164.

# DYNAMICS OF RANDOM BOOLEAN NETWORKS.

Didier PELLEGRIN

Laboratoire TIM3, Institut IMAG
BP68
38402 SAINT MARTIN-D'HERES

## 1) Introduction

Random boolean networks have been studied in order to modelize complex systems composed of many parts. S.Kauffman (1970) founds his choice of such models upon a comparison between genetic nets and networks of randomly interconnected binary elements. He gives important statistical results for nets with connectivity k=1, 2, 3, N (number of elements in the net).

For the last years, two-connected nets have been studied in detail : "the behavior of randomly interconnected deterministic nets in which each element received just two inputs from other elements is biologically reasonable" (Kauffman 1970 ). The dynamics of such networks is characterized by the existence of a stable core, a good approximation of which may be computed through the concept of forcing domain introduced by F. Fogelman (1985).

In this paper we continue the study of F.Fogelman on random boolean networks of connectivity 2.

## 2) Definitions

### Network:

A boolean network is defined by a mapping $F : \{0,1\}^n \mapsto \{0,1\}^n$. A net with connectivity k has $2^{2^{**}k}$ possible boolean functions, thus 16 in interconnectivity 2.

These 16 mappings are classified (fig.1) :

constant functions : 0 and 1                    (0,15)

transfer mappings   : (or forcing mapping in one variable) with only one input, they transfer it or reverse it. (3,5,10,12)

forcing mappings    : for one value (forcing value) of each input their value (forced value) is independent of the other input.

(1,2,4,7,8,11,13,14)

XOR, ⇔        : they always depend on their two inputs.

We choose a regular connection graph : a plane lattice with connection between nearest neighbours, and a periodic structure of connection, at the boundaries we connect symetrical nodes (fig2).

NATO ASI Series, Vol. F20
Disordered Systems and Biological Organization
Edited by E. Bienenstock et al.
© Springer-Verlag Berlin Heidelberg 1986

<u>Dynamics</u> :

Starting from random initial conditions, the network computes its new state <u>in parallel</u> : all cells change states at the same time. F. Fogelman (1985  ) and F. Robert (to appear) compare parallel with sequential iterations.

A limit cycle C associated with a given initial condition, distinguishes a stable part (SP(C)) from an oscillating part  (OP(C)) :

SP(C) = { stable elements during the limit cycle}.

OP(C) = { oscillating elements during the limit cycle}.

**Stable core** = ∩ SP(C)
            C∈𝓒

                            𝓒 is the set of limit cycles.

**Oscillating core** = ∩ OP(C)
            C∈𝓒

The definitions of cores are obviously not constructive : it is impossible to compute the dynamics for all initial conditions.

We then use two methods for estimating the cores, particulary the stable core :

i) **simulation** of large numbers of dynamics.

ii) a mathematical tool : **forcing domain**.

### 3) Simulations

For 1000 initial conditions, limit cycles have been computed and for each limit cycle, the stable part and the oscillating part have been determined. For a given network, figure 3 shows the number of times each node belonged to oscillating parts, figure 4 has the same signification for an other network. We note two different behaviors :

<u>figure 3</u> : we see 59 different numbers and the two particular 0 (stable core) and 1000 (oscillating core) appear unstable when the number of initial conditions simulated increases : 5 cells oscillated 1 time and 4 cells oscillated 999 times.

<u>figure 4</u> :  we see only 4 numbers 0, 466, 543, 1000, and we can think that here we have a good approximation of the cores.

These  two examples show the difficulty to compute cores by using only simulations.

### 4) A mathematical tool : forcing domain.

The concept of forcing mappings introduced by S. Kauffman (1972) is used by F. Fogelman to compute an approximation of stable core (1985)  : the forcing domain.

The forcing graph of F is a partial graph of G  (connection graph)  composed of those arcs i-j which are forcing (if the origin i falls into its forced value then the extremity j will also fall into its forced value) The <u>definition of the forcing domain</u> is based upon an extension of forcing circuits with three recursive rules :

■ extremity of forcing arcs
■ automata with two stable inputs
■ extremity of connection arcs with the origin in its non-forced value which is forcing value of the extremity.

F. Fogelman proved that it provides a good approximation of the stable core ( for parallel and sequential iterations), but some simulations show three problems in this method :

i) some forcing circuits are not always stable : forcing circuits without forcing inputs may oscillate (Fogelman 1985).

ii) the extension of forcing circuit appears sometimes not sufficient : if a mapping has a constant input equal to its non-forced value,it is equivalent to a transfer map.

iii) there are isolated cells in stable cores : example in figure 5.

## 5) conclusion

We propose a <u>method for studying the dynamics of Random Boolean Network</u> :

I) Compute forcing domain with a modification :
  1) rule out of forcing circuits without forcing inputs
  2) add a fourth recursive rule: transform the mappings into their equivalent

II)Then consider a partial network with constant inputs :
  1)this partial network is obtained from the original network by taking out the forcing domain.
  2)<u>search for frustrated circuits</u> to compute the oscillating core.
  3)<u>search for steps</u> : where a step is a set of elements with a common behaviour with respect to a large number of initial conditions. We visualize steps in figure 3 and 4 by a same number. The "level" of a step represents the probability of oscillating ( ex : a non-frustrated circuit oscillates with a probability 7/8 and the simulation of 5000 initial conditions gives a level of 4431/5000)
  4) isolated points are taken care of with some formal iterations.

| code | 0 | 1 | 2 | 3 | 4 | 5 | 6 | 7 | 8 | 9 | 10 | 11 | 12 | 13 | 14 | 15 |
|---|---|---|---|---|---|---|---|---|---|---|---|---|---|---|---|---|
| name | cont | nor | ⇒ | $\overline{T}_2$ | ⇐ | $\overline{T}_1$ | xor | nand | and | ⇔ | $t_1$ | ⇐ | $t_2$ | ⇒ | or | taut |
| forced value | 0 | 0 | * | 0 | * | * | 1 | 0 | * | * | 1 | * | 1 | 1 | | |
| forcing value n°1 | 1 | 0 | * | 1 | * | * | 0 | 0 | * | * | 1 | * | 0 | 1 | | |
| forcing value n°2 | 1 | 1 | * | 0 | * | * | 0 | 0 | * | * | 0 | * | 1 | 1 | | |

figure 1

```
  0    0   49   49  826  839    0    0    0  575  567  567    0    0    0    0
120  382  381   49  774  839  467  467  575  575  320  320    0  784  784    0
120  120  381  382  495  467  467    0  575  575  849  895  784  784  784    0
587  587  381  381  495  495  467  467  575  575  574  838  875  531  784  587
587    0  381  381  382  495  476  465  384  574  574  876  876    0    0  587
587  587  294  294  382  476  402  402  402  402    1  876  876  876    0  587
483  587  708  294  649  518  694  392  180  374  758  756    1    1    0  159
139    0  247    0    0  518  518  518  994  934  857  857    1    1  284  284
233  278  278    0    0  518  518  518  993  992  857  857    0    0  284  284
  0  278  278  229  517  517    0    0  987  992  857  161  248    0  139  139
  0    0  999  999  517  517  713  350  987  992    0  161  248  248  248  240
 10   10  999  999  517    0  350  350  987  741 1000  248  248  248  248  240
 10   10    0    0    0  350  350    0    0 1000 1000 1000    0  248  248  248
  0    0    0    0    0  350  350  350    0 1000 1000 1000    0    0  248  248
  0    0    0    0    0  350    0    0    0    0 1000 1000    0    0    0  248
  0    0   49   49    0    0    0    0    0    0  567  567    0    0    0    0
```

Figure 3

```
1000    0    0    0    0    0    0    0    0    0    0    0    0    0    0 1000 1000
1000    0    0    0 1000    0    0    0    0    0    0    0    0    0    0  466 1000
1000    0    0 1000 1000 1000    0    0    0    0    0    0    0  466  466  466 1000
1000    0    0 1000 1000 1000    0    0    0    0    0    0  466  466  466    0    0
   0    0    0 1000    0 1000    0    0    0    0    0    0    0    0  466    0    0
1000    0    0    0    0    0    0    0    0    0    0    0  543  543 1000 1000
1000 1000    0    0    0    0    0    0    0    0    0  543  543  543 1000 1000
1000 1000  543  543    0 1000 1000 1000    0    0    0  543  543  543  543 1000
1000  543  543    0    0    0 1000 1000 1000    0    0    0  543  543 1000 1000
1000 1000  543    0    0    0 1000 1000    0    0    0    0  543  543    0 1000
   0 1000    0    0    0    0 1000 1000    0    0    0    0  543  543    0 1000
   0    0    0    0    0    0 1000    0    0    0    0  543  543    0    0    0
   0    0    0    0 1000 1000    0    0    0    0    0    0    0    0    0    0
   0    0    0 1000 1000 1000 1000 1000 1000    0    0    0    0    0    0    0
   0    0    0    0 1000 1000 1000 1000 1000    0    0    0    0    0    0    0
1000    0    0    0 1000    0    0    0    0    0    0    0    0    0 1000    0
```

Figure 4

Figure 2

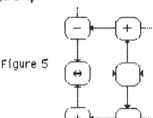

Figure 5

## Bibliography

F. FOGELMAN-SOULIE (1985) : " Contributions à une théorie du calcul sur réseaux", thesis Grenoble.

S. KAUFFMAN (1970 ) : " Behaviour of randomly constructed nets". In Towards a theoretical biology Ed. C.H. Waddington, vol3, Edinburg University Press

S. KAUFFMAN (1979) : " Assessing the probable regulatory structures and dynamics of metazoan genome. Kinetic logic ". In Lecture notes for Biomathematics. Ed. R. Thomas, 29, 30-61. Berlin Springer Verlag.

S. KAUFFMAN (1972) : "The organization of cellular genetic control systems". Lecture on Mathematics in the Life Sciences, Vol 3, American Mathematical society, Providence, RI, 63_116.

F.ROBERT (1985) : "Discrete iterations" Springer Verlag, to appear.

# RANDOM FIELDS AND SPATIAL RENEWAL

## POTENTIALS

J. DEMONGEOT & J. FRICOT
IMAG - TIM3
BP 68
38402 ST MARTIN D'HERES CEDEX - FRANCE

Abstract :

By using an approach similar to that used for Markov random fields, we propose a spatial version of renewal processes, generalizing the usual notion in dimension 1. We characterize the potentials of such renewal random fields and we give a theorem about the presence of phase transition. Finally, we study the problem of the sampling of renewal fields by means of a random automaton, we show simulations and discuss the stopping rules of the process of sampling.

## 0. Introduction :

The concept of cellular automata was introduced in the deterministic case by J. VON NEUMANN (1). The underlying idea was that systems consisting of many identical components interconnected in a uniform and regular way, each obeing simple local rules, could exhibit a highly complex overall behavior. In the stochastic case, the local rules are random and the asymptotic behavior of networks of such automata is more complex ; nevertheless, the probability measures ruling asymptotic states of these automata are in many cases remarkable : they remember the properties of local rules, in particular they have a spatial Markovian character with the same range as the local rules. Hence, we can use random automata to simulate probability measures on a set of spatial states in a Markovian way ; this possibility is very interesting in order to build and study epidemiological models (2,3), neuromimes (see these Proceedings) and image processing techniques (4). For the sake of simplicity, we will only consider binary automata with two states 0 and 1, transition rules expressing interactions between individuals, neurones or pixels...

We define firstly random automata networks and associated random fields ; then we study particular fields : we recall main properties of Markov fields and we give corresponding properties for the new notion of spatial renewal fields. After we build random automata networks and spatial birth and death processes having the same renewal field as asymptotic state measure. Such an automata network gives a way to simulate the renewal field ; hence, we exhibit simulations and propose stopping rules for the simulation process.

NATO ASI Series, Vol. F20
Disordered Systems and Biological Organization
Edited by E. Bienenstock et al.
© Springer-Verlag Berlin Heidelberg 1986

# 1. Random automata networks :

## Definition 1.1.

A <u>random automaton</u> A is defined by the quintuplet $(Q, Y, Z, F, G)$, where :

- Q is the state space, Y the set of inputs and Z the set of outputs

- F is a function from $Y \times Q^2$ to $[0,1]$ verifying :

$$\forall \, y \in Y, \, \forall \, q \in Q, \, \sum_{q' \in Q} \, F(y,q,q') = 1$$

$(F(y,q,q'))$ denotes the probability to go from state q to state $q'$, when the input is y)

- G is a function from $Q \times Z$ to $[0,1]$ verifying :

$$\forall \, q \in Q, \, \sum_{z \, \in \, Z} \, G(q,z) = 1$$

$(G(q,z)$ denotes the probability to observe the output z, when the state is q)

## Definition 1.2.

A <u>random automata network</u> N is defined by the triplet $(X, R, C)$, where :

- X is the set of cells (or sites) of the network

- R is a function from X to set of all random automata A

$(R(x)$ is the automaton associated to the site x of X)

- C is a function from $X \times Z^X \to Y$

$(C(x,f)$ is the connexion map of the network N, denoting the input received in x, when the output in all sites of X is described by the function f)

We will study in the following only binary random automata, for which we will choose :

$Q = Z = \{0,1\}$

$X \subset \mathbb{Z}, \, |X| < +\infty$

$Y = Q^X \equiv P(X)$, the identification of Y to the set of subsets of X being made in order to simplify formulas.

Then, we can define for all  x in X , D in Y and q in Q :

$$F_x (D, q, 1) = \frac{e^{U(D \cup \{x\})}}{e^{U(D \cup \{x\})} + e^{U(D \setminus \{x\})}} \quad ,$$

where  U is a _potential_ , i.e. function from

Y to $\mathbb{R}$ verifying : U $(\emptyset)$  = 0

We have also  :

$$\forall\, x \in X , \quad \forall\, q \in Q , \quad G_x (q,q) = 1$$
$$\text{and } \forall\, f \in \{0,1\}^X , \quad C ( x,f) = \{y \in X ; \quad f(y) = 1\}$$

Definition 1.3.

A binary  _random field_  on X  is a set of binary random

variables  $\{V_x\}_{x \in X}$  ; the canonical measure $\mu$  of this field

is defined  on   (Y, B (Y)), where B (Y) denotes the  borelian

$\sigma$ -algebra generated by the cylinders of Y, e.g.  by the sets of subsets

of X defined by :

$$\forall\, A \subseteq B \subset X, \; [A,B] = \{ D \subset X ; \; D \cap B = A \}$$

From the  Hammersley  – Clifford theorem (5) , we prove easily that

the canonical  measure $\mu$ of  each binary random field on X  can be

defined by a  potential  U such that :

$$\forall\, A \subset X , \quad \mu ( [A, X ] ) = \frac{e^{U(A)}}{Z (X)} \quad ,$$

where the  normalizing constant  Z (X)   verifies  :

$$Z (X) = \sum_{B \subset X} e^{U(B)}$$

$\mu$  is  called the  _Gibbs measure_  associated to the potential U  and
the corresponding random field is called the  field associated to
the random automata network  whose potential is U.

We will give examples of such fields in the following.

## 2. Markov random fields

### Definition 2.1.

A binary random field is a <u>Markov field</u>, if its canonical measure $\mu$ verifies :

i) $\mu$ is strictly positive on B (Y)

ii) $\forall x \in X$ , $\forall A \subset X$ , $\mu([\{x\}, X] | [A, X]) = \mu([\{x\}, X] | [A \cap \partial x \times X])$,

where $\partial x$ denotes a neighbourhood of x , for example the set of the nearest neighbours of x or any neighbourhood defined from a subset D of X by : $\partial x = (D+X) \setminus \{x\}$ , where $D+x = \{y+x\}_{y \in D}$ .

We will call <u>range</u> of the Markov field, we have defined above, the diameter of D, the distance d on $\mathbb{Z}^d$ being the Manhattan distance

$$( d (x,y) = \sum_{i=1}^{d} | x_i - y_i | ) \ .$$

A Markov field has two main properties :

1) we can characterize its potential U by an <u>interaction</u> V with the following dual relations :

$$\forall A \subset X , V (A) = \sum_{D \subset A} ( -1)^{|A \setminus D|} U(D)$$

$$U(A) = \sum_{D \subset A} V (D)$$

Then we have :
   $\mu$ is the measure of a Markov field, if and only if V(A)= 0 , for every subset A of X , which does not verify :
   $x,y \in A \Rightarrow x \in \partial y$

2) let us suppose that $X_n$ increases to $\mathbb{Z}^d$ as n goes to infinity and define boundary conditions on $\partial X_n = \underset{x \in X_n}{\cup} \partial x \setminus X_n$ ,

by considering the conditional measure :

$$\mu_{F_n} (.) = \mu_n (. | [F_n , \partial X_n]) \quad , \text{ where } \mu_n$$

is the measure of the field defined on $X_n \cup \partial X_n$.

Then, if $\mu_{F_n}$ tends to same measure $\mu$ on $(P(\mathbb{Z}^d), B(P(\mathbb{Z}^d)))$
for any sequence of $F_n$'s in the $\partial x_n$'s , we say that the potential U
does not have a phase transition.

We have for example the following result due to Ruelle (6) :
if $\partial x$ is the nearest neighbourhood and if U is translation and
rotation invariant, then there exists a phase transition only if

$$V(\{x\}) + 2\nu \ V(\{x,y\}) = 0 \ , \text{ where } y \in \partial x.$$

## 3. Renewal random fields

Our aim is to define a generalization of the classical renewal
processes in dimension 1 ; for this purpose, we will define the
projection $\pi(x,A)$ of the point $x$ on the subset A by :

$$\pi(x,A) = \{ y\in A \ ; \ d(x,y) = d(x,A)\}, \text{ where }$$

$d(x, A) = \inf_{z \in A} d(x,z)$. After, we will define a renewal field in the

same way as a Markov field by using a conditional property for $\mu$ .

## Definition 3.1.

A binary random field is a renewal field , if its canonical measure $\mu$
verifies :

i) $\mu$ is strictly positive on B (Y)

ii) $\forall \ x \in X \ , \ \forall \ A \subset X \ ,$ $\dfrac{\mu \ ([\{x\} , X]| \ [A, X])}{\mu \ [\{x\}, X \ ]| \ [\pi(x,A),S(x,A)])}$ =

where $S(x,A)$ denotes the sphere centered in $x$ whose radius is
$d(x,A)$ :

$$S(x,A) = \{ y\in X \ ; \ d(x,y) \leqslant d(x,A)\}$$

We have, like for the Markov fields, two main properties expressed
by the following theorems (cf. (7) for the proofs).

## Theorem 3.1.

The measure $\mu$ is the canonical measure of a renewal random field, if
and only if the interaction V of its potential U verifies :

$$\exists \ j \in \mathbb{R} \ ; \quad \text{diam (A)} > 1 \quad \Rightarrow \quad V(A) = (1)^{|A|} j$$
$$\text{and diam (A)} \leqslant 1 \quad \Rightarrow \quad V(A) \text{ takes any real value}$$
$$\text{except for} \quad V(\phi) = 0.$$

## Theorem 3.2.

U is the potential of a renewal random field, if and only if there are
a real number $j$ and a Markov field having as nearest neighbours
potential $U'$ such as :

$$\forall \ A \subset X \ ; \ A \neq \emptyset, \ U(A) = U'(A) - j$$

For the interactions, we have :

$$V(\{x\}) = V'(\{x\}) - j \text{ and } V(\{x,y\}) = V'(\{x,y\}) + j, \text{ where } x \in \partial y$$

## Theorem 3.3.

If V is the potential of a renewal random field and if V is translation
and rotation invariant, then we have a phase transition, only if :

$$V ( \{x\} ) + 2\nu \ V (\{x,y\} ) = (2\nu -1)j$$

4.  Birth and death processes

Definition 4.1.

A birth and death process on Y  with continuous time is  a family $\{V(t)\}_{t \in \mathbb{R}}$ of set valued random variables verifying the following property :  between   t and   t+dt, we can only

- remain at the same state  $V (t) \subset X$
- take a point $x \notin V(t)$ with probability $\beta (x,V(t) ) dt$

$(V (t+dt) = V(t) \cup \{x\} )$
- take away a point $x \in V(t)$ with probability
$\delta (x,V(t) )dt$        $( V(t+ dt) = V(t)\backslash\{x\} )$

Theorem 4.1.

If    $\beta (x,A ) = \beta (x, \pi (x,A))$   and $\delta (x,A) = \delta (x, \pi (x,A))$ for every $A \subset X$   and if the birth and death process is time reversible

$(\beta (x,A)= \delta (x, A \cup \{x\} )$ , if $x \notin A)$,
then the process has a unique invariant measure $\mu$  , which is the canonical measure of a  renewal random field and its interaction verifies :

$$\forall \ A \subset X \ ; x \in A, \qquad e^{V(A \cup \{x\})- V(A)} = \beta (x,A) /\delta(x,A).$$

Because of the possibility to associate this renewal field to a random automata network, we can relate the birth and death process to an automata network, which is a discrete process realizing the continuous process, these two process having the same invariant measure.

We build the corresponding automata network by choosing the probability functions $F_x$  in a sequential way as follows :

- let us define a walk on X represented by the sequence
$\{ x_i \}_{i= 1,\ldots,n}$ , where $n = | X |$

- then , at each step of order kn+i, we have :

$$\forall \ A \subset X \ , \forall q \in Q, F_{x_i} ( A, q, 1 ) = \frac{e^{U (A \cup \{ x_i \} )}}{e^{U (A \cup \{ x_i \} )} + e^{U (A \backslash \{ x_i \} )}}$$

and   if $j \neq i$,    $F_{x_j} ( A,q,q ) = 1$

Then we can consider the automata network as a set valued Markov chain, by retaining uniquely the states of the network at steps o,n,2n,... , kn,...

Theorem 4.2.

The Markov chain associated to the sequential automata network above has a transition matrix  M verifying :

i)  $M =\{m_{D,E}\}_{D,E \subset X}$   ,

where $m_{D,E} = \prod\limits_{i=1}^{n} F_{x_i}(D_i, q, u_i)$

with $D_1 = D$

$\vdots$

$D_i = (D \setminus ((D \setminus E) \cap A_i)) \cup ((E \setminus D) \cap A_i)$,

where $A_i = \{x_1, \ldots, x_i\}$

$\vdots$

$D_n = E$,

and $u_i = 1$, if $x_i \in E$

$u_i = 0$, if $x_i \notin E$

ii)  the Gibbs measure $\mu$ defined by the same potential U as F is the unique left eigenvector of M , i.e. $\mu$ is the unique invariant measure of the Markov chain.

## 5.  Simulations

The sequential automata network above allows to simulate samples of the Gibbs measure $\mu$ and has the same asymptotic behaviour as the  birth and death process . The main problem of the simulation process is : how to choose stopping rules indicating the end of the transient ?
A first idea is to use the following theorem

## Theorem  5.1.

The Kolmogorov-Sinaï entropy of the Markov chain defined above is defined by :

$$H = -\sum\limits_{D \subset X} \mu(D) \sum\limits_{E \subset X} m_{D,E} \log(m_{D,E})$$

If $\delta$ denotes the  Kullback's distance , we have, for any initial measure $\nu$ on Y :

$$\delta(\nu M^k, \mu) < e^{-kH}$$, for k  sufficiently large

But it is practically impossible to calculate H when X is a  30 × 30 box of $\mathbb{Z}^d$.

Hence, the second idea is to follow  certain indicators  of convergence to the asymptotic behaviour like :

$X_i$ = total number of 1's  in the configuration obtained at the step i of the Markov chain

$Y_i$ = number of 1's in the left part of X
$Z_i$ =     "          "      right  "      "

$V_i$ ($V'_i$) = number of sites whose state changes from 0 to 0 between
        steps i - 1 and i (from 0 to 1).

$W_i$ ($W'_i$) = same definition as above by changing the roles of 1 and 0.

Because of the ergodic character of the Markov chain, which has a
unique measure $\mu$ , the temporal means like :

$$< X>_i \ = \ \frac{1}{i} \ \sum_{j=1}^{i} X_j$$

are converging almost surely to the corresponding spatial
expectations (8).

A relative convergence criterion $\dfrac{<X>_i \ - \ <X>_{i-1}}{< X>_i}$

allows us to decide the end of the transient. We can reinforce this
stopping rule by calculating with a Monte Carlo method the entropy H
after the transient, by using in the formula of H only the D's and
transitions $D \to E$ empirically observed in a sequence of 100 steps
after the transient for example .

6 .  Generalizations

1) We can choose randomly the potential U (like in the models of spin
glasses).
2) We can use $\mathbb{R}^d$ as index set of the field instead of $\mathbb{Z}^d$, by introducing
Holley-Stroock techniques.
3) We can define the mixed Markov renewal random field   by the following properties :
i) $\mu$ is a Markov field
ii)$\forall \ A \subset X, \ x \not\in A$, verifying : $\forall \ z,y \in A \cup \{x\}, y \in \partial z$, then we have :
   $\mu([\{x\},X]|[A,X]) \ = \ \mu([\{x\},X]|[\pi(x,A),S(x,A)])$
     The simulations showed in the Appendix correspond to such a mixed
field.
4) We can change the definition of $\pi(x,A)$ :
if    $\pi(x,A)_r = \{y \in A \ ; \ d(x,y) \leqslant d(x,A)+r\}$    , we define a renewal random
field of order r by replacing $\pi(x,A)$ by $\pi(x,A)_r$.
5) We can use a temporary "immunization" in the random automata network,
if automata in each sites remain in the state 0 during at less I steps,
where I is the minimal duration of the immunization. In this case, it is
very difficult to study theoretically the invariant measure of the network, but this last generalization is perhaps the more realistic approach
in the epidemiological models for example.

7.  Conclusion

     The tools presented above give the possibility of simulation in
many cases, for which we can suppose that a set valued birth and death
process or a Gibbsian measure play an important role : after a short
transient phase in general (30 to 100 iterations of the Markov chain
associated to the automata network, that is about $5 \ 10^5$ iterations
of the sequential automaton), we can simulate the asymptotic behaviour
of the studied process and hence predict or verify theoretical or experimental assumtions. The specific use in image processing (4,9) is a
last reason justifying a systematic study of particular (Markov, renewal, ...) random fields and of their associated automata networks.

## References

(1) Von Neumann, J. (1966). Theory of self reproducing automata, A.W. Burks ed., University of Illinois Press.
(2) Demongeot, J. (1983). Coupling of Markov processes and Holley's inequalities for Gibbs measures. Proc. of the IXth Prague Conference, 183-189, Academia, Prague.
(3) Demongeot, J. (1985). Random automata and random fields. Dynamical systems and cellular automata, 99-110, Academic Press, New York.
(4) Geman, S. and Geman, D. (1983). Stochastic relaxation, Gibbs distributions and the Bayesian restoration of images, preprint Brown University.
(5) Hammersley, J.M. and Clifford, P. (1971). Markov fields on finite graphs and lattices, unpublished.
(6) Ruelle, D. (1972). On the use of small external fields. Ann. of Physics 69, 364-374.
(7) Fricot, J. (1985). Champs aléatoires de renouvellement, Thesis, Grenoble.
(8) Goldstein, S. (1981). Entropy increase in dynamical systems. Isr. J. of Maths 38, 241-256.
(9) Grenander, U. (1983). Tutorial in pattern theory, preprint Brown University.

## APPENDIX
### 1) Simulation of a mixed field

Such a field verifies : $V(A) = 0$, if A does not verify : $\forall\, x,y \in A$, $x \in \partial y$ ; $V(A) = (-1)^{|A|}$, if A verifies this property and if $\text{diam}(A) > 1$; and $V(A)$ takes any value in the other cases.

The simulation below is done for a sequence of description of the 30x 30 box X spirally twisting ; the field is of order 2 and the potential U is attractive, translation and rotation invariant with interaction V defined by : $V(\{x\}) = \text{Log } 0.01$; $V(\{x,y\}) = - \text{Log } 0.23$, if $d(x,y)=1$; $j=-\text{Log } 0.1$.

$$X_0 = 72$$

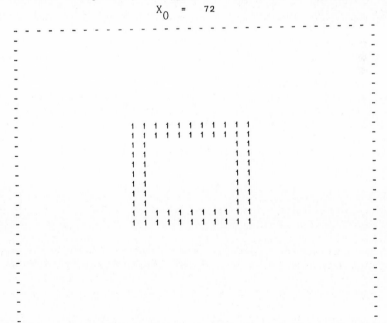

step 1        $X_1 = 98$

$V_1 = 785$
$V_1^* = 43$
$W_1 = 17$
$W_1^* = 55$

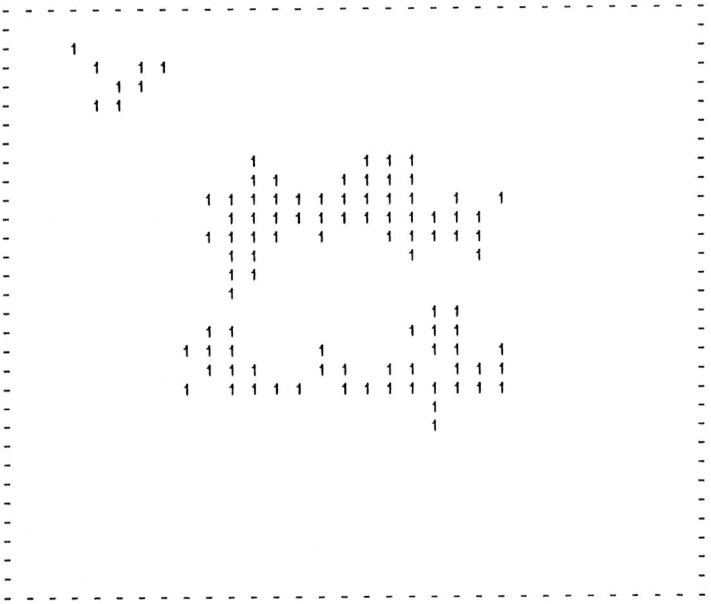

2              125

727
75
48
50

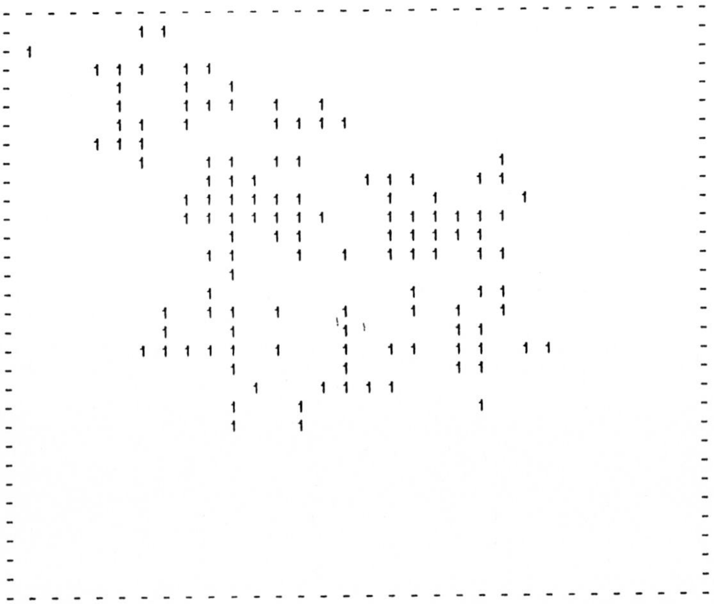

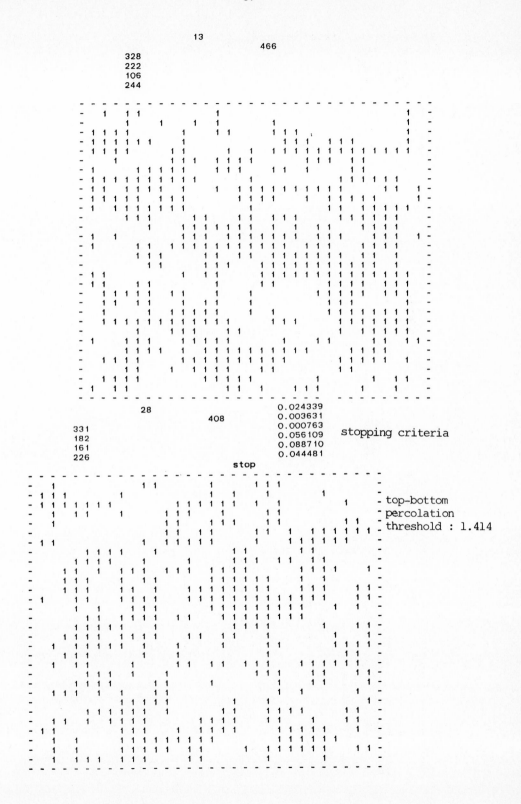

stopping criteria

stop

top–bottom
percolation
threshold : 1.414

2) Same field, but of [1] order 4    161

```
713
115
26
46
```

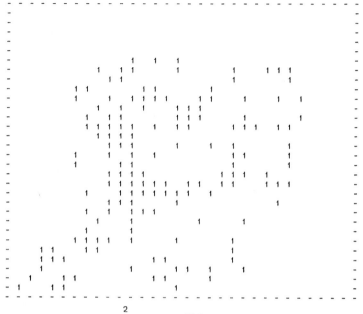

2                234

```
575
164
91
70
```

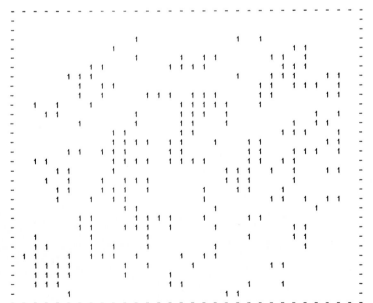

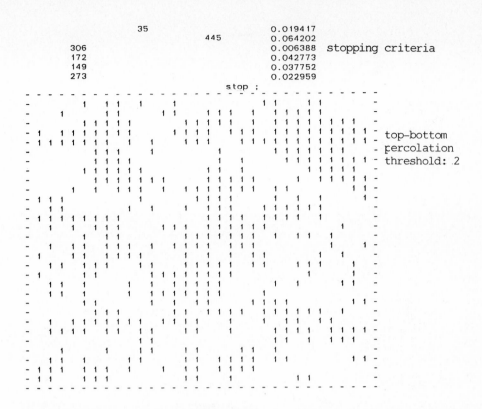

35                  0.019417
445    0.064202
306    0.006388  stopping criteria
172    0.042773
149    0.037752
273    0.022959

stop :

top–bottom
percolation
threshold: .2

3) Same field, but with immunization (I=3) and with an other initial condition

100

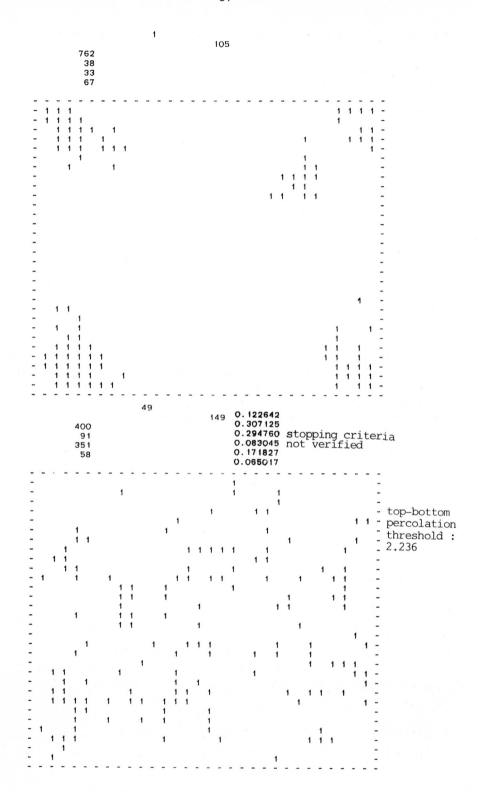

top–bottom
percolation
threshold :
2.236

stopping criteria
not verified

# LYAPUNOV FUNCTIONS AND THEIR USE IN AUTOMATA NETWORKS

**F. FOGELMAN SOULIE**, LDR, 1 rue Descartes, 75 005 PARIS
and UER de Mathématiques, Informatique, Université de Paris V.

## 1- INTRODUCTION

In this paper, we introduce the concept of **Lyapunov function** to study the dynamical behavior of automata networks. This notion, classical in continuous dynamical systems, has proved very useful for discrete systems as well.

Algebraic invariants for threshold networks [6] have long been used to characterize their dynamical behavior under parallel or sequential iterations. The concept of energy, analogous to the energy of a spin glass, was introduced [9] later on in this context and extended to the general family of positive automata [5] to show the existence of fixed points.

We review here some results which can be obtained through the definition of an adequate Lyapunov function. In the general case of boolean networks, the **entropy** of the network (as defined in [17]) plays this role: it allows us to characterize the limit set of the iteration, to give a bound on the transient and to predict the limit cycles and periods. Unfortunately, the complexity for computing the entropy makes its use hopeless in practical applications. The **energy** defined, for threshold and majority networks, by analogy with [9] proves much more tractable, since it depends only on a limited number of states. It allows to give conditions for fixed points, bounds for the transients and characterization of the attractivity of the fixed points. Applications of such properties can be made in pattern recognition : we provide an example in the recognition of words.

In §2, we give definitions (see also [13]), in §3 study the entropy of a boolean network, in §4 the energy of threshold networks, to conclude in §5.

## 2- DEFINITIONS

**Definition 1:** A *boolean mapping* in n variables is a mapping $f: \{0,1\}^n \rightarrow \{0,1\}$. A *boolean network* of n elements is a mapping $F: \{0,1\}^n \rightarrow \{0,1\}^n$, where $F=(f_1,...f_n)$, and all $f_i$ are boolean mappings in n variables.

A boolean network can be viewed as a set of interconnected elementary processors (see fig.1). Boolean networks have been extensively studied ([1],[10],[13]).

**Definition 2:** Let $F=(f_1,...f_n)$ be a boolean network of n elements. If all $f_i$ depend on k≤n variables at most, then network F is said of *connectivity* k. The

NATO ASI Series, Vol. F20
Disordered Systems and Biological Organization
Edited by E. Bienenstock et al.
© Springer-Verlag Berlin Heidelberg 1986

*connection graph* of F is the graph $\mathcal{C}=(X,\mathcal{W})$ such that $X = \{1,...,n\}$ and $(i,j) \in$ XxX is in $\mathcal{W}$ iff $f_j$ really depends on variable $x_i$:

　ie: $(i,j) \in \mathcal{W} \Leftrightarrow \exists x \in \{0,1\}^n : f_j(x_1,...,x_i,...,x_n) \neq f_j(x_1,..., \rceil x_i,...,x_n)$
　where $\rceil 0 = 1$ and $\rceil 1 = 0$.

If the n elements of network F lie in some topological space (for example $\mathbf{R}$ or $\mathbf{R}^2$ ), then we will say that F is *locally connected* or displays local -or short range- interactions if k is small with respect to n ( for example $k \leq 4$) and each element i has its function $f_i$ depending on k *neighbouring* elements. This is for example the case in cellular networks (cf [14]).

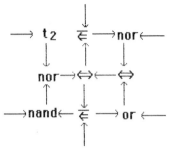

### Figure 1
This figure shows a boolean network with connectivity 2 and regular connections. Its mapping F is given by: the table of its components $f_i$ and the connection graph of the network. Connections are between nearest neighbours, and drawn on a torus.

**Definition 3:** Let $F=(f_1,...f_n)$ be a boolean network of n elements and $x^0 \in \{0,1\}^n$. We define different *iteration modes* on F with initial condition $x^0$ by $x^0 = x(0)$ and $x(t) \in \{0,1\}^n$, the state of F at time t, as:
- *parallel iteration* : $\forall t \geq 0$, $x(t+1) = F[x(t)]$
- *sequential iteration* associated to the permutation $\pi$ of $\{1,...,n\}$:
$\forall t \geq 0$, $x(t+1) = F_\pi[x(t)]$
- *block-sequential iteration* associated to a partition $(I_k)_{k=1...p}$ of $\{1,...,n\}$, ordered (which means that: $\forall x \in I_i, \forall y \in I_j$ , $i<j \Rightarrow x < y$) as: $x(t+1) = F_{(I_k)}[x(t)]$
- *random iteration*, with visiting strategy $(i_t)_{t \geq 0}$ (where $(i_t)_{t \geq 0}$ is a random process with values in $\{1,...,n\}$): $\forall t \geq 0$, $x(t+1) = F_{(i_t)}[x(t)]$.
(see [13] for full definitions).
**Remarks:**
1- the block sequential iteration associated to a partition $(\{k\})_{k=1...n}$ is the sequential iteration associated with the natural permutation.
2- the block sequential iteration associated to the trivial partition $\{1,...,n\}$ is the parallel iteration.
3- a random iteration with periodic visiting strategy of period n is a sequential iteration.

|  | Parallel | | | Sequential | | | Block sequential | | |
|---|---|---|---|---|---|---|---|---|---|
| **Cores** | 80 | 80 | 80 | 80 | 0 | 80 | 80 | 0 | 80 |
|  | 72 | 80 | 80 | 80 | 80 | 80 | 39 | 80 | 80 |
|  | 80 | 80 | 80 | 80 | 80 | 80 | 80 | 80 | 80 |
| **Stable values** | * | * | * | * | 0 | * | * | 0 | * |
|  | * | * | * | * | * | * | * | * | * |
|  | * | * | * | * | * | * | * | * | * |

**Limit cycles**

Parallel iteration: period 7.

$$
\begin{array}{ccc}
1 & 0 & 0 \\
0 & 1 & 1 \\
1 & 0 & 0 \\
\end{array}
\;\text{--->}\;
\begin{array}{ccc}
0 & 0 & 0 \\
0 & 0 & 1 \\
1 & 1 & 1 \\
\end{array}
\;\text{--->}\;
\begin{array}{ccc}
0 & 1 & 1 \\
0 & 0 & 0 \\
0 & 0 & 1 \\
\end{array}
\;\text{--->}\;
\begin{array}{ccc}
1 & 0 & 0 \\
1 & 1 & 1 \\
1 & 0 & 0 \\
\end{array}
$$

284      15      193      316

$$
\begin{array}{ccc}
1 & 1 & 1 \\
1 & 1 & 1 \\
0 & 0 & 1 \\
\end{array}
\;\text{<---}\;
\begin{array}{ccc}
0 & 0 & 1 \\
0 & 0 & 0 \\
0 & 1 & 1 \\
\end{array}
\;\text{<---}\;
\begin{array}{ccc}
0 & 0 & 0 \\
0 & 1 & 1 \\
1 & 1 & 1 \\
\end{array}
\;\text{<---}
$$

505      67      31

Sequential iteration (natural permutation): period 4.

$$
\begin{array}{ccc}
1 & 0 & 0 \\
0 & 1 & 0 \\
1 & 1 & 1 \\
\end{array}
\;\text{--->}\;
\begin{array}{ccc}
0 & 0 & 1 \\
0 & 1 & 1 \\
0 & 1 & 1 \\
\end{array}
\;\text{--->}\;
\begin{array}{ccc}
1 & 0 & 0 \\
0 & 0 & 0 \\
0 & 0 & 0 \\
\end{array}
\;\text{--->}\;
\begin{array}{ccc}
0 & 0 & 1 \\
1 & 0 & 0 \\
1 & 0 & 1 \\
\end{array}
$$

279      91      256      101

(<-------------------------------------------------------------------)

Block-sequential iteration (line by line): period 8.

$$
\begin{array}{ccc}
0 & 0 & 0 \\
0 & 0 & 0 \\
1 & 0 & 1 \\
\end{array}
\;\text{--->}\;
\begin{array}{ccc}
0 & 0 & 1 \\
0 & 1 & 1 \\
1 & 1 & 1 \\
\end{array}
\;\text{--->}\;
\begin{array}{ccc}
1 & 0 & 1 \\
0 & 0 & 1 \\
0 & 0 & 1 \\
\end{array}
\;\text{--->}\;
\begin{array}{ccc}
1 & 0 & 0 \\
0 & 0 & 0 \\
1 & 0 & 0 \\
\end{array}
$$

5      95      329      260

$$
\begin{array}{ccc}
1 & 0 & 0 \\
0 & 0 & 1 \\
1 & 0 & 1 \\
\end{array}
\;\text{<---}\;
\begin{array}{ccc}
1 & 0 & 1 \\
0 & 0 & 1 \\
1 & 0 & 0 \\
\end{array}
\;\text{<---}\;
\begin{array}{ccc}
0 & 0 & 1 \\
0 & 0 & 1 \\
0 & 0 & 1 \\
\end{array}
\;\text{<---}\;
\begin{array}{ccc}
0 & 0 & 0 \\
0 & 1 & 1 \\
1 & 1 & 1 \\
\end{array}
$$

269      332      73      31

## Figure 2

The figure shows the results of different iterations run on the network of figure 1: parallel, sequential with the natural permutation and block sequential (line by line). We have computed the limit cycles for 80 different initial conditions set at random. We show the resulting "core" ([1]) of the network: each number gives the number of times the element has been oscillating in the limit cycle. For those points which were always stable (value 0), we give the value in which they were stable. These figures differ for the different iteration modes, hence the limit cycles are different.

We also give, for each iteration mode, a limit cycle represented by the succession of states in it.

**Definition 4:** Let $F: \{0,1\}^n \rightarrow \{0,1\}^n$ be a boolean network and $F_d$ be an iteration mode on F ($F_d = F$, $F_\pi$, $F_{(lk)}$ or $F_{(it)}$). A *trajectory* of F associated to iteration $F_d$, starting at $x^0$ is a sequence $[x(t)]_{t \geq 0}$ in $\{0,1\}^n$ such that:

$$\forall t \geq 0, \; x(t+1) = F_d[x(t)] \qquad \text{and } x(0) = x^0$$

The *iteration graph* of F associated to iteration $F_d$ is a graph $\mathcal{G} = (S, \mathcal{U})$ such that S is the state space ($S=\{0,1\}^n$) and $(x,y) \in S \times S$ is in $\mathcal{U}$ iff $y=F_d(x)$. A *limit cycle* of iteration $F_d$ is a circuit of $\mathcal{G}$ and its *period* is the number of vertices in the circuit. A limit cycle of period 1 is called a *fixed point*. (see [1],[13]).

A *Lyapunov function* of the dynamics defined by $F_d$ is any function H monotonic on the trajectories: $\forall\, t \geq 0, H[x(t+1)] \leq H[x(t+1)]$
  (or $\forall\, t \geq 0, \qquad H[x(t+1)] \geq H[x(t+1)]$ )

This definition generalizes the definition of Lyapunov functions for continuous dynamic systems. In the following, we will only consider decreasing Lyapunov functions, which does not introduce any restriction.
Figure 2 shows various iterations for a boolean network.

**Remark:** All iteration modes have the same fixed points, but they may have different limit cycles.

As $\{0,1\}^n$ is a finite space, all deterministic iterations (ie non random) are ultimately periodic: $\forall\, x^0 \in \{0,1\}^n$, $\exists t(x^0), p(x^0) \in \mathbb{N}:\forall\, t \geq t(x^0), x[t+p(x^0)\,] = x(t)$
  $\forall\, t \leq t(x^0)$, or $\forall q: 0 < q < p(x^0), \qquad x(t+q) = x(t)$
where $x(t)$ is the trajectory starting at $x^0$.
  $t(x^0)$ is called the *transient length* before entering the limit cycle $[x(t(x^0)),x(t(x^0)+1),...,x(t(x^0)+p(x^0)-1)]$ of length the *period* $p(x^0)$.

## 3- ENTROPY IN BOOLEAN NETWORKS.

**Definition 5:** Let $F: \{0,1\}^n \rightarrow \{0,1\}^n$ be a boolean network and let p be a probability distribution on the state space $S = \{0,1\}^n$. The *entropy* of network F associated to p is the function E defined by:
  $E(p) = -\sum_{s \in S} p_s \log_2 p_s \qquad$ where $p_s$ is the probability of state $s \in \{0,1\}^n$.

**Theorem 1:** Let $p^0$ be an initial probability distribution on S, $F: \{0,1\}^n \rightarrow \{0,1\}^n$ be a boolean network and $F_d$ be an iteration mode on F. $F_d$ defines a trajectory $p(t)$ with: $p(0) = p^0$ and $p_s(t) = \text{Proba}[x(t) = s]$.
  Then: $\forall\, t \geq 1, E[p(t)] \leq E[p(t-1)] \qquad$ and $\quad E[p(t)] = E[p(t-1)] \quad$ iff
  $\forall\, s \in S$, either $F_d^{-1}(s) = \emptyset$, or $\exists!\ \sigma \in F_d^{-1}(s) : p_\sigma(t-1) \neq 0$.
  This theorem shows that entropy E is a Lyapunov function for the dynamics on p, it can also be used to study the dynamics of network F.

**Theorem 2:** Let $p^0$ be an initial probability distribution on S, $F: \{0,1\}^n \rightarrow \{0,1\}^n$ be a boolean network and $F_d$ be an iteration mode on F (deterministic). Let $\mathfrak{R}(t)$ be the set $\{s \in S : p_s(t) \neq 0\}$, ie the set of "live" elements at time t. Then $[\mathfrak{R}(t)]_{t \geq 0}$ is a decreasing sequence (with respect to inclusion) of subsets of S, whose intersection $\mathfrak{R} = \cap_{t \geq 0} \mathfrak{R}(t)$ is non-empty: it is the *limit set* of the iteration, it contains all the states which are part of a limit cycle. Furthermore, E(t) is convergent and its limit $E_\infty$ is such that: $\exists t^* \geq 1$:

$$E_\infty = - \sum_{s \in \mathfrak{R}} p_s(t^*) \log_2 p_s(t^*)$$

This theorem thus allows to characterize those states which are part of limit cycles. (see fig.3).

**Corollary 1**: If s and $\sigma$ are two states in $\mathfrak{R}$ such that: $F_d(s) = \sigma$, then:

$\forall \ t \geq t^*, \ p_\sigma(t+1) = p_s(t)$.

*Proof*: obvious, because: $\forall \ s' \in F_d^{-1}(\sigma), s' \neq s \Rightarrow s' \notin \mathfrak{R} \Rightarrow p_{s'}(t+1)=0$.

This corollary allows to find out the structure of limit cycles and periods from the distribution $(p(t))_{t \geq t^*}$ (see fig.3).

**Theorem 3**: Let $p^0$ be an initial probability distribution on S, F: $\{0,1\}^n \rightarrow \{0,1\}^n$ be a boolean network and $F_d$ be an iteration mode on F (deterministic). Then the transient length of $F_d$ is bounded below:

$\forall \ x^0 \in \{0,1\}^n, \ t(x^0) \leq n/ \ \varepsilon_F$

with $\varepsilon_F = \text{Min}\{|E_s(t)|/ \ t \geq 1, s \in S: F_d^{-1}(s) \cap \mathfrak{R}(t-1)$ has 2 elements at least$\}$

and $E_s(t) = \sum_{\sigma \in F_d^{-1}(s)} p_\sigma(t-1) \log_2 [ p_\sigma(t-1) / \sum_{\tau \in F_d^{-1}(s)} p_\tau(t-1)]$

$(E(t)-E(t-1) = \sum E_s(t))$.

*Proofs* for theorems 1-3 were given in [2] for parallel and sequential iterations. They can easily be extended to block sequential and random iterations using the same arguments, which rely on the reduction occurring in the state space when one applies function F. (see fig.3).

**Corollary 2**: If $\exists \ t > 0$ such that E(t)=0, then there exists a unique fixed point and no limit cycle.

*Proof*: obvious: $\exists ! \ s \in S: p_s(t) = 1$ and $\forall \ \sigma \neq s, \ p_\sigma(t) = 0$ : s is the fixed point.

**Example**: Homogeneous boolean networks of mappings XOR and $\Leftrightarrow$ have been studied ([1],[6]). It has been shown that their dynamics can be represented by the equation: $x(t+1) = Ax(t) + b$     (in $\mathbb{Z}/2$)
where x(t) is the state of the network at time t, A is the incidence matrix of the network ($a_{ij} = 1$ iff (i,j) is an arc of the connection graph $\mathcal{C}$, ie (i,j) $\in W$ ), and $b_i = 0$ if $f_i$ is XOR, $b_i = 1$ if $f_i$ is $\Leftrightarrow$.

Then, it is easy to see that: $\forall \ t \geq 0, \ x(t+1) = A^{t+1} x(0) + [ A^t + ... + A + I] b$ and thus that there exists a unique fixed point when A is nilpotent (ie $\exists r: A^r = 0$).

This result can also be proved by making use of the entropy:

$\forall \ s \in S, \ p_s(t) = \sum_{\sigma \in S} \text{proba}[x(t)=s/x(t-1) = \sigma] \ \text{proba}[x(t-1) = \sigma]$

**Figure 3:** Entropy of a boolean network: For the network of fig.1,we have computed the probability distribution p(t),starting from a uniform distribution p(0) on state space, for the parallel and sequential (with natural permutation) iterations. In figure 3-1, we show the variation in time of sets $\mathfrak{R}(t)$ (ie. elements s such that $p_s(t) \neq 0$). The limit set $\mathfrak{R}$ appears as the set of those elements which remain "alive" forever; the probability distribution is then periodic and the limit cycles can be determined by applying corollary 1. The entropy is decreasing and convergent to $E_\infty$ (fig. 3-2). The maximum transient length is equal to the time t* to reach $E_\infty$ and is bounded by $n/e_F$ (fig.3-3).

**Figure 3-1:** variation of the probability distribution in time.
– parallel iteration.

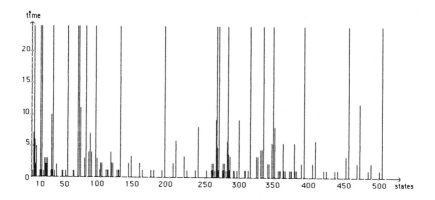

The maximum transient length is 13 and p then becomes periodic:

| states/t | 13 | 14 | 15 | 16 | 17 | 18 | 19 | 20 | 21 |
|---|---|---|---|---|---|---|---|---|---|
| 4 | .074 | .078 | .031 | .023 | .074 | .078 | .031 | .023 | .074 |
| 13 | .059 | .089 | .066 | .031 | .031 | .027 | .094 | .183 | .051 |
| 15 | .023 | .008 | .016 | .047 | .027 | .019 | .019 | .023 | .008 |
| 31 | .027 | .019 | .019 | .023 | .008 | .016 | .047 | .027 | .019 |
| 52 | .031 | .027 | .094 | .184 | .051 | .059 | .089 | .066 | .031 |
| 67 | .047 | .027 | .019 | .019 | .023 | .008 | .016 | .047 | .027 |
| 69 | .051 | .059 | .089 | .066 | .031 | .031 | .027 | .094 | .184 |
| 79 | .031 | .031 | .027 | .094 | .184 | .051 | .059 | .089 | .066 |
| 93 | .023 | .074 | .078 | .031 | .023 | .074 | .078 | .031 | .023 |
| 129 | .027 | .094 | .184 | .051 | .059 | .089 | .066 | .031 | .031 |
| 193 | .019 | .023 | .008 | .016 | .047 | .027 | .019 | .019 | .023 |
| 268 | .089 | .066 | .031 | .031 | .027 | .094 | .183 | .051 | .059 |
| 271 | .094 | .184 | .051 | .059 | .089 | .066 | .031 | .031 | .027 |
| 284 | .008 | .016 | .047 | .027 | .019 | .019 | .023 | .008 | .016 |
| 316 | .019 | .019 | .023 | .008 | .016 | .047 | .027 | .019 | .019 |
| 335 | .031 | .023 | .074 | .078 | .031 | .023 | .074 | .078 | .031 |
| 349 | .184 | .051 | .059 | .089 | .066 | .031 | .031 | .027 | .094 |
| 393 | .078 | .031 | .023 | .074 | .078 | .031 | .023 | .074 | .078 |
| 457 | .066 | .031 | .031 | .027 | .094 | .184 | .051 | .059 | .089 |
| 505 | .016 | .047 | .027 | .019 | .019 | .023 | .008 | .016 | .047 |

This gives the following limit cycles (n°1 underlined above is shown in fig.2):
  n°1:  284_ 15_ 193_316 _ 31 _67_505_(284): period 7
  n°2:  13_69_349_271_129_52 _79_457_268_(13):period 9
  n°3:  4_ 93_335_393_(4): period 4.

–Sequential iteration.

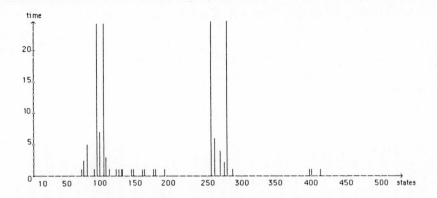

The maximum transient length is thus 8 and p then becomes periodic:

| states/t | 8 | 9 | 10 | 11 | 12 | |
|----------|-----|-----|------|------|------|--|
| 91 | .375 | .125 | .250 | .250 | .375 | 1 limit cycle: period 4 |
| 101 | .250 | .250 | .375 | .125 | .250 | **101...279... 91...256** |
| 256 | .250 | .375 | .125 | .250 | .250 | (underlined) |
| 279 | .125 | .250 | .250 | .375 | .125 | exhibited in fig. 2. |

Figure 3-2: Entropy

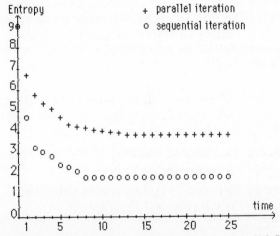

+ parallel iteration
o sequential iteration

For the parallel iteration: $E_\infty$ = 3.949 and for the sequential: $E_\infty$= 1.906 .

Figure 3-3: Transient length.
For the parallel iteration, we find: $e_F$ = .0039, hence the transient length is bounded by $9/e_F$ = 2304.
Indeed the actual maximum transient length 13 is ≤ 2304: the bound is <u>very</u> broad.
For the sequential iteration $e_F$ = .0313, the bound is $9/e_F$ = 288 and we have 8 ≤ 288.

$$\Rightarrow p_S(t) = \sum_\sigma p_\sigma(t-1) \Rightarrow \quad p_S(t) = \sum_z p_z(0)$$

$$\{\sigma : s = A\,\sigma + b\} \qquad\qquad \{z : s = A^t\, z + [\, A^{t-1} + ... + A + 1]\, b\}$$

If $A$ is nilpotent, then $\forall\ t \geq r: A^t = 0, \Rightarrow p_S(t) = \sum_z p_z(0)$

$$\{z : s = [\, A^{r-1} + ... + A + 1]\, b\}$$

$$\Rightarrow \qquad p_S(t) = \quad\begin{array}{ll} 1 & \text{if } s = [\, A^{r-1} + ... + A + 1]\, b \\ 0 & \text{otherwise} \end{array}$$

$$\Rightarrow \forall\ t \geq r, \qquad E(t) = 0$$

and thus there exists a unique fixed point : $s^* = [\, A^{r-1} + ... + A + 1]\, b$

```
6 6 9 6 6 6 6 9     1 1 0 0 1 1 1 0        B D 0 0  0 0 0 D
6 9 9 9 9 9 6 6     0 0 1 1 1 0 0 0        E C E 0  0 0 0 0
6 6 9 9 6 9 6 6     1 0 0 1 1 0 1 1        0 D B D  0 0 0 0
9 9 6 6 9 9 9 9     0 0 1 1 0 1 0 1        0 0 E C  E 0 0 0
9 9 9 9 6 9 9 6     0 1 1 1 0 1 0 0   A=   0 0 0 D  B D 0 0
6 9 6 6 9 9 9 9     1 0 0 1 1 1 1 1        0 0 0 0  E C E 0
9 6 9 9 6 9 6 9     1 1 1 1 0 1 0 1        0 0 0 0  0 D B D
9 9 6 6 9 9 9 6     1 1 1 1 0 1 0 1        E 0 0 0  0 0 E C
```

with:

```
         01000001     00000000    00000000     10000000    00000000
         00000000     10100000    01000000     00000000    00000000
         01010000     00000000    00000000     00100000    00000000
         00000000     00101000    00010000     00000000    00000000
   B =   00010100  C= 00000000 D= 00000000 E=  00001000 O= 00000000
         00000000     00001010    00000100     00000000    00000000
         00000101     00000000    00000000     00000010    00000000
         00000000     10000010    00000001     00000000    00000000
```

### Figure 4

This figure shows a boolean network $F: \{0,1\}^n \to \{0,1\}^n$, n=8x8, with regular connections (as in fig.1) with mappings XOR (6) and $\Leftrightarrow$ (9) only: left. For 80 different initial conditions, set at random, the network always ended up in a fixed point (shown in the middle). The incidence matrix $A$ of the network (right) is nilpotent: $A^8 = 0$. (see also [8]).

**Definition 6:** A boolean mapping $f: \{0,1\}^n \to \{0,1\}$ is *forcing* in its i-th variable iff:

$$\exists\ x^*_i \in \{0,1\}, \exists\ v^*_i \in \{0,1\} : \forall\ y \in \{0,1\}^n, y_i = x^*_i \Rightarrow f(y) = v^*_i$$

Variable i is called a *forcing variable* of mapping f, $x^*_i$ the *forcing value* of variable i, $v^*_i$ the *forced value* of mapping f for variable i.

This notion, first introduced by Kauffman [11], has proved very useful to characterize the stable elements in the iteration on F [2],[12].

**Definition 7:** Let $F : \{0,1\}^n \to \{0,1\}^n$ be a boolean network of connectivity k=2 and let $\mathcal{C} = (X, \mathcal{W})$ be its connection graph. We define recursively two labellings $\ell : X \to \{0,1\}$ of the nodes of $\mathcal{C}$ and $v : \mathcal{W} \to \{0,1,*\}$ of the arcs of $\mathcal{C}$ as follows:

(1) if $f_j$ is forcing in 2 variables $i_1$ and $i_2$ (ie. $(i_1,j),(i_2,j) \in \mathcal{W}$), let $v*_j$ be its unique forced value and $x*_i$ be the forcing value of $f_j$ for $i = i_1,i_2$ ( $x*_i =*$ if $f_j$ is a constant mapping).

Then we take: $\mathcal{L}(j) = v*_j$ and $v(i,j) = x*_i$ for $i = i_1,i_2$.

(2) if $f_j$ is not forcing, $\mathcal{L}(j)$ and $v(.,j)$ are not defined.

(3) if $f_j$ is forcing in 1 variable $i$ and if $\mathcal{L}(i)$ has already been defined and $\mathcal{L}(j)$ has not, then we take: $\mathcal{L}(j) = f_j[\mathcal{L}(i)]$ and $v(i,j) = \mathcal{L}(i)$

At the end of the process, $\mathcal{L}$ is defined on a subset Y of X and $v$ on a subset $\mathcal{W}'$ of $\mathcal{W}$. Then, arc $(i,j)$ in $\mathcal{W}'$ is *forcing* iff: $i \in Y$ and $\mathcal{L}(i) = v(i,j)$ or $v(i,j) = *$ (we then set $v(i,j) = \mathcal{L}(i)$). Graph $\mathcal{F} = (Y, \mathcal{W}_f)$ is called the *forcing graph* of F iff $\mathcal{F}$ is the subgraph of $\mathcal{C}$ such that $(i,j) \in \mathcal{W}$ is in $\mathcal{W}_f$ iff $(i,j)$ is forcing. Any circuit of graph $\mathcal{F}$ is called a *forcing circuit* of graph $\mathcal{C}$.

It has been shown [2] that a forcing circuit C would generally fall into a fixed value $v*_C$ under either parallel or sequential iterations. This value is called the stable value of the circuit and is –in first approximation– defined as: $v*_C(i) = \mathcal{L}(i)$, for all points $i$ on C (see [2] for details and also [11] and [12] in this volume).

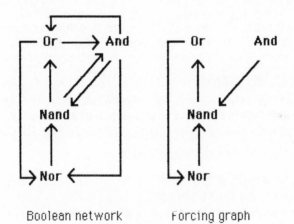

Boolean network          Forcing graph

### Figure 5

The figure shows a boolean network F: $\{0,1\}^4 \rightarrow \{0,1\}^4$ and its forcing graph. Circuit C=(Or, Nor, Nand) is forcing and its probability $P_C$ to be in its forcing value is, by computing the 3 first terms in theorem 4: $P_C = 1/8 + 1/2 + 3/10 + ... = 37/40+ ...$ We thus see that this circuit will almost certainly fall into its stable value, which is indeed the result found in simulations.

**Theorem 4:** Let F : $\{0,1\}^n \rightarrow \{0,1\}^n$ be a boolean network of connectivity k=2, C be a forcing circuit of length k and stable value $v*_C$.

Let $P_C$ = Proba[ $\exists\, t \geq 0 : \forall\, i \in C$, $x_i(t) = \boldsymbol{v}^*_C(i)$ ], where $x(t)$ is a trajectory of the parallel iteration on F. Then:

$$P_C = 1/2^k + \sum_{t \geq 1} \frac{\displaystyle\sum_{i,j:\, \boldsymbol{v}^*_C \in A_{ij}} \text{Proba}[e_{ij}(t-1)]}{\displaystyle\sum_{i,j} \text{Proba}[e_{ij}(t-1)]} [1-1/2^k] \prod_{\tau=0}^{t-2} [1 - \frac{\displaystyle\sum_{i,j:\, \boldsymbol{v}^*_C \in A_{ij}} \text{Proba}[e_{ij}(\tau)]}{\displaystyle\sum_{i,j} \text{Proba}[e_{ij}(\tau)]}]$$

where: $e_{ij}$ is the event: exactly i elements of C are in their stable value $\boldsymbol{v}^*_C(i)$ (there are $C_k^i$ such events, indexed by $j = 1,...,C_k^i$ )

$A_{ij} = \{v \in \{0,1\}^k : x,y \in \{0,1\}^{n-k} : (v,x) = F(x_{ij},y)\}$ ($x_{ij}$ is the state vector of those elements of C which are in their stable value $\boldsymbol{v}^*_C$ in event $e_{ij}$).
*Proof:* see [2].

Figure 5 shows this probability for a given boolean network.

## 4- ENERGY IN THRESHOLD NETWORKS

**Definition 7:** Let $F : \{0,1\}^n \rightarrow \{0,1\}^n$ be a boolean network. F is a *threshold network* iff all its components $(f_1,...,f_n)$ are *threshold mappings* :

$$\forall\, i(1,...,n),\ \forall\, x \in \{0,1\}^n,\quad f_i(x) = \begin{cases} 1 & \text{if } \sum_j a_{ij} x_j - b_i \geq 0 \\ 0 & \text{otherwise} \end{cases}$$

where $A=(a_{ij})$ is a real nxn matrix and $b=(b_i)$ is a real n-vector. $a_{ij}$ are called the *synaptic weights* or interaction coefficients, $b_i$ are the *thresholds*.

$f_i$ is called a *strict threshold mapping* iff:

$$f_i(x) = \begin{cases} 1 & \text{if } \sum_j a_{ij} x_j - b_i > 0 \\ 0 & \text{if } \sum_j a_{ij} x_j - b_i < 0 \end{cases}$$

It is easy to show that any threshold mapping is equivalent to a strict threshold mapping ([3]) by changing the thresholds.

**Definition 8:** Let $F : \{0,1\}^n \rightarrow \{0,1\}^n$ be a boolean network. F is a *majority network* iff all its components $(f_1,...,f_n)$ are *majority functions*:

$$\forall\, i \in \mathbf{J}_0,\ \forall\, x \in \{0,1\}^n,\quad f_i(x) = \begin{cases} 1 & \text{if } \sum_{j \neq i} a_{ij} x_j - b_i > 0 \\ x_i & \text{if } \sum_{j \neq i} a_{ij} x_j - b_i = 0 \\ 0 & \text{if } \sum_{j \neq i} a_{ij} x_j - b_i < 0 \end{cases}$$

$$\forall\, i \in \mathbf{J}_1,\ \forall\, x \in \{0,1\}^n,\quad f_i(x) = \begin{cases} 1 & \text{if } \sum_{j \neq i} a_{ij} x_j - b_i > 0 \\ 1-x_i & \text{if } \sum_{j \neq i} a_{ij} x_j - b_i = 0 \\ 0 & \text{if } \sum_{j \neq i} a_{ij} x_j - b_i < 0 \end{cases}$$

It is easy to see that any majority function is equivalent to a threshold function, with the same coefficients $a_{ij}$, and $a_{ii}$ such that:

$$\forall i \in J_0, a_{ii} \geq 0 \qquad\qquad \forall i \in J_1, a_{ii} < 0$$

(see also [5]).

**Definition 9:** Let $F : \{0,1\}^n \to \{0,1\}^n$ be a threshold or majority network. We define the *energy* of the network for the different iteration modes as follows:

- sequential, block sequential or random iterations: $\forall x \in \{0,1\}^n$,

$$E(x) = -1/2 \sum_{i=1}^{n} x_i \sum_{j \neq i} a_{ij} x_j + \sum_{i=1}^{n} (b_i - a_{ii}) x_i \text{ (threshold)}$$

$$E(x) = -1/2 \sum_{i=1}^{n} x_i \sum_{j \neq i} a_{ij} x_j + \sum_{i=1}^{n} b_i \, x_i \text{ (majority)}$$

- parallel iteration: $\forall x \in \{0,1\}^n$, $\quad E[x(t)] = \xi[x(t), x(t-1)]$

with: $\quad \xi[u,v] = \sum_{i=1}^{n} u_i \sum_{j=1}^{n} a_{ij} v_j + \sum_{i=1}^{n} b_i [u_i + v_i]$ (threshold)

**Remark:** this energy coincides with the energy of a spin glass with field **b**, as defined for example in [9] only in the case of sequential, block sequential and random iterations. The energy for the parallel iteration does not have any physical meaning in this context.

**Theorem 5:** Let $F : \{0,1\}^n \to \{0,1\}^n$ be a threshold network.
   Then: $\forall t \geq 0$, $x(t+1) \neq x(t) \Rightarrow E[x(t+1)] < E[x(t)]$
- for any sequential iteration, where **A** is a symmetric matrix, with non negative diagonal elements.
- for any block-sequential iteration associated to the ordered partition $(I_k)$, where **A** is a symmetric matrix with its blocks $A_k = (a_{ij})_{i,j \in I_k}$ non negative definite on the set $\{-1,0,1\}$, for all k.
- for the parallel iteration, where **A** is a symmetric matrix.
(in each case, E is the energy defined for that iteration mode).

   If F is a majority network, then:
- for any sequential iteration, if **A** is a symmetric matrix:
   $\forall t \geq 0,\quad x(t+1) \neq x(t) \qquad\qquad \Rightarrow \qquad E[x(t+1)] \leq E[x(t)]$
Furthermore, if $\exists k : x_k(t+1) \neq x_k(t)$, then:
   $k \in J_0 \qquad\qquad\qquad \Rightarrow \qquad E[x(t+1)] < E[x(t)]$
   $k \in J_1$ , and there is no tie for $k \Rightarrow E[x(t+1)] < E[x(t)]$
(where $x_k(t)$ is the state of the network when, starting from time t, k only has changed state).

   This theorem thus shows that the energy is a Lyapunov function of the network for the different iteration modes. Note that the energy depends only on

**Figure 6**: iterations on threshold and majority networks.
This figure shows the results of iterations run on threshold networks (fig.6-1) with non symmetric matrix **A** and on majority networks with symmetric matrix **A**, with diagonal not fully non negative (fig.6-2). Networks with symmetric matrix **A** and non negative diagonal only have fixed points for sequential iterations (fig7), but otherwise they may have long periods.

**Figure 6-1**: threshold network with non symmetric integer matrix **A** ($a_{ij}=+1$, marked +, or $-1$, marked $-$), regular local connections and size n=10x10.

```
  -   -   -   +   +   +   +   -   -   -
-*- -*- +*- -*- +*- +*+ -*- +*+ +*- -*-
  -   -   +   -   +   -   -   -   -   -
  +   -   +   -   -   +   -   +   +   +
+*+ -*- -*+ +*+ +*+ +*- -*- -*+ +*- +*-
  -   -   +   +   _   +   +   +   -   +
  +   +   +   -   +   +   +   -   +   -
-*- -*- -*+ -*- +*- +*+ -*- -*- -*+ +*+
  -   -   -   +   +   -   -   -   +   -
  +   -   -   -   +   -   +   +   +   +
-*- -*- +*- -*+ -*+ +*- +*+ -*+ +*+ +*-
  -   +   -   +   +   +   -   -   +   +
  +   +   +   +   -   +   -   +   +   +
+*- +*+ -*- -*+ +*+ -*- +*+ +*+ -*+ -*+
  +   -   +   +   -   -   -   +   +   +
  -   -   +   +   +   -   -   +   +   +
+*- +*+ -*+ -*+ +*+ -*+ -*- +*- -*- -*-
```

| 0 | 0 | 0 | 0 | 0 | 0 | 16 | 16 | 16 | 0 |
|---|---|---|---|---|---|----|----|----|---|
| 0 | 0 | 0 | 0 | 0 | 0 | 0 | 0 | 16 | 0 |
| 0 | 0 | 0 | 0 | 0 | 0 | 0 | 0 | 0 | 0 |
| 0 | 0 | 16 | 16 | 16 | 0 | 0 | 0 | 0 | 0 |
| 0 | 0 | 16 | 0 | 16 | 16 | 0 | 0 | 0 | 0 |
| 16 | 16 | 0 | 0 | 0 | 0 | 0 | 0 | 0 | 16 |
| 16 | 16 | 16 | 0 | 0 | 0 | 14 | 0 | 0 | 0 |
| 0 | 0 | 16 | 0 | 14 | 10 | 14 | 0 | 0 | 0 |
| 16 | 0 | 16 | 16 | 14 | 0 | 0 | 0 | 0 | 0 |
| 16 | 0 | 0 | 0 | 0 | 0 | 16 | 16 | 0 | 16 |

Parallel iteration (same notations as in fig 2 with 16 initial conditions). The average period is 24.00.

```
  +   -   +   +   -   -   -   +   +   +
  +   +   -   +   +   +   -   -   -   +
+*- +*+ -*- -*+ +*+ -*- +*+ +*+ -*+ -*+
  +   -   +   +   -   -   -   +   +   +
  +   +   -   +   +   +   -   -   -   +
-*- -*- -*+ -*- +*- -*+ +*+ -*- +*+ -*+
  -   -   -   +   +   -   -   -   +   -
  +   +   +   -   -   +   +   +   -   -
-*- -*+ +*- +*+ -*- +*+ +*+ +*- -*- -*-
  -   +   -   +   +   +   -   +   -   -
  +   +   +   -   -   +   +   -   +   -
+*- +*+ +*- -*- +*- -*- +*- +*- -*- +*-
  -   +   -   +   +   -   -   +   -   +
```

Sequential iteration (with natural permutation). Average period is 12.00.

| 0 | 0 | 0 | 0 | 0 | 0 | 16 | 16 | 16 | 0 |
|---|---|---|---|---|---|----|----|----|---|
| 0 | 0 | 0 | 0 | 0 | 0 | 0 | 0 | 16 | 0 |
| 0 | 0 | 0 | 0 | 0 | 0 | 0 | 0 | 16 | 0 |
| 0 | 0 | 16 | 16 | 16 | 0 | 0 | 0 | 0 | 0 |
| 0 | 0 | 16 | 0 | 16 | 16 | 0 | 0 | 0 | 0 |
| 16 | 16 | 0 | 0 | 0 | 0 | 0 | 0 | 0 | 0 |
| 16 | 16 | 16 | 0 | 0 | 0 | 4 | 0 | 0 | 0 |
| 0 | 0 | 16 | 0 | 4 | 0 | 4 | 0 | 0 | 0 |
| 16 | 0 | 16 | 16 | 4 | 0 | 0 | 0 | 0 | 0 |
| 16 | 0 | 0 | 0 | 0 | 0 | 16 | 16 | 0 | 16 |

**Figure 6-2**: this figure shows a majority network, of size n=8x8, with symmetric integer matrix **A** (same conventions as in fig. 6-1 for the connections), and diagonal not fully non-negative: diagonal elements are shown in matrix D. Elements in $J_0$ were all found stable (in 0 or 1 as shown on the fig.), elements in $J_1$ sometimes oscillate (marked 3 in the fig.).

$$
D=\begin{pmatrix}
-0.5 & 0.9 & 0.1 & -0.2 & -0.2 & 0.2 & 0.6 & 1.0 \\
-1.0 & -0.8 & -0.1 & -0.9 & 0.1 & -1.0 & 0.8 & 0.0 \\
0.7 & -0.4 & -0.9 & -0.4 & -0.2 & 0.8 & 0.6 & -0.8 \\
-0.8 & 0.0 & -0.9 & 0.3 & -0.5 & 0.3 & 0.9 & 0.1 \\
0.5 & -0.3 & -0.1 & 0.0 & -0.5 & 0.8 & -0.4 & 0.9 \\
0.4 & -0.5 & 0.4 & 0.4 & -0.6 & 0.8 & 0.7 & 0.4 \\
1.0 & 0.0 & 0.3 & -1.0 & -0.7 & 0.1 & 0.0 & -0.2 \\
0.9 & 0.3 & -0.2 & -1.0 & 1.0 & -0.6 & 0.1 & -0.9
\end{pmatrix}
$$

```
*-*+*-*+*+*+*-*+     1 0 0 3 0 0 0 1
-  + + -  + -  - -
*+*+*+*+*-*+*+*-     0 0 0 3 0 1 1 3
-  + + -  + + + -
*+*+*+*-*+*-*+*+     1 0 0 3 3 1 1 1
+  - - - - + + +
*-*+*+*+*+*-*-*-     0 1 1 1 1 1 0 1
-  - + + + - + +
*+*+*+*+*-*-*-*-     1 3 1 1 1 0 1 0
+  - + - - + + +
*-*+*+*+*-*-*-*-     1 3 1 1 1 0 1 0
-  + + - - - + -
*+*+*+*+*-*-*+*-     1 1 1 3 0 1 3 3
-  - - - - - + +
*+*-*-*+*+*+*-*+     0 0 3 3 1 3 1 3
```

one (or two for the parallel iteration) state, **contrary** to the entropy defined in §3, which depends on the whole distribution of states. The complexity for computing the energy is thus much lower than for the entropy.

**Corollary 3**: Under the conditions of theorem 5:
- all sequential, random and block sequential iterations on threshold networks only have *fixed points*. Furthermore, the transient length of any trajectory is bounded below :

$$\forall x^0 \in \{0,1\}^n, \ t(x^0) \le [1/\varepsilon][1/2 \sum_{i,j=1}^{n} |a_{ij}| + \sum_{i=1}^{n} |b_i|]$$

with $\varepsilon = \text{Min} \{|E[x(t)]-E[x(t+1)]|/ \ x^0 \in \{0,1\}^n, \ x(t+1) \ne x(t)\}$

Moreover, if **A** is an integer matrix $(a_{ij} \in \mathbb{Z})$: $t(x^0) \le \sum_{i,j} |a_{ij}| + 2\sum_i |b_i|$

if **A** has its elements in $\{-1,0,1\}$ $(a_{ij} \in \{-1,0,1\})$: $\qquad t(x^0) \le 3nv$

where $\forall$ i, card $\{ j \in \{1,...,n\} : a_{ij} \ne 0\} \le v \quad (v \le n)$
- the parallel iteration on a threshold network only has limit cycles of period 1 or 2. Furthermore, the transient length of any trajectory is bounded below :

$$\forall x^0 \in \{0,1\}^n, \ t(x^0) \le [1/\varepsilon][1/2 \sum_{i,j} |a_{ij}| + \sum_i |b_i|]$$

with $\varepsilon = \text{Min} \{|\xi[F^2(x),F(x)]-\xi[F(x),x]|/ \ x \in \{0,1\}^n, \ x \ne F^2(x)\}$

Moreover, if **A** is an integer matrix or if **A** has its elements in $\{-1,0,1\}$, the bounds are as before.
- for all sequential and random iterations of majority networks, in any limit cycle, all elements i with i in $J_0$ are stable, all elements i with i in $J_1$ change states only in case of a tie.

This corollary is very similar to the theorem previously proved on boolean networks by makink use of the entropy: it shows the power of the general concept of **Lyapunov functions** in the study of the dynamical behavior of a discrete dynamical system. Figure 6 shows examples of iterations on threshold and majority networks.

It is also possible to make use of the energy to study the associative memory capabilities of threshold networks ([9]).

**Definition 10**: Let F: $\{0,1\}^n \rightarrow \{0,1\}^n$ be a threshold network, with interaction matrix **A** and threshold vector **b**. Let $\mathcal{X}_S = \{x^S\}_{S \in S}$ be a family of states in $\{0,1\}^n$. F is an *associative* $\mathcal{X}_S$ *memory* iff:
- all $x^S$ are fixed points of the sequential iterations on F.
- all $x^S$ are attractive in their neighborhoods: $\forall s \in S, \ \exists V_S \ c\{0,1\}^n : x^S \in V_S$
and: $\forall x(0) \in V_S, \ \exists T \ge 0: \forall t \ge T, \ F_\pi^t [x(0)] = x^S$.

According to this definition, the memorized patterns, the $x^S$, are stable in the recognition process (the iteration), and any pattern sufficiently close to an $x^S$ will be recognized as this $x^S$. (see fig.7).

**Figure 7**: associative memory to retrieve part of our bibliography.

The figure shows the results of a sequential iteration run on a threshold network F=(**A,b**) with **A** computed by the Hopfield rule (right) and by a linear rule: $a_{ij} = \sum_s x_i^s \cdot x_j^s$, with $x \in \{-1,+1\}^n$ (left). The threshold **b** is chosen so as to satisfy the first condition in theorem 6.

The network is of size n=104, each word is coded by concatenating the binary values of the ASCII codes of its letters. We memorized the 4 words: FOGELMAN, GOLES, HOPFIELD, KAUFFMAN. But a network of this size is capable of memorizing about 8 words. Words in **bold** are fixed points.

Obviously, authors in our bibliography are not (linearly) independent. Nethertheless, our system allows to retrieve words several bits away from the original word (one letter different is equivalent to at most 8 bits different). This shows that the second condition (sufficient for attractivity in the first neighborhood) in theorem 6 is not very restrictive: we did not impose it and the system still work well. Performances are much improved by working in $\{-1,+1\}^n$ (see [16]).

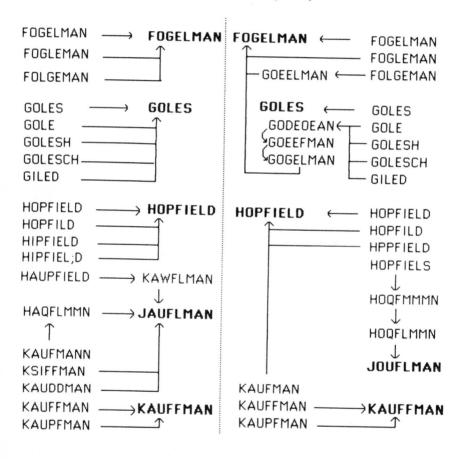

**Theorem 6**: If **A** is a symmetric matrix, with non negative diagonal elements, if **A**, **b** and $\mathbf{x}_S$ satisfy the following conditions: $\forall s \in S, \forall i,j$:

- $\quad grad_i E(x^S) \cdot (x_i^S - \exists x_i^S) \leq 0$
- $\quad [grad_j E(x^S) + (x_i^S - \exists x_i^S)] \cdot [x_j^S - \exists x_j^S] \leq 0$

then F is an associative $\mathbf{x}_S$ memory.

The first condition in theorem 6 can be written as a condition on thresholds when the $a_{ij}$ are given: $\forall i(1,...,n)$:

$$Max\{\textstyle\sum_{j\neq i} a_{ij} x_j^S \: / s \in S: x_i^S=0\} < b_i \leq Min\{\textstyle\sum_{j\neq i} a_{ij} x_j^S \: / s \in S: x_i^S=1\}.$$

In the case where the $a_{ij}$ are chosen as in Hopfield ([9])

$$\forall \; i,j \; (1,...,n), \qquad\qquad a_{ij} = \textstyle\sum_{s\in S} (2 x_i^S - 1)(2 x_j^S - 1)$$
$$a_{ii} = 0$$

The theorem gives a **necessary and sufficient condition** which ensures that the patterns are fixed points attractive in their first neighborhood. This condition holds whatever the patterns: in particular they do **not** need to be independent or orthogonal as was requested in the original Hopfield's work. This is made possible by choosing appropriately the thresholds $b_i$, which may eventually become non zero, if S is too large (see [4],[15] and [16] for more details on this).

Simulations show that, in general, conditions on the thresholds (necessary for the stability of the patterns) are sufficient to ensure their attractivity in a neighborhood much larger than the first one. This feature may be used in practical systems of pattern recognition: for example, Figure 7 shows an application of this for the memorization and retrieval of words.


# 5- CONCLUSION

We have shown on various examples of automata networks -boolean, threshold, majority- that by using a Lyapunov function, it was possible to characterize the dynamical behavior of the network under various iteration modes. Different networks have different Lyapunov functions (energy and entropy for example) and, for a given network, the Lyapunov function may differ among the different iteration modes, thus leading to eventually different dynamical behavior (fixed points only and period 2 for example). Furthermore, the complexity for computing the Lyapunov function may widely differ, depending on the network under study.

We do not know of any general Lyapunov function, valid for all networks and easily computed: the problem remains open, for the majority of the networks, to design an appropriate Lyapunov function.

# 6- REFERENCES

[1]  F. FOGELMAN SOULIE: Contributions à une Théorie du Calcul sur Réseaux. *Thèse, Grenoble* 1985.

[2]  F. FOGELMAN SOULIE: Parallel and Sequential Computation on Boolean networks. *Theoret. Comp. Sc.,* to appear.

[3]  F. FOGELMAN, E. GOLES, G. WEISBUCH: Transient Length in sequential Iterations of Threshold Functions. *Disc. Appl. Math.*, 6, pp 95-98, 1983.

[4]  F. FOGELMAN SOULIE, G. WEISBUCH:Random Iterations of Threshold Networks and Associative Memory. Submitted to *SIAM J. on Computing.*

[5]  E. GOLES CHACC: Positive Automata Networks. (this volume).

[6]  E. GOLES CHACC: Comportement Dynamique de Réseaux d'Automates. *Thèse, Grenoble* 1985.

[7]  E. GOLES CHACC, F. FOGELMAN SOULIE, D. PELLEGRIN: Decreasing Energy Functions as a Tool for Studying Threshold Networks. *Disc. Appl. Math.*, to appear.

[8]  H. HARTMAN, G.Y. VICHNIAC: Inhomogeneous Cellular Automata. (this volume).

[9]  J.J. HOPFIELD: Neural Networks and Physical Systems with Emergent Collective Computational Abilities. *Proc. Nat. Acad. Sc; USA*, vol. 79, pp 2554-2558, 1982.

[10] S.A. KAUFFMAN:Behaviour of Randomly Constructed Genetic Nets. In "Towards a Theoretical Biology". Ed. C.H. Waddington, *Edinburgh Univ. Press*, vol.3, pp 18-37, 1970.

[11] S.A. KAUFFMAN: Boolean Systems, Adaptive Automata, Evolution. (this volume).

[12] D. PELLEGRIN: Dynamics of Random Boolean Networks. (this volume).

[13] F. ROBERT:Basic Results for the Behaviour of Discrete Iterations. (this volume).

[14] G. VICHNIAC: Cellular Automata Models of Disorder and Organization. (this volume).

[15] G. WEISBUCH, D. d'HUMIERES: Determining the Dynamic Landscape of Hopfield Networks. (this volume).

[16] G. WEISBUCH, F. FOGELMAN SOULIE: Scaling laws for the Attractors of Hopfield Networks. *J. Physique Lett.*, 46, pp L623-L630, 1985.

[17] S. WOLFRAM: Statistical Mechanics of Cellular Automata. *Rev. of Modern Physics*, 55-3, pp 601-645, 1983.

# POSITIVE AUTOMATA NETWORKS (*)

## E. GOLES CH.

Dept. Matematicas, Esc. Ingenieria, U. de Chile
Casilla 5272, Correo 3, Santiago, Chile
and TIM3, CNRS

## 1 INTRODUCTION

Automata networks have been introduced as a modeling tool in several fields: neurophysiology, Fogelman(1985), Hopfield(1982), Mc Culloch(1943), selfreproducing cellular arrays, Von Neumann(1966), group dynamics, Goles(1985a) and, more recently, simulations on spin glass structures and other physical systems, Demongeot, Goles, Tchuente(1985), Fogelman(1983,1985), Goles (1985a).

In most of those applications, the study of the automata dynamics was made by computer simulations and the theoretical results were very few. This last aspect is due, principally, to the hard combinatorial analysis which is needed in order to handle the discrete nature of the problem: finite set of states, discrete cellular array, discrete time evolution, etc.

In this paper we present a mathematical framework for the analysis of a large class of networks: *the positive automata networks*, Goles(1985b,c). As a particular case of such networks we find the neural models, McCulloch, Pitts (1943) and majority functions analogous to spin glasses, Goles(1985b), Goles, Tchuente (1983). Our basic idea is to determine the local transition functions that allow associating to a given network a monotonic or Lyapunov function similar to the spin glass energy, Fogelman, Goles, Weisbuch (1983), and Goles (1985a).Through this tool, we determine the dynamical behaviour of such networks; i.e cycle lengths, convergence speed, etc.

## 2 SOME EXAMPLES AND DEFINITIONS

An *automata network* is a collection of modules, each with the same time scale, interconnected by sending the output of each module through several lines and connecting some of these to the input lines of others modules. An output may thus lead to any number of inputs, but an input may only come from at most one output. Also, we shall assume that input-output relations between two modules are symmetric; i.e if module i has module j as an input then module j has module i as an input.

(*) I am grateful for partial support by Fondo Nacional de Ciencias (Chile)

Each module can be in a finite number of states; i.e: $Q = \{0,1, ....,p-1\}$. The state at time t of each module depends only on the states at time t-1 of its input modules. The control of each module is made according to a *positive local function* that takes into account the states of the neighbouring modules at time t-1. As an example, let us take the set of states $Q=\{-1, +1\}$ and the cellular array of n=4 modules:

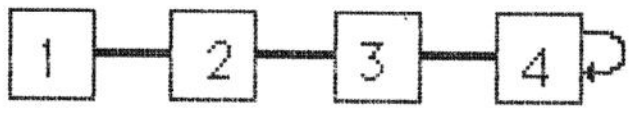

fig. 1

The local functions associated to each module are the following:

Module 1 : $1(x_2)$

Module 2 and 3 : $1(x_{i-1} + x_{i+1})$     $(x_1, x_2, x_3, x_4) \in \{-1,+1\}^4$

Module 4 : $1(x_3 + x_4)$

Where         $1(u) = \begin{cases} -1 & \text{if } u<0 \\ +1 & \text{otherwise} \end{cases}$         is a threshold function

We are interested in the parallel evolution of this network; i.e :

$$x(t+1) = \begin{bmatrix} x_1(t+1) \\ x_2(t+1) \\ x_3(t+1) \\ x_4(t+1) \end{bmatrix} = \begin{bmatrix} 1(x_2(t)) \\ 1(x_1(t) + x_3(t)) \\ 1(x_2(t) + x_4(t)) \\ 1(x_3(t) + x_4(t)) \end{bmatrix} \qquad \begin{array}{l} t= 0, 1, 2, .... \\ x(0) \in \{-1,+1\}^4 \end{array}$$

The dynamics diagram associated with the initial vector (1-1-1-1) is the following:

$$\underset{x(0)}{1\text{-}1\text{-}1\text{-}1} \mapsto \underset{x(1)}{-11\text{-}1\text{-}1} \mapsto \underset{x(2)}{1\text{-}11\text{-}1} \mapsto \underset{x(3)}{-11\text{-}11} \mapsto \underset{x(4)}{1\text{-}111} \mapsto \underset{x(5)}{-1111} \mapsto \underset{\substack{x(6) \\ \text{fixed point}}}{1111}$$

fig. 2

If we define, for this example, the operator :

$$E(u,v) = -( u_1v_1 + u_2(v_1 + v_3) + u_3(v_2 + v_4) + u_4(v_3 + v_4)) \; ; \; u, v \in \{-1, +1\}^4$$

we have :

$$E(x(1),x(0)) = E(x(2),x(1)) = E(x(3),x(2)) = E(x(4),x(3)) =$$

$$E(x(5),x(4)) = E(x(6),x(5)) = -5 \quad \text{and} \quad E(x(6),x(6)) = -7$$

Clearly the network evolves towards a stable state where all the modules take the same value. Furthermore the operator E decreases along the trajectory and stable states, $(-1-1-1-1)$ and $(1111)$, are global minima of E.

Physically, we can say that neighbouring modules try to take the same state in order to minimize operator E. On the other hand, the fact that E is not strictly decreasing rises a problem for the computation of the transient length of the network. In order to solve this problem, we can slightly change the local functions of the network without changing its dynamics and by this method we determine a strictly decreasing operator E :

Module 1 : $1(x_2)$ ;  Modules 2 and 3 : $1(x_{i-1} + x_{i+1} + 1)$ ;  Module 4 : $1(x_3 + x_4 + 1)$

It is easy to see that the dynamics of this network is the same as in the previous model (see fig.2), but now, if we take the following expression for E :

$$E(u,v) = -(u_1v_1 + u_2(v_1 + v_3) + u_3(v_2 + v_4) + u_4(v_3 + v_4) + u_2 + v_2 + u_3 + v_3 + u_4 + v_4)$$

It is strictly decreasing along the trajectory; i.e :

$$E(x(t),x(t-1)) = -2(t-1) - 1 \qquad \text{for} \quad t = 1, \dots, 6$$

and $E(x(6),x(6)) = -13$ is the global minimum of E

## 3 THRESHOLD NETWORKS DYNAMICS

In a more general context, let us suppose that we have n modules with local transition functions :

$$\text{Module i} : 1( \Sigma_{j=1,n} a_{ij}x_j ) \qquad 1 \leq i \leq n$$

where $x_j \in \{-1, +1\}$ and $A = (a_{ij})$ is an integer symmetric nxn matrix.

$$-\|A\| - 2\Sigma_{i=1,n} b_i \leq E(u,v) \leq \Sigma_{i=1,n} b_i - n, \text{ for any } u,v \in \{-1,+1\}^n$$

where $\|A\| = \Sigma_{i,j=1,n} |a_{ij}|$

From the previous properties we conclude that this class of automata networks verifies:

$P \leq 2$  (the network only has fixed points or cycles of length two)

$T \leq (\|A\| + 3 e - n)/2$, where $e = \Sigma b_i$ is the number of even lines in matrix A

Furthermore, the first example proves that the bound is reached (see fig. 2).

The previous results are very general and other proofs are available, Goles(1985a) Poljak(1984). The goal of our approach is to associate to the network a *Lyapunov function* that takes into account the deep properties of local functions; i.e: for any u,v belonging to $\mathbb{R}$ :

$(1(u)-1(v))u \geq 0$     and

$(1(u)-1(v))u = 0$     iff   $1(u) = 1(v)$ or $u = 0$

Functions verifying these properties are called *strictly positive functions*, Goles(1985a,b).


## 4. POSITIVE NETWORKS

In this paragraph we generalize the previous results to a large class of networks, the *positive networks*. The basic idea is that we are able to study dynamics on a network if we can define a *Lyapunov function*. Hence we characterize local funcions that allow determining a Lyapunov operator.

Let Q be a subset of $\mathbb{R}^p$ and f be a function from $\mathbb{R}^p$ into Q. We shall say that f is *positive* iff :

$\langle f(u)-f(v), u \rangle \geq 0$     for any u,v in $\mathbb{R}^p$

and *strictly positive* if f also verifies :

$\langle f(u)-f(v), u \rangle = 0$ iff $f(u) = f(v)$ or $u = 0$, where $\langle x, y \rangle = \Sigma_{i=1,p} x_i y_i$

We shall study the parallel evolution of this network; i.e :

$$x_i(t+1) = 1( \Sigma_{j=1,n} a_{ij} x_j(t) ) \quad t=0,1, \ldots ; \quad x(0) \in \{-1,+1\}^n$$

Similarly to the previous example, since the entries $a_{ij}$ are integers, we can change the precedent network for the following, which has the same dynamics:

$$x_i(t+1) = 1( \Sigma_{j=1,n} a_{ij} x_j(t) + b_i ) \quad t=0,1, \ldots ; \quad x(0) \in \{-1,+1\}^n$$

where $b_i = \begin{cases} 0 & \text{if } \Sigma_{j=1,n} a_{ij} \text{ is odd} \\ 1 & \text{otherwise} \end{cases}$

Clearly, since $\{-1,+1\}^n$ is a finite set, all the trajectories $\{ x(t) : t \geq 0 \}$ are ultimately periodic. Hence for every configuration $x \in \{-1,+1\}^n$ there exist numbers $p(x), t(x) \in \mathbb{N}$ called respectively the cycle and the transient lengths, such that:

$$x(t+p(x)) = x(t) \text{ for any } t \geq t(x) \quad \text{and} \quad x(t+q) \neq x(t) \text{ for any } t < t(x) \text{ and } 0 < q < p(x)$$

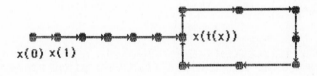

We denote $T = \max\{ t(x) ; x \in \{-1,+1\}^n \}$ and $P = \max\{ p(x) ; x \in \{-1,+1\}^n \}$. In order to estimate $T$ and $P$ we define :

$$E(u,v) = -\Sigma_{i=1,n} u_i \Sigma_{j=1,n} a_{ij} v_j - \Sigma_{i=1,n} b_i (u_i + v_i) ; \quad u,v \in \{-1,+1\}^n$$

Since $A$ is a symmetric matrix it is easy to prove that $E$ is a Lyapunov function associated to the network ; i.e :

1) If $x(t+1) \neq x(t-1)$ then $\Delta_t E = E(x(t+1), x(t)) - E(x(t), x(t-1)) < 0$

2) Since $a_{ij} \in \mathbb{Z}$ : $|\Delta_t E| \geq 2$

3) The operator $E$ is bounded by the following expression :

Examples of positive functions are the following:

$$f: \mathbb{R}^p \mapsto \mathbb{R}^p ; \quad f(x) = \begin{cases} x/\|x\| & \text{if } x \neq 0 \\ 0 & \text{otherwise} \end{cases}$$

$$f: \mathbb{R}^p \mapsto \{e_1, \dots, e_p\} \quad \text{where } e_i = (0,\dots,\overset{i}{\underset{\downarrow}{1}},\dots,0)$$

$$\text{and} \quad f(x_1, \dots, x_n) = e_k \quad \Leftrightarrow \quad \begin{cases} x_k \geq x_i & \text{for } i \geq k \\ \text{and} \\ x_k > x_i & \text{for } i < k \end{cases}$$

It is easy to see that for p=1 (f a real function) f is positive iff it is a threshold function, Goles(1985a, b):

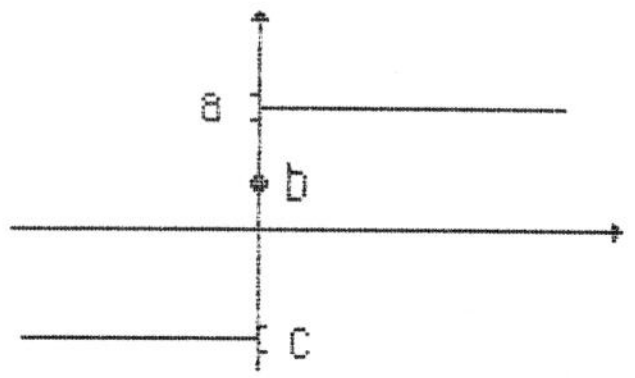

$$f(x) = \begin{cases} c & \text{if } x<0 \\ b & \text{if } x=0 \\ a & \text{if } x>0 \end{cases}$$

fig. 3

Other results for the characterization of positive functions can be found in Goles, Martinez(1985b,c).

Let us now take a network with n modules, and let $Q \subset \mathbb{R}^p$ be a finite set of states. Let us suppose that each module has as a control a strictly positive function $f_i$ from $\mathbb{R}^p$ into Q. We are interested in the dynamics of the system:

$$x_i(t+1) = f_i( A^{(i)}x(t) + b^{(i)} ) \quad \text{for } i=1, \dots, n ; t=0,1,\dots ; x(0) \in Q^n$$

where $b^{(i)} \in \mathbb{R}^p$ and $A^{(i)}$ is a npxn real matrix.

Let us define the linear operator:

$$A: Q^n \to \mathbb{R}^{pn}$$
$$x = (x_1, \dots, x_n) \mapsto \begin{pmatrix} A^{(1)} \\ \dots \\ A^{(n)} \end{pmatrix} x \quad ( A \text{ is a npxnp real matrix})$$

hence if we consider the following operator:

$$E(u,v) = -\langle u, Av \rangle - \langle b, u+v \rangle \qquad \forall u,v \in Q^n$$

we prove, Goles(1985a), that:

If A is an integer symmetric matrix and $f_1, \ldots, f_n$ are strictly positive functions then E is a *Lyapunov function*; i.e:

$$x(t+1) \neq x(t-1) \quad \Rightarrow \quad \Delta_t E = E(x(t+1),x(t)) - E(x(t),x(t-1)) < 0, \text{ for } t \geq 3$$

Hence we conclude that the cycle length of the network, P, may only be 1 or 2 and the transient length, T, is bounded by :

$$T \leq \max\{3, 2(E_M - E_m)\}$$

where $E_M = \max\{E(u,v) ; u,v \in Q^n\}$ and $E_m = \min\{E(u,v) ; u,v \in Q^n\}$

Obviously, in each particular case we can give better bounds on T. For instance, for a regular cellular array; i.e :

$$a_{ij} \in \{0,1\} \quad \text{and} \quad |V(i)| = \text{card}\{j ; a_{ij}=1\}=p \quad \text{for } 1 \leq i \leq n$$

we prove that T = O(np) ;i.e the transient length is a linear function on n, Goles (1985a). In the general case (an arbitrary real symmetric matrix), we may exhibit symmetric networks with exponential transient length,Goles, Olivos(1981).


## 5 MAJORITY NETWORKS

As an other example of positive networks we define the *local generalized majority functions* in the following way :
Let $Q = \{1, \ldots, p\}$ be the state set and $f_i$ be the local transition functions defined by:

$$f_i(x) =k \iff \begin{cases} \sum_{x_j=k} a_{ij} \geq \sum_{x_j=r} a_{ij} & \text{if } k \geq r \\ \\ \sum_{x_j=k} a_{ij} \geq \sum_{x_j=r} a_{ij} & \text{if } k < r \end{cases} \qquad \text{for } i=1, \ldots, n \text{ and } x_j \in Q$$

We may interpret this network as a society of n persons (each module): $P_1, \ldots, P_n$ and Q the set of opinions which may be assumed by any person. In this contex, $a_{ij}$

is the weight of the interaction between $P_i$ and $P_j$. If $x = (x_1, ..., x_n)$ represents the state of the society (i.e $P_i$ has opinion $x_i$), then, with respect to opinion r, $P_i$ is subjected to an influence whose total weight is $\sum \{ a_{ij} ; x_j = r \}$. The function $f_i$ can be interpreted by saying that, at time t+1, $P_i$ selects its most preferred opinion among those of maximum weight.

Generalizations of this kind of models can be found in Goles,Tchuente(1983) and Poljak(1983).

This class of functions may be studied in the positive framework previously introduced by coding the states of the network as follows:

$$i \rightarrow e_i = (0, ... ,1, ... ,0) \in \mathbb{R}^p \quad \text{for } i=1, ... ,p \quad \text{(see Goles(1985b))}$$
$$\uparrow$$
$$i$$

It is not difficult to see that the new local functions are positive and, from previous results, we conclude that, for A symmetric, the cycles are bounded by two and the transient length is small for regular arrays, Goles(1985a,b). As a particular case let us take $Q = \{ 0,1,2,3 \}$ and the regular array $\mathbb{Z} \times \mathbb{Z}$ :

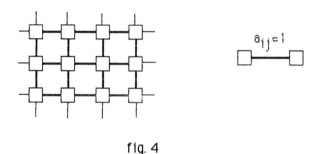

fig. 4

Some evolution patterns, for the parallel iteration on majority functions, are the following:

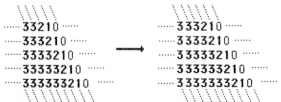

fig. 5

In fig. 5, frontier diagonals propagate at a regular speed 1.

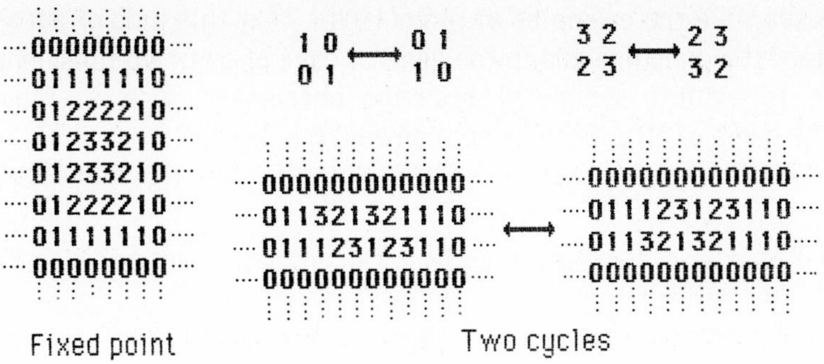

```
 ::::::::
···00000000···        1 0 ←——→ 0 1        3 2 ←——→ 2 3
···01111110···        0 1        1 0        2 3        3 2
···01222210···
···01233210···      ::::::::::::            ::::::::::::
···01233210···    ···000000000000···     ···000000000000···
···01222210···    ···011321321110···     ···011123123110···
···01111110···    ···011123123110··· ←——→ ···011321321110···
···00000000···    ···000000000000···     ···000000000000···
 ::::::::          ::::::::::::            ::::::::::::

 Fixed point                 Two cycles
```

fig. 6

Let us take now the following local function:

$$f_i(x) = \begin{cases} 0 & \text{if} \quad \sum_{j\in V(i)} x_j \leq 1 \\ 1 & \text{if} \quad \sum_{j\in V(i)} x_j = 2 \\ 2 & \text{if} \quad 3 \leq \sum_{j\in V(i)} x_j \leq 7 \\ 3 & \text{if} \quad 8 \leq \sum_{j\in V(i)} x_j \end{cases} \qquad \text{for any } i \in \mathbb{Z} \times \mathbb{Z}$$

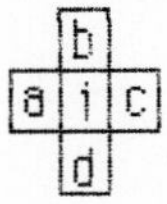

$$V(i) = \{a,b,c,d\} \subset \mathbb{Z} \times \mathbb{Z}$$

fig. 7

Functions of this kind are called *multithreshold* and, by coding, they can be studied as *positive functions*. An evolution pattern is the following:

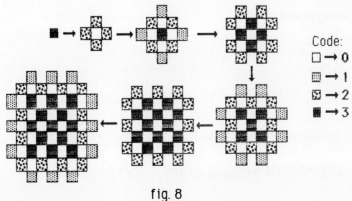

Code:
□ → 0
▦ → 1
▨ → 2
■ → 3

fig. 8

It is interesting to see, like the examples of Ulam(1970), that this kind of automata resembles a tapestry machine and, in the general case of positive functions, it is not possible to exhibit symmetry breaking phenomena such as moving configurations in the game of life, Berlekamp, Conway, Guy (1982).

# 6 ANTISYMMETRIC NETWORKS

In previous paragraphs we have only studied symmetric networks; i.e A being a symmetric matrix. Here we shall associate a Lyapunov function to a threshold antisymmetric network.

Let us suppose that we have n modules with local transition functions:

$$\text{Module } i : 1(\textstyle\sum_{j=1,n} a_{ij} x_j) \qquad \text{for } i=1, \dots, n$$

where $x_j \in \{-1, +1\}$ and $A=(a_{ij})$ is an integer antisymmetric matrix; i.e $a_{ij}=-a_{ji}$.
We shall study the parallel iteration on this network; i.e:

$$x_i(t+1) = 1(\textstyle\sum_{j=1,n} a_{ij} x_j(t))$$

We shall suppose also that $\sum_{j=1,n} a_{ij}$ is odd for any $1 \le i \le n$. This hypothesis implies that:

for any $x \in \{-1, +1\}^n$ and $i=1, \dots, n$ : $\sum_{j=1,n} a_{ij} x_j \neq 0$

Considering the function defined by:

$$E(u,v)=- \textstyle\sum_{i=1,n} u_i \sum_{j=1,n} a_{ij} v_j \qquad \text{where} \quad u,v \in \{-1, +1\}^n$$

we have the following property of E:

If there exists $k \in \{1, \dots, n\}$ such that $x_k(t+1) = x_k(t-1)$ then:

$$\Delta_t E = E(x(t+1),x(t)) - E(x(t),x(t-1)) < 0$$

In fact, since A is antisymmetric, it is easy to see that:

$$\Delta_t E = - \textstyle\sum_{i=1,n} (x_i(t+1)+x_i(t-1)) \sum_{j=1,n} a_{ij} x_j$$

since $\sum a_{ij}x_j(t) \neq 0$ for any $x \in \{-1,+1\}^n$ and $i=1, \dots, n$; we have that:

$$-(x_i(t+1)+x_i(t-1)) \sum_{j=1,n} a_{ij}x_j(t)=0 \quad \text{iff} \quad x_i(t+1) \neq x_i(t-1)$$

and

$$-(x_i(t+1)-x_i(t-1)) \sum_{j=1,n} a_{ij}x_j(t)<0 \quad \text{iff} \quad x_i(t+1) = x_i(t-1) \quad \square$$

Furthermore, from the previous property it is easy to see that the network has only cycles of length 4 and, in a way similar to paragraph 3, the transient, T, is bounded by:

$$T \leq \| A \|/2$$

Example: for the matrix A defined by:

$$A = \begin{pmatrix} 0 & 1 & -1 & -1 \\ -1 & 0 & 1 & -1 \\ 1 & -1 & 0 & 1 \\ 1 & 1 & -1 & 0 \end{pmatrix}$$

we have the unique cycle:

$$-1\ 1\ -1\ -1 \longrightarrow 1\ 1\ -1\ 1$$
$$\uparrow \qquad\qquad\qquad\qquad \downarrow$$
$$1\ -1\ 1\ 1 \longleftarrow -1\ -1\ 1\ -1$$

fig. 9

# 7 CONCLUSION

The results presented here show how the dynamical behaviour of positive automata networks closely resembles physical systems such as spin glasses. This allowed us to prove fairly general results, a situation which is rather unusual in the study of automata dynamics. I believe that deeper studies will be made possible through an extensive use of physical analogies, thus allowing to develop new results in automata networks and their applications (neural networks, pattern recognition, etc).
Finally, for positive networks it is possible to study other evolutive procedures; i.e each module depending on the last k states, Goles(85a), Goles, Tchuente (83), sequential iterations, Fogelman(83, 85), Goles(85), etc. For all these procedures we are able to exhibit Lyapunov functions which allow determining the dynamical behaviour of the network.

# REFERENCES

BERLEKAMP E.R., CONWAY J.H., GUY R.K. (1982):Winnings Ways, vol. 2, chap. 25, Academic Press

DEMONGEOT J., GOLES E.,TCHUENTE M. (1985): Cellular Automata and Dynamical Systems, Proc. Workshop Luminy, Sept. 1983, Academic Press

FOGELMAN F. (1985): Contribution à une théorie du calcul sur réseaux,Thèse d'Etat, Université de Grenoble, France

FOGELMAN F., GOLES E.,WEISBUCH G. (1983): Transient length in sequential iteration of threshold functions, Disc. Applied Maths.,Vol. 6

GOLES E., OLIVOS J. (1981): The convergence of symmetric threshold automata, Inf. and Control, Vol. 51

GOLES E., TCHUENTE M. (1983): Iterative behaviour of generalized majority functions, Math. Soc. Sciences, Vol. 4

GOLES E. (1985a): Comportement Dynamique de Réseaux d'Automates, Thèse d'Etat, Université de Grenoble, France

GOLES E. (1985b): Dynamics on positive automata networks, Theor. Comp. Sciences, to appear

GOLES E., MARTINEZ S. (1985c): Dynamics and properties on positive functions, Res. Report, Dept. Matematicas, U. Chile

HOPFIELD J.J. (1982): Neural networks and physical systems with emergent collective computational abilities, Proc. Nat. Acad. Sc. USA, 79, pp 2554-2558

Mc CULLOCH W., PITTS W. (1943): A logical calculus of the ideas immanent in nervous activity, Bull. Math. Biophysics, Vol 5

POLJAK S. (1983): On periodical behaviour in societies with symmetric influences, Combinatorica, Vol. 3, 1, 119-121

POLJAK S., TURZIK D. (1984): On pre-period of discrete influence systems,Res. Report, Dept. of Systems Ing., U. Czechoslovakia

TCHUENTE M. (1982): Contribution à l'étude des méthodes de calcul pour des systémes de type coopératif, Thèse d'Etat, Université de Grenoble, France

ULAM S.M. (1970): Some mathematical problems connected with patterns of growth of figures, Essays on Cellular Automata, Univ. of Illinois Press

VON NEUMANN J. (1966): Theory of self-reproducing automata, BURKS W. ed., Univ. Illinois Press

# DIRECTIONAL ENTROPIES OF CELLULAR AUTOMATON-MAPS

## John MILNOR

Institute for Advanced Study
Princeton N.J. 08540 U.S.A.

Consider a fixed lattice $L$ in $n$-dimensional euclidean space, and a finite set $K$ of symbols. A correspondence $\mathbf{a}$ which assigns a symbol $\mathbf{a}(z) \in K$ to each lattice point $z \in L$ will be called a *configuration*. An $n$-dimensional cellular automaton can be described as a map which assigns to each such configuration $\mathbf{a}$ some new configuration $\mathbf{a}' = f(\mathbf{a})$ by a formula of the form

$$\mathbf{a}'(z) = F(\mathbf{a}(z+v_1), \cdots, \mathbf{a}(z+v_r)),$$

where $v_1, \cdots, v_r$ are fixed vectors in the lattice $L$, and where $F$ is a fixed function of $r$ symbols in $K$. I will call $f$ the *cellular automaton-map* which is associated with the *local map* $F$. If the alphabet $K$ has $k$ elements, then the number of distinct local maps $F$ is equal to $k^{k^r}$. This is usually an enormous number, so that it is not possible to examine all of the possible $F$. Depending on the particular choice of $F$ and of the $v_i$, such an automaton may display behavior which is simple and convergent, or chaotic and random looking, or behavior which is very complex and difficult to describe. (Compare [Wolfram].)

The *topological entropy* of the map $f$ is a numerical invariant $0 \le h(f) \le \infty$ which measures the information content per unit time step in a large but finite subset $S$ of the lattice $L$ when the map $f$ is iterated over and over. (Compare [Walters].) To be more precise, suppose that we consider all possible initial configurations at time zero and iterate the cellular automaton-map $t$ times. Let $N = N(t, S)$ be the number of distinct histories which can be observed, within the set $S$, over these $t$ timesteps. Then $h(f)$ can be defined as the supremum over $S$ of the quantity $\lim_{t \to \infty} \log(N)/t$. In the case $n = 1$ this invariant is always finite. However in the higher dimensional case it is usually infinite.

This note will describe an $n$-dimensional *directional entropy function* which takes finite values, and hence measures possibly useful properties of the map $f$, even in the higher dimensional case. This directional entropy generalizes not only the topological entropy of $f$ itself, but also the entropies of compositions of $f$ with lattice translations. To see the usefulness of such compositions, consider two rather trivial examples.

**Example 1.** Let $L = Z$ be the 1-dimensional lattice of integers, and let $s$ be the right shift map $\mathbf{a} \mapsto \mathbf{a}' = s(\mathbf{a})$ where $\mathbf{a}'(i) = \mathbf{a}(i-1)$. Then the topological entropy is given by $h(s) = \log(k)$, where $k$ is the number of elements in the alphabet $K$. In fact, if we follow the history of successive configurations $\mathbf{a} \mapsto \mathbf{a}' \mapsto \mathbf{a}'' \mapsto \cdots$ for $t$ timesteps, throughout a strip of lattice points of width $w$, then we can observe exactly $N = k^{t+w-1}$ possible distinct histories. Hence $h(s) = \sup_w \lim_{t \to \infty} \log(N)/t$ is equal to $\log(k)$.

**Example 2.** Again in the 1-dimensional case, let $\sigma(\mathbf{a}) = \mathbf{a}'$ be the sum map

$$\mathbf{a}'(i) \equiv \mathbf{a}(i) + \mathbf{a}(i-1) \pmod{k}.$$

Then again the number $N$ of possible histories over a strip of height $t$ timesteps and of width $w$ lattice points is equal to $k^{t+w-1}$, hence the topological entropy is equal to $\log(k)$.

Thus topological entropy, by itself, cannot distinguish between the completely trivial shift map $s$ and the rather complicated sum map $\sigma$. However, if we compose $s$ with an arbitrary translation $s^m$ then we obtain the cellular automaton-map $s^{m+1}$ with entropy $|m+1| \log(k)$, while the composition of $\sigma$ with $s^m$ has entropy equal to $\max(m+1, -m) \log(k)$. The former is zero for $m = -1$ while the latter is always strictly positive.

NATO ASI Series, Vol. F20
Disordered Systems and Biological Organization
Edited by E. Bienenstock et al.
© Springer-Verlag Berlin Heidelberg 1986

Now suppose that we are given some probability measure $\mu$ on the space of configurations, which is $f$-invariant in the sense that $\mu(\Sigma) = \mu(f^{-1}(\Sigma))$ for any measurable set $\Sigma$ of configurations. Then in place of topological entropy $h(f)$ we can study the measure theoretic entropy $h_\mu(f)$. The definition is similar to that sketched above except that, in place of the number $\log(N)$, we must substitute $\sum -\mu_j \log(\mu_j)$, to be summed over the $N$ possible histories for $t$ timesteps on the set $S$, where $\mu_j$ is the probability of the $j$-th possible history. If this measure $\mu$ is also translation-invariant, then again we can study the entropy of the composition of $f$ with an arbitrary translation. (Compare [Sinai].) This also is usually infinite in the higher dimensional case, but can be replaced by an appropriate $n$-dimensional directional measure theoretic entropy.

One particular translation-invariant measure plays a special role. Define the *standard measure* $\mu_0$ on the space of configurations by specifying that the symbols associated with the different lattice points are independent random variables, the probability of a given symbol occurring at a given lattice point being $1/k$.

**Theorem.** *A cellular automaton-map $f$ preserves this standard measure $\mu_0$, in the sense that $\mu_0(\Sigma) = \mu_0 f^{-1}(\Sigma)$ for any measurable set $\Sigma$ of configurations, if and only if $f$ is onto, that is, if and only if for every configuration $\mathbf{a}'$ there exists a configuration $\mathbf{a}$ satisfying the equation $f(\mathbf{a}) = \mathbf{a}'$.*

In the 1-dimensional case this was proved by Blankenship and Rothaus as reported by Hedlund, and later by Amoroso and Patt. The higher dimensional case was proved by Maruoka and Kimura. (Caution: Here one is very close to some unsolvable problems. For example Yaku has shown that the question of whether there exists a solution $\mathbf{a}$ to the equation $f(\mathbf{a}) = \mathbf{a}'$ is recursively unsolvable for a general cellular automaton-map $f$ and finite configuration $\mathbf{a}'$ in dimensions greater than one.)

The definition of the $n$-dimensional directional entropy functions associated with an $n$-dimensional cellular automaton-map $f$ can be outlined as follows. Consider first a doubly infinite sequence $\cdots, \mathbf{a}(-1), \mathbf{a}(0), \mathbf{a}(1), \mathbf{a}(2), \cdots$ of configurations on the lattice $L \subset R^n$ which satisfy the equation $\mathbf{a}(t+1) = f(\mathbf{a}(t))$ for every (positive or negative) integer $t$. We will also work with the $(n+1)$-dimensional lattice $Z \times L$, embedded in the euclidean space $R \times R^n = R^{n+1}$. Thinking of each $\mathbf{a}(t)$ as a configuration on the $n$-dimensional sublattice $t \times L \subset Z \times L$, we can consider all of the $\mathbf{a}(t)$ simultaneously as horizontal slices of a configuration $\hat{\mathbf{a}}$ which is defined throughout this $(n+1)$-dimensional lattice. An $(n+1)$-dimensional configuration $\hat{\mathbf{a}}$ which can be obtained in this way will be called a *complete history* for $f$.

For any finite subset $S$ of the lattice $Z \times L$, define the (topological) *information content* $H_{\text{top}}(S)$ to be the logarithm of the number $N$ of distinct configurations on $S$ which arise as restrictions of complete histories to $S$. Similarly, if $\mu$ is an $f$-invariant and translation-invariant measure on the space of configurations on $L$, then a corresponding measure theoretic information content $H_\mu(S)$ is defined. These satisfy

$$0 \leq H_\mu(S) \leq H_{\text{top}}(S) \leq |S| \log(k),$$

where $|S|$ is the number of elements in $S$. Both the topological and measure theoretic information contents satisfy three basic properties: they are *translation invariant*, *monotone*, and *subadditive*. That is $H(S+v) = H(S)$ for any vector $v$ in the lattice, $H(S) \leq H(S')$ whenever $S \subset S'$, and $H(S \cup T) \leq H(S) + H(T)$ for any finite sets $S$ and $T$.

Given $n$ vectors $v_1, \cdots, v_n$ in the ambient vector space $R^{n+1}$, the *directional entropy* $h(v_1, \cdots, v_n)$ is now defined as follows. Let $P = P(t_1 v_1, \cdots, t_n v_n)$ be the parallelepiped with edges $t_1 v_1, \cdots, t_n v_n$ where the $t_i$ are large real numbers. We will thicken this parallelepiped, so as to obtain an $(n+1)$-dimensional set, by forming the vector sum $P+B$ where $B$ is some fixed bounded subset of $R^{n+1}$. Then $h(v_1, \cdots, v_n)$ is defined to be the supremum over $B$ of $\lim \sup H((P+B) \cap (Z \times L))/t_1 \cdots t_n$, taking the lim sup (or equivalently the lim inf) as the $t_i$ tend to infinity. This quantity is well defined and finite, and depends only on the exterior product of the vectors $v_i$.

In the 1-dimensional case, if $u$ is the unit timelike vector, then $h(u)$ coincides with the usual entropy of the map $f$. Lind and Smillie show by example that the topological entropy $h(v)$ is not necessarily continuous as a function of the vector $v$. However, if we restrict attention to "right-spacelike" vectors, then they show that this function is continuous, with $h(v+w) \leq h(v)+h(w)$. This result generalizes easily to higher dimensions, and to measure theoretic entropy. If $f$ is actually an invertible automaton-map, then [Boyle and Krieger, §2.16] show that equality holds for right-spacelike vectors, that is $h(v+w) = h(v)+h(w)$. In fact they give an effective computation of the resulting linear function. I do not know whether their results can be generalized to higher dimensions.

## REFERENCES

AMOROSO S. and PATT Y. N. (1972): Decision procedure for surjectivity and injectivity of parallel maps for tessellation structures, J. Comp. Syst. Sci. **6**, 448-464.

BOYLE M. and KRIEGER W. (to appear): Periodic points and automorphisms of the shift.

COVEN E. M. (1980): Topological entropy of block maps, Proc. Amer. Math. Soc. **78**, pp 590-594.

DEMONGEOT J., GOLES E. and TCHUENTE M. (1985): Dynamical systems and cellular automata, Academic Press.

FRIED D. (1983): Entropy for smooth abelian actions, Proc. Amer. Math. Soc. **87**, pp 111-116.

GOODWYN L. W. (1972): Some counter-examples in topological entropy, Topology **11**, pp 377-385.

HEDLUND G. A. (1969): Endomorphisms and automorphisms of the shift dynamical system, Math. Syst. Th. **3**, pp 320-375.

LIND D. and SMILLIE J. (in preparation).

MARUOKA A. and KIMURA M. (1976): Condition for injectivity of global maps for tessellation automata, Inform. and Contr. **32**, 158-162.

SINAI Ya. (to appear): On a question of J. Milnor, Comment. Math. Helv.

WALTERS P. (1975): Ergodic Theory - Introductory Lectures, Lecture Notes in Math. **458**, Springer.

WOLFRAM S. (1984): Universality and complexity in cellular automata, Physica **10D**, pp 1-35; Cellular automata as models of complexity, Nature **311**, pp 419-424.

YAKU T. (1973): The constructibility of a configuration in a cellular automaton, J. Comp. and System Sci. **7**, pp 481-496.

# 2 PHYSICAL DISORDERED SYSTEMS

# ON THE STATISTICAL PHYSICS OF SPIN GLASSES

M. Mézard [*]
Dipartimento di Fisica, Università di Roma I,
Piazzale A.Moro 2, 00185 Italy

## Foreword

These notes are intended to provide a brief introduction to the recent progresses on the physical understanding of the equilibrium properties of spin glasses (in mean field theory). In order to try and make them more accessible, I have systematically avoided all the computations, which are described in many good recent reviews (see for instance /1-4/) in which one will find also more extended references. An introduction to the replica method, from which a large part of the following results have been deduced, can be found in /4/. Parts of the following text are inspired from a "pedagogical" article I wrote in french recently /18/.

## 1 - A model of spin glasses

Spin glasses are magnetic materials in which the interaction between the magnetic moments (spins) are random and conflicting. The canonical examples are obtained from the dilution of a magnetic metal ($M_g$, Fe,...) in a noble metal (Au, Ag, Cu...) at concentrations of the order of the %. In

---

[*] On leave From laboratoire de Physique Théorique de l'E.N.S., Paris.

NATO ASI Series, Vol. F20
Disordered Systems and Biological Organization
Edited by E. Bienenstock et al.
© Springer-Verlag Berlin Heidelberg 1986

such a situation the interaction energy between the magnetic moments $\vec{m}$ and $\vec{m}'$ of two magnetic impurities, $E = - J(n)\vec{m} \cdot \vec{m}'$ depends on the distance $n$ between them with an oscillating law $J(n) = n^{-3} \cos \frac{n}{a}$ (a is a constant of the order of the interatomic distances). As the positions of the impurities are random, their couplings are either ferromagnetic ($J(n) > 0$) or antiferromagnetic ($J(n) < 0$) depending on the pair of magnetic impurities one considers. The system is frustrated: if one looks for the configuration of spins which minimizes the energy, one finds that it is impossible to minimize at the same time the interaction energies of all pairs of spins (because of the existence, e.g., of triplets of spins $m_i$, $m_j$, $m_k$ such that $J_{ij} J_{jk} J_{ki} < 0$ - frustrated loops -).

Disorder and frustration are two basic properties of spin glasses. Their consequences are important at low temperature where the allowed configurations of the system are low energy ones. Indeed there exist many configurations of the magnetic moments which are very different one from another but have energies very close to the minimal one. Many fascinating experimental observations /1/ - freezing properties, existence of very large relaxation times (which can be of the order of several days!), hysteresis effects, ....- can be traced back to these properties.

In order to understand the collective properties of such an assembly of spins with random couplings, one wants to build up and study in details a simplified model which contains nevertheless all the important characteristics of the system. This was achieved in 1975 by S.F.Edwards and P.W.Anderson who introduced a model of a spin glass based on the simple hamiltonian /5/:

$$\mathcal{H} = - \sum_{i,j} J_{i,j} \, \sigma_i \, \sigma_j \tag{1}$$

As in the Ising model of the ferromagnetism, one considers a system of N spins $\sigma_i$ on a periodic lattice, each of
which can be in either of two states $\sigma_i = + 1$ or $\sigma_i = - 1$.
But the couplings between the nearest neighbour spins $J_{ij}$ are
now random variables which can be either positive or negative. In this modelisation one forgets about random positions
and oscillating interactions but one keeps their most
important consequences: the random interactions and the
frustration. Indeed the E.A. hamiltonian is probably the
simplest one which can be written possessing these two
properties.

Confronted to such a microscopic model the first
question one wants to answer is: what are the different
possible phases of the system and their characteristics?
Usually this problem is addressed by the use of the so called
mean field approximation in which a given spin reacts to the
field created in average by all its neighbours, this field
being in turn determined selfconsistently. This
approximation, which amounts to neglecting the fluctuations
of the magnetizations on neighbouring sites, becomes exact
when there are a lot of them, i.e. in a space of large
dimension.

For spin glasses the mean field theory was defined
clearly by Sherrington and Kirkpatrick (S.K.) /6/ as the
infinite range version of the E.A. model, in which each spin
is supposed to interact with all the N-1 other ones. Even
within this approximation it turned out to be difficult to
solve this problem on which theorists have concentrated much
effort during the last ten years.

The different approaches, analytical as well as
numerical, which have been used /1/ so far build up a
coherent and clear picture of the nature of the new
thermodynamical phase which appears at low temperature: the
spin glass phase. The corner stone of our understanding is
the solution of the S.K. model by the mysterious replica

method proposed by Parisi in 1979 /7/ and the physical interpretation he gave four years later /8/. In the following I will try to describe the physical results obtained with this approach, but one should keep in mind that it has not yet been rigorously demonstrated that it is the solution of the S.K. model.

## 2 - Ergodicity breaking

In the framework of the mean field theory one finds that, besides the paramagnetic and ferromagnetic phases, the system can be in a new type of phase: the spin glass phase. This is the equilibrium phase at low temperatures when the average coupling between the impurities is small. The original and specific properties of this phase are the breaking of ergodicity, and the existence of many equilibrium states.

To understand this phenomenon, let us first give a look at the simpler case of the Ising model. At low temperature, in the ferromagnetic phase, there appears in zero magnetic field a spontaneous magnetization. Because of the symmetry by the reversal of all the spins, the density of magnetizations can be either positive or negative ( $\pm$ M (T) ): the ergodicity is broken and there exist two equilibrium states "+" and "-" (two ergodic components) which can be characterized by the value of the global magnetization.

In the spin glass phase the situation is more complex. There appear an infinite number of equilibrium states and they are not related to each other by a symmetry law. In every equilibrium state $\alpha$ the local magnetizations $m_i^\alpha$ on each site are nonzero, there is hence some local ordering. However in zero magnetic field the global magnetization per spin $M^\alpha = \frac{1}{N} \sum_i m_i^\alpha$ is of order $1/\sqrt{N}$ and vanishes in the thermodynamic limit: there is no globally preferred direction.

The difficulties of the studies of the spin glass phase originate in this fundamental property of ergodicity breaking. Indeed the use of the Boltzmann Gibbs formula to compute the thermal average $\langle O \rangle$ of a given observable $O(\sigma_i)$:

$$\langle O \rangle = \left[ \sum_{\{\sigma_j = \pm 1\}} e^{-\beta \mathcal{H}} O(\sigma_i) \right] \cdot \left[ \sum_{\{\sigma_j = \pm 1\}} e^{-\beta \mathcal{H}} \right]^{-1} \qquad (2)$$

is correct only for ergodic systems. In the spin glass phase one can compute with this formula only averages over all the equilibrium states of the system /8-10/:

$$\langle O \rangle = \sum_{\alpha} P_\alpha \langle O(\sigma_i) \rangle_\alpha \qquad (3)$$

where $\langle O(\sigma_i) \rangle_\alpha$ is the average value of the observable $O$ in the state $\alpha$, and $P_\alpha$ is the probability of this state. This is already true in the ferromagnetic case. Below the Curie temperature, in spite of the existence of two equilibrium states "+" and "-" with $\langle \sigma_i \rangle_\pm = \pm M$, a direct computation of $\langle \sigma_i \rangle$ with the Boltzmann-Gibbs formula gives (with free boundary conditions $P_+ = P_- = \frac{1}{2}$) a zero magnetization ($\langle \sigma_i \rangle = P_+ \langle \sigma_i \rangle_+ + P_- \langle \sigma_i \rangle_- = 0$) which demonstrates the necessity of modifying the Boltzmann-Gibbs prescription.

The well known solution used in the ferromagnetic case which consists in choosing one of the states, for instance by applying a small uniform magnetic field, doesn't work for spin glasses. The reason is the following: the different pure states correspond to well defined priviledged directions locally, but is is impossible to determine them at each point of the sample unless one possesses a sort of Maxwell demon! One must hence keep to Boltzmann-Gibbs averages (3) and try to deduce from them the properties of the different states $\alpha$.

3 - An order parameter function

Nevertheless it is possible to find in the spin glass phase some kind of order. Instead of trying to pick up one state one must rather compare them from a geometrical point of view. Each state is characterized by its probability $P_\alpha$ and the local magnetization on each site i: $m_i^\alpha = \langle \sigma_i \rangle_\alpha$ . A natural measure of the distance between two states $\alpha$ and $\beta$ is given by their overlap $q^{\alpha\beta}$ defined as a scalar product in phase space:

$$q^{\alpha\beta} = \frac{1}{N} \sum_i m_i^\alpha m_i^\beta \tag{4}$$

The larger the overlap, the smaller the distance between the states.

In order to characterize the differences between equilibrium states, taking into account there respective weights, one introduces the probability law, $P(q)$, that two states $\alpha$ and $\beta$ picked up at random have an overlap q:

$$P(q) = \sum_{\alpha,\beta} P_\alpha P_\beta \, \delta(q^{\alpha\beta} - q) \tag{5}$$

This distribution of distances between equilibrium states has been proposed as an order parameter for spin glasses by Parisi, who has shown how it can be computed with the replica method in the mean field theory /8/. Clearly this kind of function can be defined generally for all the systems which possess several equilibrium states, once one has identified a good measure of their distances. It also provides a criterion of characterization of the different phases, as well for disordered as for ordered systems. For instance in the Ising model in zero field this function is a simple $\delta$ peak in the paramagnetic phase (one state only, in which the local magnetization vanishes), which gives rise to two $\delta$ peaks below the Curie temperature: $P(q) = \delta(q + M^2) + \delta(q - M^2)$ (two equiprobable states with uniform magnetizations).

In the S.K. model the function $P(q)$ in zero field (averaged over the samples) presents also two $\delta$ peaks at $q = \pm q_M$, but there is moreover a whole continuum spectrum between these peaks (fig.1). There exist an infinity of states and their mutual overlaps can take all the values between $-q_M$ and $+q_M$. The order parameter is no longer a number, but a whole function $P(q)$.

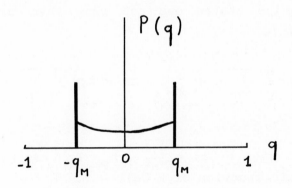

Fig.1: The order parameter function in zero field, averaged over the samples, in the S.K. model.

Although it is beyond the scope of this article to explain how one can compute this function, it is of some interest to know the starting idea.
The Maxwell demon which would pick up a given state is

replaced by a second system of spins $\{s_i\}$, exactly identical to the first one (it has the same set of couplings $J_{ij}$), which is compared to this one by the computation of the thermal average:

$$g(y) = \langle \exp \frac{y}{N} \sum_{i=1}^{N} s_i \sigma_i \rangle_{(2)}$$

(6)

in the double system in which the hamiltonian is $H^{(2)} = H(s) + H(\sigma)$. As the two systems are only very weakly coupled, each of them can be independently in any of its equilibrium states, and from (3) the thermal average $\langle \rangle_{(2)}$ decomposes as:

$$\langle \rangle_{(2)} = \sum_{\alpha,\beta} P_\alpha P_\beta \langle \rangle_{\alpha,\beta}$$

(7)

where $\langle \rangle_{\alpha,\beta}$ is the thermal average in which $\{s\}$ is in state $\alpha$ and $\{\sigma\}$ in state $\beta$. Using specific properties of the equilibrium states one can show that in the infinite volume limit /8/:

$$\langle \exp \frac{y}{N} \sum_i s_i \sigma_i \rangle_{\alpha,\beta} = \exp \frac{y}{N} \sum_i m_i^\alpha m_i^\beta$$

(8)

which proves that $P(q) = \sum_{\alpha,\beta} P_\alpha P_\beta \exp(y \, q^{\alpha\beta})$ is the characteristic function of $P(q)$.

By this procedure which consists in comparing two copies of the same system, one obtains $P(q)$ (which is an information on the different states) from a standard computation of a Boltzmann Gibbs average (6), which can be approached from any method of statistical mechanics. The function of fig.1 is the result of the mean field approximation.

# 4 - Ultrametricity: order in the disorder

The geometrical analysis of the space of equilibrium states which has already provided the order parameter function can be pushed further and exhibits a rich structure /11/. One finds that the probability P ($q_1$, $q_2$, $q_3$) that, picking up three states at random, their mutual overlaps be $q_1$, $q_2$, $q_3$ is nonvanishing in only two types of cases:
- Either $q_1 = q_2 = q_3$, which corresponds to an equilateral triangle in the space of states,
- Or two overlaps are equal, e.g. $q_1 = q_2$, and the third is $q_3 > q_1$, which corresponds to an isoceles triangle in which the third side is smaller than the other two.

Spaces which possess such a property are called ultrametic. One of the possible representations of an ultrametric space is given in fig.2. It is a kind of genealogical tree, in which the extremity of each branch is a state. The overlap between two states depends only on the level in the tree at which one must get back in order to find their closest common ancestor. When the temperature of a spin glass is lowered below the critical temperature, the different equilibrium states are obtained by a process of bifurcations represented by such a tree.

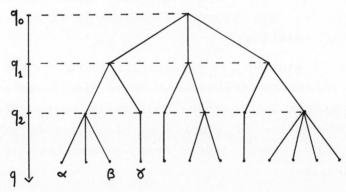

Fig.2: A representation of the ultrametric space of equilibrium states. The extremity of each branch represents a state. For the three states $\alpha$, $\beta$, $\gamma$, one has: $q^{\alpha\gamma} = q^{\beta\gamma} = q_1$; $q^{\alpha\beta} = q_2 > q_1$.

## 5 - Some properties which depend on the sample

A big surprise in the analysis of the replica solution of the S.K. model was the fact that some properties depend on the sample one considers! /11-12/.

In a given sample the (random) positions of the magnetic impurities are fixed and do not vary during the experiments. Consequently the couplings $J_{ij}$ in (1) must be choosen once for all (with a given probability law), they are "quenched" which means that they don't vary with the external parameters such as temperature and magnetic field, they depend only on the sample.

Most of the properties, and particularly the extensive thermodynamic quantities, are nevertheless independent of the particular realization of the couplings, and hence of the sample.
A simple argument which explains this consists in cutting the sample into a large number N of pieces, all macroscopic. Neglecting the surface terms, the free energy for instance is equal to the sum of the free energies of the pieces which are independent random variables. The free energy of the sample, sum of N independent random variables, equals N times the average value of one of these variables, which proves that the free energy per spin doesn't fluctuate: it is a so called "self averaging" quantity.

This kind of argument is the justification for the usual procedure in the statistical mechanics of disordered systems consisting in computing averages relatively to the disorder (here the distribution of couplings) of extensive physical quantities: this gives indeed the properties of each macroscopic sample.

Although the above argument doesn't apply to the S.K. model because the infinite ranged interactions prevent from neglecting "surface" effects, one can show that the conclusion remains valid: free energy, internal energy, magnetization are sample independent. More surprising is the fact that some properties fluctuate from sample to sample even in the thermodynamic limit, they are non self averaging. This is the case for the distribution of the weights of the states, and hence of the order parameter function P (q) defined in (5). The function P (q) plotted in fig.1 is averaged over the samples.

The computation of the fluctuations of P (q) /11,14,17/ shows in fact that this P (q) averaged has only a little probability of being realized on one given sample!

The situation is rather unwieldy since the quantities one can compute are averages over the samples of averages over the states (see (3)) of a given observable. Fortunately it can be shown that many properties, including some local informations such as the distribution of local magnetizations in a given equilibrium state, are not only sample independent but also state independent /13/.

In fact the non self averageness can be traced back to the weights of the equilibrium states $\alpha$ , which are related to their free energies $F_\alpha$ through /9/:

$$P_\alpha = e^{-\frac{F_\alpha}{T}} \cdot \left[ \sum_\gamma e^{-\frac{F_\gamma}{T}} \right]^{-1} \tag{9}$$

In a given sample, all the equilibrium states should have the same free energy per spin (the minimal one) $\lim_{N \to \infty} \frac{F_\alpha}{N} = F$, but the 1/N corrections $f_\alpha$ to these quantities ($F_\alpha = NF + f_\alpha$) differ from state to state and this is the source of their different weights. In (9) the value F disappears, the weights depend only on the $f_\alpha$ , and as these are terms of order 1/N, it is not surprising that the values of the weights are non self averaging.

The distribution of the weights obtained with the replica method in the S.K. model /11/ have received recently a very simple interpretation in terms of the free energies $f_\alpha$ : these are independent random variables with an exponential distribution /14/.

Imposing a cutoff $f_c$ to make the number of states finite, say equal to M, the distribution of the $f_\alpha$ of these states is up to an overall scale:

$$\mathcal{P}(f_1, \cdots, f_M) = \prod_{\alpha=1}^{M} \beta e^{\beta(f_\alpha - f_c)} \quad ; \quad f_\alpha < f_c \quad (10)$$

where $\beta$ is a parameter ( $\beta < \frac{1}{T}$ ) related to the weight of the $\delta$ functions in the average order parameter function. If one lets the cutoff $f_c$ go to infinity at the same times as M, with $Me^{-\beta f_c}$ = v kept constant, one obtains from (10) all the results on the distribution of the weights of the states.

For instance one sees clearly that the number of equilibrium states is infinite, but there exist some samples in which one of the states has a dominant weight ($P \sim 1$) at low temperatures: this is the case when there is a finite gap between the ground state and the first excited one. The infinity of the states comes from the region of large free energies, which corresponds to states having a vanishinghy small weight and thus essentially irrelevant for the thermodynamics.


6 - Conclusions


Clearly the replica method is a very powerful tool which can give many informations but is hard to dominate!

We are now on the way of a no-replica understanding of the spin glass phase in the sense that the physical content of Parisi's replica symmetry breaking Ansatz /7/ has now been found: The ultrametricity and the distribution (10) of the free energies are physical properties which are equivalent to this Ansatz.

One should mention in this respect the existence of a model, simpler than the S.K., in which one can check the results of the replica method /15/ since it is equivalent to a model of random energy levels which can be solved by alternate methods /16/. This is a very useful cousin of the S.K. model which shares with it many typical properties of the spin glass phase - infinite number of equilibrium states, fluctuations of the weights of the states /17/.....

To conclude, our physical understanding of the nature of the spin glass phase in the mean field theory has increased a lot recently. The relevance of the above results for real three dimensional spin glasses remains however an open problem. It would be nice to relate the large spectrum of relaxation times observed experimentally to some process of diffusion by hopping from one state to the other in an ultrametric space.

There is still a long way to go, but certainly the mean field theory of spin glasses is one of the most elaborate model in the statistical physics of disordered systems, which has generated in ten years many important concepts in this field.

References

1    "Heidelberg colloquium on spin glasses", J.L.Van Hemmen
     and I.Morgenstern editors. Lecture notes in Physics
     Vol.192, Springer Verlag (1983).

2    C.De Dominicis, Heidelberg colloquium loc.cit..

3    G.Toulouse, Heildelberg colloquium loc.cit..

4    G.Parisi, lectures given at the 1982 "Les Houches" summer
     school (North-Holland). Field theory and Statistical Mechanics.

5    S.F.Edwards and P.W.Anderson, J.Phys. $\underline{F5}$ (1975) 965.

6    D.Sherrington and S.Kirkpatrick, Phys.Rev.Lett. $\underline{32}$ (1975)
     1792; S.Kirkpatrick and D.Sherrington, Phys.$\overline{\text{Rev}}$. $\underline{B27}$
     (1978) 4384.

7    G.Parisi, Phys.Rev.Lett. $\underline{43}$ (1979) 1754 and J.Phys. $\underline{A13}$
     (1980) L 115, 1101, 1887.

8    G.Parisi, Phys.Rev.Lett. $\underline{51}$ (1983) 1206.

9    C.De Dominicis and A.P.Young, J.Phys. $\underline{A16}$ (1983) 2063.

10   A.Houghton, S.Jain and A.P.Young, J.Phys. $\underline{C16}$ (1983) L375
     and Phys.Rev. $\underline{B28}$ (1983) 2630.

11   M.Mézard, G.Parisi, N.Sourlas, G.Toulouse and
     M.A.Virasoro, Phys.Rev.Lett. $\underline{52}$ (1984) 1156 and
     J.Physique $\underline{45}$ (1984) 843.

12   A.P.Young, A.J.Bray and M.A.Moore, J.Phys. $\underline{C17}$ (1984)
     L149.

13   M.Mézard and M.A.Virasoro, Rome preprint (1985), to
     appear in J.Physique.

14   M.Mézard, G.Parisi and M.A.Virasoro, J.Physique Lett. $\underline{46}$
     (1985) L 217.

15   D.Gross and M.Mézard, Nucl.Phys. $\underline{B240}$ /FS12/ (1984) 431.

16   B.Derrida, Phys.Rev.Lett. $\underline{45}$ (1980) 79 and Phys.Rev. $\underline{B24}$
     (1981) 2613.

17   B.Derrida and G.Toulouse, J.Physique Lett. $\underline{46}$ (1985)
     L223.

18   In "Images de la Physique 1985", CNRS ed.

# SYMBOLIC COMPUTATION METHODS
# FOR SOME SPIN GLASSES PROBLEMS

Bernard LACOLLE

LABORATOIRE TIM3  B.P. 68
38402 S$^t$ Martin d'Hères Cédex
FRANCE

## INTRODUCTION

The purpose of this paper is the computation of exact expressions (symbolic expressions) of two fundamental functions used in Statistical Physics. In the first part we study *the partition functions* of finite two-dimensional and three-dimensional Ising models . These partition functions can be expressed with polynomials and we want to compute all the coefficients of these polynomials <u>exactly</u> . In the second part we deal with *the free energy* of some regular two-dimensional Ising models . We write these functions in the form of double integrals and we carry out the calculation of the kernels of these integrals .

Computation of partition functions and free energy have justified a lot of works because many interesting informations may be obtained from their exact expressions . For the partition functions there exist some results in two dimensions but the sizes of the systems are very small (Ono(1968),Suzuki(1970),Katsura(1971),Abe(1976)...) . In the first part of this lecture we can extend the results from 6x6  (36 sites ) systems up to 12x12 systems (144 sites) . In three dimensions there exist results for the perfect Ising model only (Pearson(1982)) . We present algoritms valid for spin glasses problems . In the same way , in the second part of the lecture , we present computational techniques for the free energy problem . We can obtain exact solutions for a large class of regular models . We give applications of the method to some well known models but our aim is to present the use of a *symbolic computation language*  (REDUCE) as an *investigation tool* for the Physicist .

The aim of this paper is to present computational techniques and not to give physical considerations about the numerical results .

NATO ASI Series, Vol. F20
Disordered Systems and Biological Organization
Edited by E. Bienenstock et al.
© Springer-Verlag Berlin Heidelberg 1986

# PART I : COMPUTATION OF PARTITION FUNCTIONS

## 1 The two-dimensional rectangular lattice with Ising Spins

The physical problem is a rectangular lattice with m rows and n sites per rows . These sites are to be occupied by two kinds of spins ($+1$ or $-1$) . Interactions exist only between nearest neighbours .

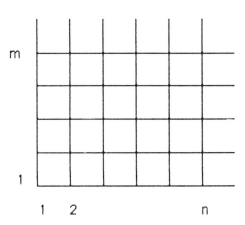

$J_{ij}$( $= +$ or $-1$) is the coupling constant for the neighbour pair $(i,j)$ and $\sigma_i$ (resp. $\sigma_j$) the value of the spin on the site $i$ (resp. $j$) . The energy of interaction for the neighbour pair $(i,j)$ is given by :

$$- J_{ij} \sigma_i \ \sigma_j$$

The vector : $\sigma = (\sigma_i) \in \{-1, +1\}^{m \times n}$ , is called a *spin configuration* and we associate with the spin configuration $\sigma$ , the quantity :

$$E(\sigma) = - \sum_{(i,j)} J_{ij} \sigma_i \sigma_j$$

(where the sum runs over all the pairs of nearest neighbours on the lattice) . In this work the partition function will be denoted by :

$$Z(\beta) = \sum_{\sigma} e^{-\beta E(\sigma)}$$

(where $\Sigma$ implies the summations over all the possible values of $\sigma$ : $2^{mn}$ values of $\sigma$) . Sometimes we replace the previous expression by :

$$P(e^{-\beta}) = (e^{-\beta})^{2mn} Z(\beta)$$

where $P(e^{-\beta})$ is a polynomial in $z = e^{-\beta}$ , with integer coefficients .

## 2   COMPUTATION of  P  ( or Z )  : The main problems .

The enumeration of all the states is too expansive . With this method the computation of  Z has been carried out for a 6x6 system only . For greater systems the coefficients of the polynomials become very large integers and they cannot be represented with the  data types  of usual programming languages .

## 3   MATRIX   EXPRESSION   of   Z   :   The   transfer   matrix   method   (Kramers(1941),Kaufman(1949))

We associate with the $i^{th}$ and $(i+1)^{th}$ row of the lattice , a square matrix $C_i$. The horizontal entries of $C_i$ are the states of spins in the $i^{th}$ row and the vertical entries of  $C_i$ are the states of spins in the $(i+1)^{th}$ row . We define   $C_i(k,l)$ by :

$$C_i(k,l) = e^{-\beta E_{kl}}$$

where $E_{kl}$ is the sum of:
1) the interactions between the spins of the  $i^{th}$ row ( in the state k)
2) the interactions between the spins of the $i^{th}$ row and the spins of the
  $(i+1)^{th}$ row (in the states k and l respectively) .

If we apply periodic boundary conditions in the vertical lattice direction , we have the classical result :

$$Z(\beta) = \text{Trace} (C_1 \, C_2 \, \dots\dots C_m \, )$$

The numerical computation of a product   $C_i \, C_{i+1}$ requires   $2^{3n}$ elementary operations (it is less than $2^{nxm}$   !) but our aim is the computation of this product with polynomial coefficients . It is not possible to perform the computations for matrices with polynomial coefficients . For this reason we will  reconstruct the polynomial  $Z(\beta)$ ( or $P(e^{-\beta})$ ) by an *interpolation process*. But , to avoid numerical errors , we must use interpolation over finite fields : for example $\mathbb{Z}/p\mathbb{Z}$ where p is a prime integer . For some reasons which will become clear latter , this method (*modular arithmetic*) solves the problem of the great size of coefficients .

# 4 MODULAR ARITHMETIC and the Chinese Remaindering

(Aho,Hopcroft,Ullmann(1974))

## 4.1 Modular representation of an integer

If $p_0, p_1, \ldots, p_{n-1}$ are (pairwise relatively) prime integers and

$$p = \prod_{i=0 \ldots n-1} p_i$$

then we can represent any integer $u \in \{0,1,2,\ldots,p-1\}$ uniquely by the set of residues : $u_0, u_1, u_2, \ldots, u_{n-1}$ where $u_i = u[p_i]$.

## 4.2 Converting an integer from modular notation to usual notation

We have : $p_0, p_1, \ldots, p_{n-1}$ relatively prime integers and residues $u_0, u_1, \ldots, u_{n-1}$.

We wish to find the integer $u$ ($\in \{0,1,\ldots,p-1\}$) represented by the previous residues.

Let : $c_i = p/p_i$ (the product of all the $p_j$'s except $p_i$)

$$d_i = c_i^{-1}[p_i] \text{ ( i.e. } d_i c_i = 1[p_i] \text{ )}$$

Then we have :

$$u = \sum_{i=0,1,\ldots,n-1} c_i d_i u_i \ [p]$$

## 4.3 Reconstruction of a polynomial P with integer coefficients

Let : $x_0, x_1, \ldots, x_r \in \mathbb{Z}/p_i\mathbb{Z}$ $x_k \neq x_l$ for $k \neq l$
( with $r$ greater than or equal to the degree of P and $p_i$ a prime integer ) .

Let : $y_j = P(x_j)[p_i]$
$j = 0,1,\ldots,r$
$y_j \in \{0,1,\ldots,p_i-1\}$

There exist an unique polynomial $P_i(z)$ ( with its coefficients in $\mathbb{Z}/p_i\mathbb{Z}$ )
so that : $P_i(x_j) = y_j \; [p_i]$
$$j = 0,1,...,r \quad d^\circ P_i <= r$$

We can reconstruct these polynomials with a classical algorithm of numerical analysis .**The coefficients of the polynomial $P_i$ are the residues (modulo $p_i$) of the coefficients of the polynomial P** .So, with a good choice of p and $p_i$ , it is possible to reconstruct the polynomial P .

**Conclusion** :    The main operation will be the numerical computation of $P(z = e^{-\beta})$ with z belonging to $\mathbb{Z}/p_i\mathbb{Z}$ .

# 5  A MATRIX FORMULATION OF THE PROBLEM

## 5.1 Notations

We note : $1 = \begin{pmatrix} 1 & 0 \\ 0 & 1 \end{pmatrix}$   $s = \begin{pmatrix} 1 & 0 \\ 0 & -1 \end{pmatrix}$   $c = \begin{pmatrix} 0 & 1 \\ 1 & 0 \end{pmatrix}$

Denote the Kronecker product : $A \otimes B = (B(i,j) A)$ and :

$$\mathbf{J} = 1 \otimes 1 \otimes \ldots \ldots \otimes 1 \otimes \ldots \ldots \otimes 1$$
$$S_r = 1 \otimes 1 \otimes \ldots \ldots \otimes s \otimes \ldots \ldots \otimes 1$$
$$C_r = 1 \otimes 1 \otimes \ldots \ldots \otimes c \otimes \ldots \ldots \otimes 1$$
$$\qquad\qquad r^{th} \text{ factor} \qquad n^{th} \text{ factor}$$

for $r = 1, 2, ..., n$

## 5.2 Periodic boundary conditions in the vertical lattice direction

We have  the expressions (Kaufman (1949)) :

$$Z(\beta) = \text{Trace} (A_1 B_1 A_2 B_2 \ldots \ldots A_m B_m)$$
$$A_k = \overset{n}{\underset{i=1}{\Pi}} (\text{ch}(\beta J_k^r) \; \mathbf{J} + \text{sh}(\beta J_k^r) \; S_r S_{r+1})$$

$$B_k = \prod_{i=1}^{n} \left( \exp(\beta J_k^{\overset{v}{r}}) \; \mathbf{J} + \exp(-\beta J_k^{\overset{v}{r}}) \; C_r \right)$$

where $J_k^{\overset{o}{r}}$ is the value of the interaction between the $r^{th}$ site and the $(r+1)^{th}$ site in the $k^{th}$ row and $J_k^{\overset{v}{r}}$ the value of the interaction between the $r^{th}$ sites in the $k^{th}$ and $(k+1)^{th}$ rows.

If we deal with $P(e^{-\beta}) = (e^{-\beta})^{2mn} Z(\beta)$ we replace $A_k$ by $e^{-\beta}A_k$ and $B_k$ by $e^{-\beta} B_k$ and we consider the variable $z = e^{-\beta}$ ( or $z = e^{-2\beta}$ ).

Let : $\quad S_r S_{r+1} = T_r$ for $r = 1, 2, ..., n-1$

$$\Gamma_i = \begin{matrix} \alpha_{n-1} & \alpha_{n-2} & & \alpha_1 & \beta_n & \beta_{n-1} & & \beta_1 \\ T_{n-1} & T_{n-2} & ...... & T_1 & C_n & C_{n-1} & ...... & C_1 \end{matrix}$$

with $\alpha_i , \beta_i \in \{0,1\}$ and $\overline{i = \alpha_{n-1}\alpha_{n-2}...\alpha_1\beta_n\beta_{n-1}...\beta_1}$ (binary representation).

If we note $\overline{\alpha}$ the negation of $\alpha$ we have very simple product rules :

$$\begin{matrix} \alpha_{n-1} & \alpha_{n-2} & & \alpha_r & & \alpha_1 & \beta_n & \beta_{n-1} & & \beta_1 \\ T_{n-1} & T_{n-2} & ...... & T_r & ...... & T_1 & C_n & C_{n-1} & ...... & C_1 \end{matrix} T_r =$$

$$(+ \text{ or } -) \begin{matrix} \alpha_{n-1} & \alpha_{n-2} & & \overset{-}{\alpha_r} & & \alpha_1 & \beta_n & \beta_{n-1} & & \beta_1 \\ T_{n-1} & T_{n-2} & ...... & T_r & ...... & T_1 & C_n & C_{n-1} & ...... & C_1 \end{matrix}$$

$$\begin{matrix} \alpha_{n-1} & \alpha_{n-2} & & \alpha_1 & \beta_n & \beta_{n-1} & & \beta_r & \beta_1 \\ T_{n-1} & T_{n-2} & ...... & T_1 & C_n & C_{n-1} & ......C_r...... & C_1 \end{matrix} C_r =$$

$$\begin{matrix} \alpha_{n-1} & \alpha_{n-2} & & \alpha_1 & \beta_n & \beta_{n-1} & & \overset{-}{\beta_r} & \beta_1 \\ T_{n-1} & T_{n-2} & ...... & T_1 & C_n & C_{n-1} & ...... C_r ...... & C_1 \end{matrix}$$

The product $A_1B_1A_2B_2......A_mB_m$ can be expressed as :

$$\prod_{i=1}^{2mn} ( a_1 \, \mathbf{J} + b_1 \, \mathbf{V}_1 )$$

The matrices $\mathbf{V}_1$ are $\mathbf{T}_r$ or $\mathbf{C}_r$ matrices , $a_1$ , $b_1$ are scalar and the product can be easily evaluated on the base matrices $\Gamma_i$ (Lacolle (1984)) .


## 5.3 Free boundary conditions in the vertical lattice direction

The expression of Z becomes :

$$V^TA_1B_1A_2B_2......A_mV$$

where V is the vector of $\mathbf{R}^{2^n}$ with all the components equal to 1 . In the same way we use the basis of vectors (Lacolle(1984)) :

$$V^T (T_{n-1}^{\alpha_{n-1}} \quad T_{n-2}^{\alpha_{n-2}}......T_1^{\alpha_1} )$$


## 6 A FEW WORDS ABOUT THE COMPLEXITY OF THESE ALGORITHMS
(Lacolle (1984))

**Periodic boundary conditions** : The total amount of elementary operations for the computation of P grows like $m^2n^2 \, 2^{2n}$ .

**Free boudary conditions** : The total amount of elementary operations for the computation of P grows like $m^2n^2 \, 2^n$ .

**Polynomial algorithms** : There are polynomial algorithms for the two-dimensional problems (Barahona(1980)) but the designing of a computer program for these algorithms is not easy .

## 7 NUMERICAL RESULTS (for more details see Lacolle(1984)).

We give two typical examples of partition functions :

### Periodic two-dimensional Ising model 8x8

(CPU : 1500 seconds on a 3033 IBM computer)

Coefficients of $Z(\beta)$ with the variables $z = e^{-\beta}$. If the degree is not a multiple of 4, then the coefficient is equal to 0. We have the same coefficients for negative degrees.

| | | |
|---|---|---|
| DEGREE 128 : 2 | DEGREE 60 : 8582140066816 | |
| DEGREE 124 : 0 | DEGREE 56 : 36967268348032 | |
| DEGREE 120 : 128 | DEGREE 52 : 149536933509376 | |
| DEGREE 116 : 256 | DEGREE 48 : 564033837424064 | |
| DEGREE 112 : 4672 | DEGREE 44 : 1971511029384704 | |
| DEGREE 108 : 17920 | DEGREE 40 : 6350698012553216 | |
| DEGREE 104 : 145408 | DEGREE 36 : 18752030727310592 | |
| DEGREE 100 : 712960 | DEGREE 32 : 50483110303426544 | |
| DEGREE 96 : 4274576 | DEGREE 28 : 123229776338119424 | |
| DEGREE 92 : 22128384 | DEGREE 24 : 271209458049836032 | |
| DEGREE 88 : 118551552 | DEGREE 20 : 535138987032308224 | |
| DEGREE 84 : 610683392 | DEGREE 16 : 941564975390477248 | |
| DEGREE 80 : 3150447680 | DEGREE 12 : 1469940812209435392 | |
| DEGREE 76 : 16043381504 | DEGREE 8 : 2027486077172296064 | |
| DEGREE 72 : 80748258688 | DEGREE 4 : 2462494093546483712 | |
| DEGREE 68 : 396915938304 | DEGREE 0 : 2627978003957146636 | |
| DEGREE 64 : 1887270677624 | | |

### Ising Model 12x12 with 5% of negative bonds (with free boundary conditions) (CPU : 1700 seconds on a 3033 IBM computer )

We have no place to write all the coefficients of $Z(\beta)$ in $z=e^{-\beta}$. We give the largest coefficients in order to give an idea of their size.

| | | |
|---|---|---|
| DEGREE | 4 : | 10647394629240737858721129737272093500070436 |
| DEGREE | 2 : | 10894604599170712949764296857122613058800492 |
| DEGREE | 0 : | 10978285657878906413343517649219523904 15568 |

# 8   THE THREE DIMENSIONAL CUBIC LATTICE WITH ISING SPINS

The three dimensional problem is much more difficult . There exist some results about a 4x4x4 system (Pearson(1982)) . But these methods take into account the translation invariance , the cubic symmetry ...Therefore they cannot be used for a three dimensional spin glass .

## 8.1 Computational Method for a three dimensional spin glass

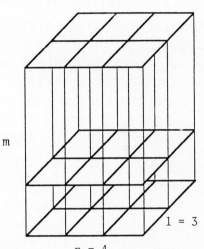

The cubic lattice : **lxnxm** is the superposition of **m** slices of **nxl** sites .We must apply free boundary condition in the vertical lattice direction . We use the transfer matrix method as we have done for the two-dimensional problem .

We consider the basis of vectors :

$$s^{\alpha_1} \otimes s^{\alpha_2} \otimes \ldots \ldots \otimes s^{\alpha_{ln}}$$

where :   $\alpha_i \in \{0,1\}$   $s^0 = \begin{pmatrix} 1 \\ 1 \end{pmatrix}$   $s^1 = \begin{pmatrix} 1 \\ -1 \end{pmatrix}$

The problem is always the computation of a product :

$$v^T ( \prod_{l=1...L} (a_l \mathbf{J} + b_l \mathbf{V}_l) ) V$$

V is the vector ($\in R^{2^{nl}}$ ) with all the components equal to one ,$a_l$ and $b_l$ are scalar and the vectors $\mathbf{V}_l V$ have very simple expressions with the base matrices .We do not give more details . The computational rules of the previous product are similar to the two-dimensional case (Lacolle(1984)) .

## 8.2 Complexity : The total amount of elementary operations for the computation of the partition function grows like $m^2 n^2 l^2 \, 2^{ln}$ and there is no polynomial algorithm (Barahona(1980)) .

## 8.3 Numerical results

### The 3x3x20 perfect Ising Model ( $2^{180}$ states . CPU : 900 seconds on a 3033 IBM computer )

We do not give all the coefficients of $Z(\beta)$ with the variable $z = e^{-\beta}$ .
In order to show the size of the coefficients we give the largest ones .

DEGREE 5  58751042827573490667622156997950511400561734712007092
DEGREE 3  59927514882490904992270566534304997602630982466607276
DEGREE 1  60524664054113422404873574636296722279029344622884008

### A 4x4x4 Spin Glass   with 6% of negative bonds and free
boundary conditions ( CPU : 4000 seconds on a 3033 IBM computer )
Coeficients of $Z(\beta)$ with the variable $z = e^{-\beta}$ . The coefficients of odd degrees are equal to zero and we have the same coefficients for negative degrees . The degree of each coefficient is written before it .

| | | |
|---|---|---|
| 126 2 | 82 3759611806 | 38 8917310280638986 |
| 124 6 | 80 8246151390 | 36 14487440534384674 |
| 122 20 | 78 17931674850 | 34 23026561410692462 |
| 120 82 | 76 38655653400 | 32 35778924196732746 |
| 118 256 | 74 82591244728 | 30 54310563694206838 |
| 116 800 | 72 174849197750 | 28 80485349516892072 |
| 114 2382 | 70 366653130452 | 26 116375398398281142 |
| 112 6710 | 68 761260494524 | 24 164086931926208242 |
| 110 18620 | 66 1564241260836 | 22 225492698533112166 |
| 108 49976 | 64 3179488818812 | 20 301880001859569784 |
| 106 130458 | 62 6389467411840 | 18 393547754253785552 |
| 104 333230 | 60 12687567602696 | 16 499412404466667466 |
| 102 834370 | 58 24879036291004 | 14 616703894275196226 |
| 100 2052884 | 56 48144435797986 | 12 740840559179617648 |
| 98 4965874 | 54 91878383001432 | 10 865558978358692368 |
| 96 11836054 | 52 172789036151000 | 8 983338402666010486 |
| 94 27839450 | 50 319978267432702 | 6 1086103582211950428 |
| 92 64683010 | 48 583014627693346 | 4 1166125854665225814 |
| 90 148633854 | 46 1044329160315750 | 2 1216986536950534826 |
| 88 338026654 | 44 1837525172007056 | 0 1234435851567174472 |
| 86 761345602 | 42 3173249625150622 | |
| 84 1699308608 | 40 5373898176625762 | |

# PART II : SYMBOLIC COMPUTATION OF THE FREE ENERGY OF SOME REGULAR TWO-DIMENSIONAL SPIN GLASSES

## 1 INTRODUCTION

On the square lattice , with the notation :

$$E(\sigma) = - \sum_{(i,j)} J_{i,j} \, \sigma_i \, \sigma_j \quad \text{and } J_{ij} = 1$$

$$Z(\beta) = \sum_\sigma e^{-\beta E(\sigma)}$$

L. Onsager established in 1944 the famous result :

$$f(\beta) = \lim_{n \mapsto \infty} \frac{1}{n^2} \log Z(\beta) = \log 2 + \frac{1}{8\pi^2} \int_0^{2\pi}\int_0^{2\pi} \log(ch^2 2\beta - sh2\beta(cos\omega + cos\varphi))d\omega\,d\varphi$$

In this lecture we call $f(\beta)$ the free energy ; it is not exactly the physical definition of this function . After L.Onsager a lot of different methods have been developped to establish this previous expression . Now , let us assume that $J_{ij}$ takes only two values +1 or -1 . Some interesting spin glass models may be represented by regular models wher $J_{ij}$ are periodically distribued ( The Villain model , the B model , the C model ,the layered models...) The solutions of these models ( and some other models ) are well known (Bryskin(1980),Longa(1980),Villain(1977),Hoever(1981), Wolff(1983)) .The first results have been obtained with combinatorial methods and most recently with a representation in terms of fermions operators . With combinatorial methods the free energy is expressed as a double integral and the kernel of this integral is a determinant $\Delta$ . The size of this determinant is related to the size of the *unit cell* of the periodical model . For the Villain model the size of determinant is 8x8 , for the C model the size is 16x16 ... But for other models , with larger unit cells , the computation of the determinant is too dificult .

To make this evaluation possible :
1) We introduce a more compact expression of the previous determinant $\Delta$ with the *spinor analysis* formalism introduced by B.Kaufman (1949).
2) We use the symbolic programming language REDUCE (for a descrition of this language see Hearn (1983)) for computing the expressions of these determinants .

We want to show that the Kaufman method associated with a symbolic computation language is an *efficient interactive tool* to obtain the exact expression of free energy .This method does not replace other ones which give more compact solutions but we want to present an *investigation tool* for the Physicist .

## 2 SPINOR ANALYSIS (Boerner(1970),Kaufman(1949)) .

We summarize the problem in a few words .The partition function of the Ising model is the trace of a power of the transfer matrix V . It has been shown that V is a $2^n$ -dimensional "spin representative" of a 2n-dimensional rotation O .The eigenvalues of V are related to the eigenvalues of O .The periodicity of interactions on the lattice makes the matrix O cyclic.

### 2.1 General notations

We consider a model with qr rows and pr sites on a row with the same notations as in the first part of this paper. On a row the values of interactions are distribued with the period p . On a column the interactions are distribued with the period q .The partition function can be expressed by :

$$Z_{m,n} = \text{Trace} ( A_1^{(n)} B_1^{(n)} \ldots\ldots A_m^{(n)} B_m^{(n)} ) = \text{Trace} (C^{(n)})^r$$

with m = qr , n = qr and :

$$C^{(n)} = A_1^{(n)} B_1^{(n)} \ldots\ldots A_q^{(n)} B_q^{(n)}$$

We want to obtain :

$$\lim_{r \mapsto \infty} (1/pqr^2) \log \text{Trace} ( C^{(pr)} )^r$$

We can set up a rotation $\tilde{D}^{(pr)}$ corresponding to $C^{(pr)}$ and , after many calculations , the previous limit becomes (Lacolle(1984)) :

$$-\log 2\text{sh}\beta + \frac{1}{2} \frac{1}{pq} \lim_{n \mapsto \infty} \frac{1}{4\pi r} \int_0^{2\pi} \log |\det( \tilde{D}^{(pr)} - \mathfrak{I} e^{i\varphi} )| \, d\varphi$$

where $\mathfrak{I}$ is the identity matrix .

## 2.2 Construction of $\tilde{D}^{(pr)}$ . Illustration with the case q=1 .

We illustrate the method in the case q=1 but it is not more difficult to deal with the case q ≠ 1 . However if we consider horizontal layers , the particular case q=1 (with the periodicity considerations of the section 2.3) leads to a final result with 2x2 matrices .

We note the values of $\beta\, J_{ij}$ by $\overset{v}{\beta_k}$ and $\overset{o}{\beta_k}$ and we repeat periodically on the lattice the unit cell :

All these models ca be reduced to the case : $\beta_i = \beta$ and we have :

$$\tilde{D}^{(pr)} = \tilde{A}^{(pr)}\ \tilde{B}^{(pr)}$$

$$\tilde{A}^{(pr)} = \begin{pmatrix} ch2\beta & 0 & 0 & \ldots & sh2\beta \\ 0 & sh2\beta & sh2\beta & \ldots & 0 \\ 0 & sh2\beta & ch2\beta & \ldots & 0 \\ \cdot & \cdot & \cdot & \cdots & \cdot \\ sh2\beta & 0 & 0 & \ldots & ch2\beta \end{pmatrix}$$

$$\tilde{B}^{(pr)} = \begin{pmatrix} b_1 & & 0 & \ldots & 0 \\ 0 & & & b_2 & \ldots & 0 \\ \cdot & & \cdot & & \cdots & \cdot \\ 0 & & 0 & 0 & \ldots & b_n \end{pmatrix}$$

with :

$$b_k = \begin{pmatrix} coth\,2\beta_k & -1/sh2\beta_k \\ & \\ -1/sh2\beta_k & coth2\beta_k \end{pmatrix} \qquad \beta_k = \overset{v}{\beta_k}$$

## 2.3 Row periodicity

Now we take into account the row periodicity which makes the matrices $\tilde{A}^{(pr)}$ and $\tilde{B}^{(pr)}$ block cyclic. We denote the blocks:

$$a_1 = \begin{pmatrix} ch2\beta & 0 & 0 & . & . & . & 0 \\ 0 & ch2\beta & sh2\beta & . & . & . & 0 \\ 0 & sh2\beta & ch2\beta & . & . & . & 0 \\ . & . & . & & . & . & . & 0 \\ 0 & 0 & 0 & . & . & . & ch2\beta \end{pmatrix}$$

$$a_2 = \begin{pmatrix} 0 & 0\,0\ldots 0 \\ . & . & . & \ldots \\ sh2\beta & 0\,0\ldots 0 \end{pmatrix}$$

$$\delta = \begin{pmatrix} coth2\beta_1 & -1/sh2\beta_1 \\ -1/sh2\beta_1 & coth2\beta_1 \\ & & . \\ & & & coth2\beta_p & -1/sh2\beta_p \\ & & & -1/sh2\beta_p & coth2\beta_p \end{pmatrix}$$

The final result is:

$$f(\beta) = -\log 2 + \frac{1}{2} - \frac{1}{8p\pi^2} \int_0^{2\pi}\int_0^{2\pi} \log\{sh^p 2\beta |det((a_1\delta - \Im e^{i\varphi}) + a_2\delta e^{i\omega} + a_3\delta e^{-i\omega})|\}\,d\omega\,d\varphi$$

where $\Im$ is the identity matrix (for more details see Lacolle(1984)).

## 2.4 USE of the Symbolic language REDUCE (for a description of *Reduce* see Hearn(1983))

We have no place to give any details about the use of *REDUCE*.

The coefficients of matrices $a_1, a_2, a_3, \delta$ (sh2$\beta$, ch2$\beta$,...) and the variables $e^{i\omega}, e^{i\varphi}$ are *symbolic expressions*. Using the symbolic language *REDUCE* we can perform all the matrix operations (multiplications, evaluation of determinant,...) with these symbolic expressions. We give to the system the simplification rules ( $\cos^2 + \sin^2 = 1$ ,...) and we have a *symbolic expression* of the result .

**Example 1 :**

We consider the model constructed by the repetition of the cell :

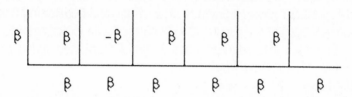

With the notations $x = $ sh2$\beta$ , $\varphi = $ phi , $\omega = $ w the developped expression of the argument of the logarithm (2.3) is given in the form :

```
16*sh(x)**10*(2*cos(2*phi) - 3) + 32*sh(x)**9*
( - 2*cos(3*phi) + 5*cos(phi)) + 8*sh(x)**8*(6*cos(4*phi)
+ cos(2*phi) - 45) + 16*sh(x)**7*( - cos(5*phi)
- 9*cos(3*phi) + 51*cos(phi)) + 2*sh(x)**6*(cos(6*phi)
+ 44*cos(4*phi) - 124*cos(2*phi) + cos(w) - 450)
+ 16*sh(x)**5*( - cos(5*phi) - 5*cos(3*phi) + 89*cos(phi))
+ 4*sh(x)**4*(10*cos(4*phi) - 96*cos(2*phi) - 243)
+ 1024*sh(x)**3*cos(phi) + 32*sh(x)**2*( - 5*cos(2*phi) - 14)
+ 256*sh(x)*cos(phi) - 64
```

**Example 2 :**

We can carry out the computation for the model constructed by the repetition of the cell :

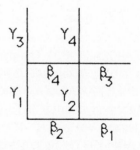

We obtain a developped expression of the kernel with the eight parameters $\beta_1, \beta_2, \beta_3, \beta_4, \gamma_1, \gamma_2, \gamma_3, \gamma_4$ .For this expression see Lacolle (1984) .

# BIBLIOGRAPHY

Y.ABE,S.KATSURA (1976) : Phase transition and distribution of zeros of the partition functions for an antiferomagnetic Ising model and for a hard core lattice gas . J.Phys.Soc. Japan,Vol. 40,642

A.V.AHO,J.E.HOPCROFT,J.D.ULLMANN (1974) : The design and analysis of computer algorithms . Addison Wesley

F.BARAHONA (1980) : Application de l'optimisation combinatoire à certains problèmes de verres de spins : complexité et simulation. Thèse Docteur Ingénieur Grenoble

H.BOERNER (1970) : Representations of groups with special considerations for the needs of modern physics . North Holland Publishing Company

V.V.BRYSKIN,A.Y.GOLTSEV,E.E.KUDINOV (1980) : Some exact results for the 2D Ising model with regular disposition of frustated squares . J. Phys. C : Solid State Phys. ,13

A.C.HEARN (1983) : Reduce user's manual The Rand Corporation Santa Monica April 1983

P.HOEVER,W.F.WOLFF,J.ZITTARTZ (1981) : Random layered frustration models Z.Physik B, Condensed matter 41

S.KATSURA,Y.ABE,.M.YAMAMOTO (1971) : Distribution of zeros of the partition functions of the Ising model . J.Phys.Soc. Japan ,Vol. 30,347

B.KAUFMAN (1949) : Crystal Statistics II . Partition function evaluated by spinor analysis . Phys.rev. Vol 76 ,Number 8

H.A.KRAMERS,G.H.WANNIER (1941) : Statistics of the two-dimensional ferromagnet Part I . Phys.Rev. 60

B.LACOLLE (1984) : Sur certaines méthodes de calcul de la Physique Statistique . Thèse Mathématiques Universite De Grenoble

L.LONGA,A.M.OLES (1980) : Rigourous Properties of the two-dimensional Ising model with periodically distibued frustration . J.Phys. A 13

S.ONO,Y.KARAKI,M.SUZUKI,C.KAWABATA (1968) : Statistical thermodynamics of finite Ising Model I . J. Phys. Soc. Japan Vol.25 N° 1

L.ONSAGER (1944) : Cristal Statistics I . A two-dimensional model with an order-disorder transition . Phys.rev. 65 ,Number 3-4

R.B. PEARSON (1982) :The partition function of the Ising Model on the periodic 4x4x4 lattice Phys.rev. B (3),26

M.SUZUKI,C.KAWABATA,S.ONO,Y.KARAKI,M.IKEDA (1970) : Statistical thermodynamics of finite Ising Model II . J. Phys. Soc. Japan , Vol.29 , N° 4

J.VILLAIN (1977) : Spin glass with non random interactions J. Phys. C : Solid State Phys. , Vol. 10 , N° 4

W.F. WOLFF , J. ZITTARTZ (1983) : Spin glasses and frustration models : Analytic results . Lecture Notes in Physics 192

# ORDER AND DEFECTS IN GEOMETRICALLY FRUSTRATED SYSTEMS

R. MOSSERI* and J.F. SADOC**

*Laboratoire de Physique des Solides, CNRS, F-92195 Meudon Principal Cedex

**Laboratoire de Physique des Solides, Université Paris-Sud, F-91405 Orsay

## I - THE CURVED SPACE MODEL

In many physical (and perhaps biological) systems, there is a contradiction between the local order (local building rule) and the possibility for this local configuration to generate a perfect tiling of the euclidean space. It is the case, for instance, for amorphous metals where long range order is absent while the local configuration is rather well defined with the presence of five fold (icosahedral) symmetry[1]. Our aim is to provide a definition for the notions or "order" and "defects" in these materials. In a first step the underlying space geometry is modified in order for the local configuration to propagate freely (without "frustration"). This is often possible by allowing for curvature in space (either positive or negative). The obtained configuration in space is called the "Constant Curvature Idealization" of the disordered material. Let us take a simple example. In 2D a perfect pentagon tiling of the euclidean plane is impossible (fig. 1-a). It is however possible to tile a positively curved surface, a sphere, with pentagons leading to a pentagonal dodecahedron. Let us try now to densely pack spheres in 3D (this is an approximation to amorphous metals structures). Four spheres build a regular tetrahedron, but a perfect tetrahedral packing is impossible because the tetrahedron dihedral angle is not a submultiple of $2\pi$ (fig. 1-b). A perfect tetrahedral packing becomes possible

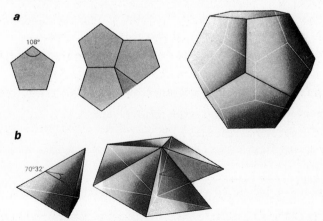

*Fig. 1 : geometrical frustration in 2 and 3 Dimensions*

NATO ASI Series, Vol. F20
Disordered Systems and Biological Organization
Edited by E. Bienenstock et al.
© Springer-Verlag Berlin Heidelberg 1986

if space is curved. 600 regular tetrahedra fill the (hyper) surface of a hypersphere S3[*]. The obtained configuration is called the polytope {3,3,5} using standard notations[2]. It contains 120 12-fold coordinated vertices with perfect icosahedral local configuration and each polytope edge shares 5 regular tetrahedra.

## II - OTHER CURVED SPACE TEMPLATES

### II-1. Tetracoordinated non crystalline structures

In covalent structures like aSi or aGe, the local configuration is almost perfectly described as an atom at the center of a tetrahedron with its four neighbours at the tetrahedron vertices. This local order can lead to numerous regular structures in either euclidean and curved space. It is thus necessary to add some details at a slightly larger scale in order to completly define the local configuration. The "local building rule" must include informations about the ring configuration such as :
  - the number of edges in the ring
  - the twist of the rings
  - the existence, or absence, of cages with faces defined by the rings.

With this in mind, several tetracoordinated polytopes have been described[3] :
  - the polytope {5,3,3} , the dual of the {3,3,5} which is a packing of 120 dodecahedra on S3. This polytope is characterized by the occurence of cages (the dodecahedra) and the oddness nature of its rings (pentagons). It is a template for clathrate like structures and possibly amorphous ice.
  - the polytope "240", which has 240 vertices, each one being surrounded by 18 twisted-boat hexagons. This polytope medium range order is very similar to that encountered in older euclidean model proposed for aSi and aGe, but here this configuration propagates throughout the 3D space without defects. Its theoretical electronic properties have been calculated and compared with experiments[4].
  - Another tetracoordinated polytope has been described[3] whose local order is intermediate between that of polytope {5,3,3} and "240".

### II-2. Curved cholesteric structures

J. Sethna[5] has recently given a solution to the blue phase

---

* S3 can be embedded in the 4D euclidean space with equation : $x_1^2 + x_2^2 + x_3^2 + x_4^2 = R^2$. Note that only 3 of 4 coordinates are independant, S3 being a 3D (curved) manifold.

frustration problem using a spherical space. In these structures, cholesteric "linear" molecules are nearly aligned with each other, but the molecular interaction leads to a prefered small twist angle between the molecular axes. It is not possible to generate such "double twisted" director field in euclidean space. Such configuration is possible on S3 by taking the undirected tangent field to a bundle of Clifford parallel lines which fill S3.

## III - MAPPING ON EUCLIDEAN SPACE AND DEFECTS

In order to describe realistic structures, one has to map the above described templates back to euclidean space. The required geometrical transformations will define the topological defects of the euclidean structure. Metrical distortions are also generated during the mapping. In the following we describe one type of defect called a disclination.

Disclinations are created by cutting the structure and adding (or removing) a wedge of material between the two lips of the cut. Figure 2 shows the relation between disclinations, curvature and ring parity for 2D disclination point defects. In 3D disclinations are linear and also carry curvature. A flat structure (on the average) can thus be obtained from the polytope by introducing a suitable amount of disclination lines carrying negative curvature. The type of such line defects compatible with the poly-tope symmetry has been classified using homotopy theory[6]. The underlying geometry of the flattened model is better described as a corrugated 3D manifold[7]. Positively curved regions, where the local order is similar to the template polytope, are called the "ordered regions". Negatively curved regions, where disclinations are located, are the loci of defects.

Now different kind of structures are associated with different arrangement of the disclinations lines.

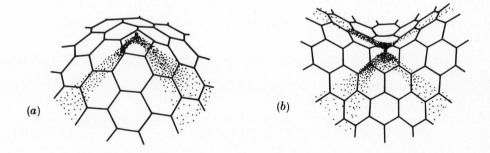

(a)          (b)

Fig. 2 : The relation between disclination , curvature and ring size in 2D. The starting configuration is a planar hexagonal network : a) a positive disclination point located at the pentagon center b) a negative disclination point located at the heptagon center.

- Periodic network of disclination lines[8]. It corresponds to complex crystalline structures like the Frank-Kasper metallic phases or crystalline clathrate with large unit cells.

- Hierarchical structure of defects[9]. The obtained non crystalline flat models belong to a larger family of recurrent structures like the Penrose tilings for instance. They could provide good models for the recently discovered icosahedral "quasi crystals"[10].

- Disordered array of line defects. It corresponds to the case of amorphous materials (and low temperature liquids probably). The disorder is no more supposed at an atomic range but rather at the range of disclination interdistances : giving the disclination network completely define the whole structure apart from small atomic displacements.

## References

1) J.F. SADOC, J. Non Cryst. Sol. 44 (1981)1. J.F. SADOC, R. MOSSERI, in"Topological disorder in condensed matter", Sol. St. Sciences 46 (Springer Verlag).
2) H.S.M. COXETER, Regular polytopes (Dover 1973).
3) J.F. SADOC, R. MOSSERI, Phil. Mag. B45 (1982) 467.
4) D. DI VINCENZO, R. MOSSERI, M.H. BRODSKY, J.F. SADOC, Phys. Rev. B 29 (1984) 5934
Same authors, submitted to Phys. Rev. B.
5) J.P. SETHNA, Phys. Rev. Lett. 51 (1983) 2198.
6) D. NELSON, M. WIDOM, Nucl. Phys. B 240 [FS 12] (1984) 113.
7) J.P. GASPARD, R. MOSSERI, J.F. SADOC, Phil. Mag. B 50 (1984) 557.
8) J.F. SADOC, J. Phys. Lett. 44 (1983) L.707.
9) R. MOSSERI, J.F. SADOC, J. Phys. Lett. 45 (1984) L-827.
Same authors, submitted to J. de Physique.
10) D. SCHECHTMAN, I. BLECH, D. GRATIAS, J.W. CAHN, Phys. Rev. Lett. 53 (1984) 1951.

# 3 FORMAL NEURAL NETWORKS

# COLLECTIVE COMPUTATION WITH CONTINUOUS VARIABLES

J. J. Hopfield
Divisions of Chemistry and Biology, California Institute
of Technology, Pasadena, CA 91125 and AT & T Bell Telephone
Laboratories, Murray Hill, NJ 07974

D. W. Tank
AT & T Bell Telephone Laboratories
Murray Hill, NJ 07974

The model on which collective computation in neural networks was originally shown to produce a content addressable memory was based on neurons with two discrete states (1) (or levels of activity) which might be thought of as "0" and "1". That model involved an asynchronous updating of the state of each neuron, and could be closely related to aspects of magnetic Ising systems. The role of exchange in the magnetic system is played by the connections between the neurons in the biological system.

In this lecture we discuss the behavior of such a system with neurons (or equivalently, amplifiers) having a continuous response. There are two important reasons for extending the analysis into this domain. First, real neurons have input-output relations which may be more reasonably modeled on a continuous basis than on an all-or-none basis. Only those properties which are present when the neural system is modeled with continuous variables might reasonably be attributed to a biological system. Fortunately, most of the interesting properties of the Ising system persist with minor alteration when the variables become continuous. Second, if we were to attempt to capture some of the properties of the idealized model in an electrical circuit made to function like the biological one, the system would have smooth (though perhaps rapid) responses of its amplifiers, and should be modeled as a set of differential equations rather than as a discrete algorithm acting in discrete time.

One of the major questions of the neurobiology of higher animals is how

NATO ASI Series, Vol. F20
Disordered Systems and Biological Organization
Edited by E. Bienenstock et al.
© Springer-Verlag Berlin Heidelberg 1986

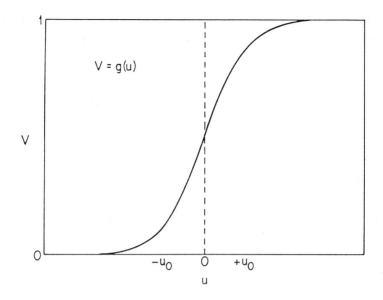

Fig.1 Firing frequency versus input for a neuron displaying action potentials.

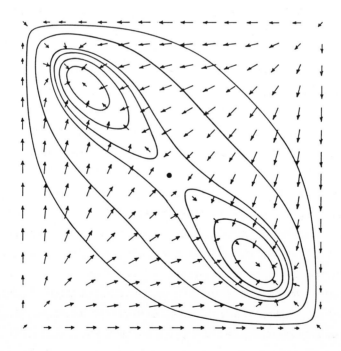

Fig.2 The flow field and E for a bistable circuit of 2 amplifiers (flip-flop).

they get the enormous computational power which they seem to display. The present analysis indicates that the large connectivity, the reciprocal connectivity, and the analog nature of the system all play important roles. The system functioning in a smooth and analog mode performs complex computations much more effectively than it does in the 1-0 mode of operation.

The output of a typical neuron in the brain of a mammal consists of series of stereotyped "spikes" (action potentials) each about 1 millisecond long. If a neuron has a strong input, then these spikes are generated at a high rate. If the input is weak or absent, then the spikes are produced at a very low rate. The mean rate of the generation of spikes as a function of the input is shown in Fig. 1. In some systems (e.g., signal processing early in the auditory system) the timing of individual spikes is an essential part of signal processing. But away from primary sensory areas in brain, it is generally believed that the individual action potential or its exact timing is unimportant. The short time average or running integral of excitation is often taken to be the effective input to a cell, and its frequency of producing action potentials the effective output (2). The situation is thought to be similar to that in electric circuit analysis, where the electrical current is taken as a continuous variable in spite of the fact that electric charge is actually carried by discrete electrons. The electrons introduce a noise which can be added to the usual analysis. We will thus adopt a smooth input-output relation in circuit theory. Similarly, noise due to the discrete nature of action potentials can be added to the present analysis, and under some circumstances even creates new computational possibilities.

We consider a set of neurons i all having gain functions $V_i = g(u_i)$, where $u_i$ is the input of neuron or amplifier i, $V_i$ is the output of neuron i, and the function g is a bounded and monotonic input-output relation like that shown in Fig. 1. The inputs to any neuron i arise from an external contribution $I_i$ and a contribution from the outputs of other neurons to which it is connected. A synaptic weight $T_{ij}$ describes the extent to which the input of neuron i is driven by the output from neuron j. In electrical circuit terms, $T_{ij}$ is the electrical conductance of a connection between the output of j and the input of i. In neurobiology, it would be viewed as the synaptic weight of the j-to-i connection.

In both neurobiology and circuits, the device i has a characteristic input capacitance and input resistance. The equations of motion of the input then have the general form

$$C_i \frac{du_i}{dt} = - \frac{u_i}{R_i} + \sum_j T_{ij} V_j + I_i$$

In order to obtain interesting results, it is essential that the elements of T can have both signs. This is most simply accomplished in biology, for both excitatory and synaptic connections can be made. In the former case the output from neuron j will drive i to produce more output, and in the latter case the output of j will drive i toward producing less output. A similar property can be constructed easily with amplifiers having both a normal and an inverting output. The connection matrix is then constructed from positive conductances connected to the normal output (for a positive $T_{ij}$) or to the inverting output (for negative $T_{ij}$). The net input impedance $1/R_i$ seen at the input of an amplifier is the result of the parallel input impedance of the amplifier itself and the conductance network which drives it. For simplicity, all the $R_i$ will be taken as identical, as will the values of $C_i$.

Simple as these equations appear, there can be no general theory of the behavior of the solutions. The computational behavior of a Cray computer could be duplicated by an appropriate choice of the matrix T. In order to obtain a tractable set of equations, we limit outselves to the special case of T being a symmetric matrix. Even this restricted form of T can perform powerful computations.

In several areas of the brain, reciprocal circuits of the sort implied by a symmetric T matrix have been observed (3). However, in many areas, the circuitry is more complex. The grey matter of the cerebral cortex is sheet-like, and each sheet has many layers. While it is often true that the output of layer A may dominantly drive layer B, and that layer B may chiefly drive layer C there are also many pathways in the other direction. Within a few of such layers, a reciprocal path will then exist from C or D back to A. Reciprocal connections between different cortical areas are very common (4). Thus while biology may not literally implement a direct reciprocal matrix, it has all the properties necessary to implement a system which is equivalent to a symmetric matrix. One may thus expect evolution to make use of the interesting computational properties which symmetric connections may produce, as well as use the much greater - but less tractable analytically - possibilities that can result when symmetry is not required.

The importance of a symmetric matrix (slightly weaker conditions would suffice) is that there is then a Lyapunov function for the system (5). The function E defined by

$$E = -\frac{1}{2} \sum_i \sum_j T_{ij} V_i V_j - \sum_i I_i V_i + \sum_i \int^{V_i} \frac{g^{-1}(V) dV}{R_i}$$

can be easily shown to be locally minimized by the equations of motion, i.e.,

$$dE/dt \leq 0, \quad dE/dt = 0 \longrightarrow dV_i/dt = 0 \text{ for all } i$$

The third term in E goes to $+ \infty$ when one or more of the $V_i$ approach their maximum or minimum values. Since the other terms in the energy are bounded if $V_i$ remains finite, the above conditions ensure that the system moves to a local minimum of E and comes to rest.

Let the extrema of $V_i$ be taken as $-1$ and $1$ for simplicity. If this is not the case, a simple transformation can be made such that it is true in new variables. Then the domain of operation of the equations of motion is the N dimensional hypercube $-1 < V_i < 1$, where N is the number of neurons or amplifiers. The infinite gain limit is particularly simple, for the last term in the energy function then vanishes on the interior of the domain, and this term then has only the effect of restricting the operation of the circuit to the interior. In the infinite gain limit, the energy function can be thought of as only having the first two terms, but limited to the domain $-1 < V_i < 1$. These first two terms can have a minimum on the interior of the domain only if T is a negative definite matrix. If not - and this will usually be the case for interesting T - all minima must be at the boundary of the hypercube. If in addition there are not self terms (i.e., $T_{ii} = 0$), then the true minima all occur at corners of the hypercube. In this limit, the stable states of the system are exactly those of a set of Ising spins with an exchange matrix $T_{ij}$ and an external field $I_i$.

At finite gain, the stable minima move from the corners of the domain into the interior. As the gain is reduced, minima disappear by combining with saddle points, and in the limit of zero gain, a single stable point remains. Figure 2 illustrates the operation of the system at finite gain for the case of two neurons, connected so as to have a pair of stable points near (1,1) and at (-1,-1). The flow field of the equations of motion and the contours of the energy function are both shown. Note that while the system generally goes downward in energy, the actual equations of motion do not produce a steepest descent path. (A steepest descent path can be produced by a slightly modified circuit (6).)

It has been shown that such a system can perform as a content-addressable memory (CAM). (1) CAM is intrinsically not merely memory. To retrieve a memory from partial content intrinsically involves a computation of which memory is most similar to the partial information given. While this is a computation, it is a very simple one. When such a system is used as a CAM, it will generally store $\varepsilon \cdot N$ memories, each of length N bits, and the retrieval clue might have $\delta \cdot N$ bits fixed and the rest given at random and $\varepsilon$ and $\delta$ are positive numbers less than 1. The computation to find the correct memory is intrinsically easy, scaling like $\varepsilon \delta N^2$ if done on a conventional computer.

Much more complex calculations can be done by such a network. In the following, we examine the behavior of the network on a difficult set of problems having considerable combinatorial complexity (7). The problems of finding meaningful interpretations of the visual or the auditory world involve such great computational complexity that machine vision and machine interpretation of natural speech have remained extremely primitive. Thus it becomes interesting to examine the computational power of a small neural network to understand its computational abilities. Unfortunately there are no good computational models of how to deal with such real-world problems. Since the essence of the problems seems to be combinatorial complexity - the difficulty of having to consider many possible solutions to a problem in order to determine the best solution - we will use a well-studied artifical example of known difficulty. The Traveling Salesman Problem (TSP) we have chosen (7) is an np-complete problem of combinatorial complexity (8). The problem falls naturally on a network of neurons, and the construction by which it does so indicates how to put a broad class of difficult problems onto such a network.

The syntax in which we are going to interpret the state of the network of neurons must first be established. In ordinary programming, it is necessary to establish how to interpret the output string of bits as an answer - for example, is the computed string of output bits to be interpreted as a binary number or as a decimal number written in ASCII? In the TSP problem on n cities, we will make use of $N = n^2$ neurons (amplifiers) to solve and convey the the problem. In the final state, some of these devices will have outputs very near to "1", and the rest will be very close to "0". We choose to arrange the neurons into n groups of n each, as indicated in the table below. Each row represents a particular city. A final-state syntax is chosen in which exactly one amplifier has an output near "1" in each row and each column. The rest of the outputs will all be near zero. We will interpret the meaning of this array as the order in which the cities are to be visited in a tour. Thus if 1's are present in the array shown below, the interpretation is to visit city C first, A second, E third, and so on.

|   | 1 | 2 | 3 | 4 | 5 |
|---|---|---|---|---|---|
| A | 0 | 1 | 0 | 0 | 0 |
| B | 0 | 0 | 0 | 1 | 0 |
| C | 1 | 0 | 0 | 0 | 0 |
| D | 0 | 0 | 0 | 0 | 1 |
| E | 0 | 0 | 1 | 0 | 0 |

Any such matrix, with one 1 in each row and each column, represents a complete closed tour of all cities. There is a 2n-fold degeneracy in the description of a path corresponding to the choices of starting point and direction of tours representing a given path.

An energy function (or alternatively a set of connections and/or inputs) must be specified which will enforce this syntax. For convenience, amplifiers or neurons are given subscripts $V_{Xi}$ to represent the amplifier which will be "on" if city X is to be in position i in the tour. The limits on the outputs of all amplifiers will be taken as 0 and 1. Then the following energy function will enforce the syntax of desired answers.

$$E = A/2 \sum_X \sum_i \sum_{j \neq i} V_{Xi} V_{Xj} + B/2 \sum_i \sum_X \sum_{Y \neq X} V_{Xi} V_{Yi} + C/2 (\sum_X \sum_i V_{Xi} - n)^2$$

where A, B, and C are positive constants. All final state output matrices describing valid tours have zero energy from these terms enforcing syntax

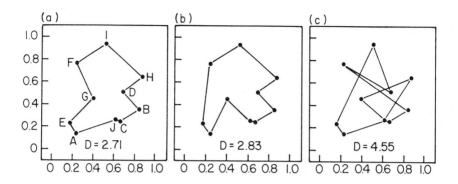

Fig.3  TSP solutions found by the network a),b), low gain; c) high gain.

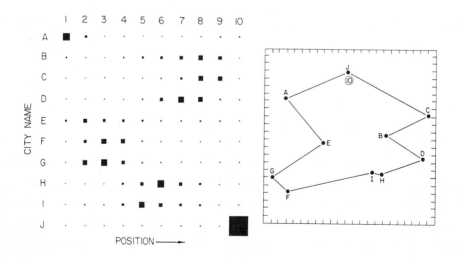

Fig.4  A stage during the analog convergence process. The linear dimensions of the squares are proportional to the outputs of the corresponding neurons. $V_{J10} = 1$.

while all other matrices have higher energy.

The successful calculation of a good path must involve adding to the problem the data – the information about the distance between cities in the particular exemplar of the problem to be solved. It must bias the n! equivalent tours (from the point of view of syntax) in such a fashion that only short tours are likely to occur as answers. Since there is empirical evidence that the minima which have the lowest E are the easiest for such networks to find (9), (deeper valleys are wider in this problem) it should be sufficient to add a term in E so that the paths which are shortest have the lowest energies. The term

$$1/2 \, D \sum_{X} \sum_{Y \neq X} \sum_{i} d_{XY} \, V_{Xi}(V_{Y,i+1} + V_{Y,i-1})$$

has such a form, for when the syntax of the answer is enforced, this term adds to the energy a number which is equal to the path length of that tour.

The shortest path now corresponds to the 2n tours which have the lowest energy. If the circuit actually found the absolute minimum of E, it would find the solution to the TSP. Since the network can be guaranteed only to find a local minimum of E, it becomes an experimental question to study how well the network does in solving the problem.

Since we do not know what the answer is, and the data in the problem is already contained in the connections, a calculation consists of starting the system somewhere in its state space and seeing what answer it finds. It would be nice to make a totally unbiased starting point. Unfortunately, the 2n-fold degeneracy of the description of all of the paths as tours produces a broken symmetry problem. Noise must be entered into the system so that it can choose a particular one of the 2n equivalent tours which describe a path, for just the same reason that the choice of the direction in which the magnetism of a ferromagnet will point when it is cooled through its phase transition depends on noise. The real world of course contains such noise, but in the artificial world of digital computers it is necessary to insert such noise. The initial $u_i$ were chosen equal to a common constant plus a small amount of random noise.

We have found in computer simulations that if the amplifiers are operated with very high gain (a very steep input-output relation in the transition region between 0 and 1), the circuit operates in an all-or-none-fashion. At any instant, most of the amplifiers will have output either very near zero or very near unity. Occasionally an amplifier will rapidly transit between an output of zero (or unity) and an output of unity (or zero). The decisions made by the amplifiers are chiefly sequential. Computing in this fashion is qualitatively similar to the operation of the original CAM algorithm (1), in which the decisions by the amplifiers were made in a stochastic order. Typical such results are shown in Fig. 3c for the case of 10 cities. The path is little better than a random path. Qualitatively different results are obtained when the gain is lower, as is shown in the paths of Fig. 3a,b. Figure 3a also is the shortest path. With appropriate parameters, the network found the best 2 (out of $1.8 \times 10^5$) paths for almost 50% of starting states.

Figure 4 shows the convergence of the network in a low gain circumstance. During this particular example, neuron J10 was held on so that city J would be definitely in position 10. The other 9 cities and their 81 neurons were free to change. At the time shown, most of the amplifiers have small outputs. The energy function was constructed with a view in mind to its meaning only when $V_{Xi} = 0$ or 1, but is being evaluated and is guiding the decision on the **interior** of the domain of operation of the circuit. At the time shown, there is no rigorous definition of the meaning of the output pattern. Yet it is qualitatively clear that some of the decision has roughly already been made. Clearly cities E, F, and G are contending for positions 2,3, and 4, though no firm decision on which will capture which has yet been made. Similarly, H and I are contending for positions 5 and 6, with no clear decision yet to be seen. A comparison of the two halves of the figure indicates that the state of the amplifiers is consistent with a rough description of a good path, and that the approximate information from one region is appropriate for other regions to act on. The analog dimension of the ouptut of the amplifiers is being used to represent a superposition of similar paths, not readily obtained in the high gain limit. All amplifiers are simultaneously progressing toward their final state, and decisions are being made in parallel rather than in series. If a 1 in position Xi means that "it is true that city X is in position i" and a 0 in the position means that "it is false that city X is in position i", we see that what is being made use of is a peculiar realm in which propositions are neither wholly true nor wholly false (until the end of the calculation). As a consequence, similar states can be simultaneously considered.

The sigmoid gain curve of finite width brings to the deterministic calculation some of the aspects of finite temperature in simulated annealing methods of solving such problems. Within the "Ising spin" (0-1) representation of the TSP problem, it would be possible to carry out a simulated annealing to search for a near-optimal solution. In such a procedure, the actual spins have only values of 0 and 1. In spin glasses of infinite range, the effective field approximation to the statistical properties of the system is often an excellent description. If the gain curve of the amplifiers is taken of the form

$$V(u) = (1 + \tanh(u/\alpha))$$

then the values of $V_i$ taken on in the stable states of the electrical network can be shown to obey the equations of the effective field approximation. This equivalence between the network analog parameters and the effective field solutions is a way of understanding the meaning of the analog values taken on by the network states. Of course, the effective field equations have many solutions, and the network picks out one particular solution during its dynamic convergence from the initial state. The actual simulated annealing procedure carried out by Kirkpatrick et al (10) for TSP was done directly with paths (the spin representation was not known) and in the path framework no effective field description was evident.

Neural network simulations have also been carried out with 30 cities and a network of 900 neurons or amplifiers. In this case, there are more than $10^{30}$ possible paths. The best solution is not found, but one of the best $10^7$ to $10^8$ paths is obtained.

In the TSP problem, there are three contributions to the biases and interconnctions which determine the dynamics. There is an input bias, the same for every amplifier or "spin", and this is equivalent to a static magnetic field in terms of a magnetic description of the system. The interconnections are from two sources. The syntactical rules for the solution contribute to the connections through terms A, B, and C in an orderly fashion. The random element in the problem, the distances between the cities, contribute only to the strengths of the interconnections $T_{ij}$, so from the point of view of spin glasses, the TSP problem has the aspect of a random exchange system. That the TSP problem is np-complete, and the ground state of a random exchange spin glass is also (11) seem united by these ideas.

Many other problems of discrete choices with combinatorial possibilities can be mapped onto a set of analog neurons in a related fashion. A particularly interesting one, for spin glass reasons, is the Euclidean match problem. A set of N cities, N even, are distributed at random in the unit square, just as in the TSP problem. In the Euclidean match case, these cities are to be grouped into N disjoint pairs, in such a way that the sum over pairs of $d_{ij}$ is as small as possible. While this problem is superficially like the TSP problem, it is only $N^3$ hard, not np.

This problem can be put into an energy function or neuronal representation in terms of 1-0 neurons by the same kind of general procedures that were used in the TSP case. A set of $N(N-1)/2$ neurons is needed, arranged as the above-diagonal elements of an NxN matrix. The neuron at position ij represents the use of a link ij. (Thus the diagonal elements are not needed, for links ii do not exist, and the other half of the matrix is not needed, for link ij and links ji are identical.) The syntax to be enforced is that there can be no more than one 1 in each row and each column. In addition, if a 1 occurs in position i,j, then no 1 should appear anywhere in column n,i or in row j,n for all n. The enforcement of these rules of syntax by means of inhibitory or "antiferromagnetic" interactions analogous to the first two syntax terms in the TSP problem can be immediately constructed of the form

$$\sum_k V_{ij}{}^*V_{ik} + V_{ij}{}^*V_{kj} + V_{ij}{}^*V_{jk} + V_{ij}{}^*V_{ki}$$

A term such as the C term in the TSP is needed to complete the syntax generation by adding a term which tends to make the total number of 1's in the matrix equal to N/2.

The data term can now be added to this problem as

$$V_{ij}{}^*d_{ij}$$

This term is minimized by keeping the sum over the bonds which are used as small as possible. The best solution to the energy function is thus the best solution to the Euclidean match problem. Note, however, that this random information term now has the form of a random field (in the usual sense of magnetism) rather than a random exchange. The belief that the calculation of the ground state of a random exchange spin glass is computationally difficult compared with a random field spin glass appears to have a parallel in our circuit designs. We have no constructions by which the TSP or other np-hard problems can be represented by circuits employing only a random _field_ data term.

The right-hand side of each of the differential equations corresponding to the energy function in the Euclidean match problem can be computed in order N steps, for the energy function is a very specialized form. As a result, unless the differential equation is "stiff", it should be possible to solve these $N^2$ differential equations in a time which scales as $N^3$. One then wonders whether the network might in this case be able to find the best solution rather than simply a very good one as in the TSP case.

Many kinds of graph problems also fit readily onto neural networks by similar constructions. Graph coloring asks, given a set of N nodes i and a set of links $l_{ij}$ connecting some of them, whether the nodes can be assigned colors from a set of k colors such that no two linked nodes will have the same color. In this case, an N by k array of neurons can be used with syntax encouraging stable states with one 1 in each column. Such states represent a putative coloring. The data term then may be taken as

$$\sum_{i,j,c} l_{ij} \, V_{ic} \, V_{jc}$$

where $l_{ij}$ is 1 if i and j are linked, and c is one of the k colors. Again, this is a random exchange problem and is np-complete. A more compact representation involving only $N\ln_2 k$ neurons has been found, but involves higher than quadratic terms in $E(V)$. [We thank Eric Baum for calling graph coloring to our attention.]

Finding a subgraph in a larger graph is a problem of interest in data base queries. The graph is described as a set of nodes A which may be connected by links of type 1 described by $l_{AB}$. A possible subgraph consists of a set of nodes a with links $l_{ab}$. The problem of finding the subgraph in the larger graph consists of asking for a correspondence a–A such that if ab are connected by a link of type $l_{ab}$, then A and B are connected by a link of the same type. A simple example of this sort consists of a set of people represented by nodes, with links of a directed type indicating (when present) a parent-offspring relation. The data base query, "are A and B second cousins" is then a subgraph-in-graph question as to whether there is a subgraph of the structure A child of c child of d child of e having offspring f having offspring g having offspring B. A matrix of neurons can be set up with rows X and columns a. The syntax allows no more than one 1 in each row, and one 1 in each column. In the example cited above, seven rows would be used, and the correspondence a-A and b-B would be forced. The five other rows would be left free to locate their own positions for 1's. The data terms consist of entries

$V_{Aa}*V_{Bb}*$(1 if graph link AB is a different type from link ab in the subgraph

*(0 if graph link AB is the same type as link ab from the subgraph)

While this problem can be put onto neural hardware, it does not fit with great grace because of the necessity to recalculate the entire connection matrix for each problem, even though the subgraph being sought might be small.

Decisions involved in nonlinear signal decoding/deconvolution can be put into this same kind of network. Consider a signal S(i) in discrete time i sampled at regular time intervals i. Suppose the signal is known to be made up of stereotype forms a(j), b(j), ..., there being many stereotype forms a, b, ..., and each lasting for a relatively short duration. The total signal then in principle has the form

$$\text{putative signal}(i) = \sum_{a,k} V_{ak} \, a(i-k)$$

where $V_{ak}$ is 1 if a stereotype pulse of type a was initiated in time interval k and is otherwise 0. The signal of each form a, b, etc. was either initiated in a time interval or was not, and if it was initiated continues with its usual stereotype time course. The nonlinear deconvolution decision consists of finding the set of $V_{ak} = 1,0$ such that

$$\sum_i (S(i) - \sum_{a,k} V_{ak} a(i-k))^2$$

is minimized. This expression is a quadratic function of the $V_{ak}$ and can therefore be taken as an energy function for building a deconvolution decision network. The form of the expression indicates that when a, b ... are known, $T_{ij}$ can be computed, and the signal S(i) enters the network as input signals to the various amplifiers $V_{ak}$.

The problem of deconvoluting the signal from an extra-cellular electrode in neurophysiology is exactly of this nature, and the different a, b are the forms of the action potential spikes from different neurons. The problem of separating about 20 roughly stereotype phonemes from a string of continuous speech is of this general nature, but made more difficult by the nonlinear effects of adjacent phoneme interactions and the fact that there are real failures of invariance between speakers, emotions, and situations.

There are many kinds of problems of considerable computational difficulty which seem to fit naturally onto simply neural networks, and for which the network should perform very effective computation when viewed on a basis of how much computation can be done by a few devices in a short time. The ten city TSP problem can be done by a set of 100 neurons or simple amplifiers. Given that the requisite amplifiers can be made from two transistors each, it is difficult to imagine a means of solving a problem of such complexity with fewer active elements. The solution can be found in a single analog convergence, which might take tens of milliseconds on neural hardware, or tens of nanoseconds in silicon hardware.

The large connectivity and analog response of the system are essential to its computational effectiveness. The large connectivity permits many parts of the data to interact in making decisions at a single site. The analog response allows all the active elements to make decisions in a self consistent fashion, rather than each individually. The speed is possible because no stochastic elements are involved - the system is simply going downhill in E, and need not do elaborate configuration averaging to sense which way to move. Instead, the analog values of neuronal state carry important information about collections of possible final states which are 'being simultaneously considered.'

Connectivity as complete as that used here is present in mammalian neuroanatomy only for groups of a few thousand cells. The computational abilities described here might refer to what a very small patch of cortex can carry out. Aspects of hierarchy must be added to describe the nature of processing done on a larger scale. And the larger scale connectivity is certainly not reciprocal. System aspects such as continuing to compute - to change state - instead of simply converging are obviously also absent in the simple model described. In neurobiology the model can at most be viewed as directly relevant to the processing done by small groups of cells, and as a metaphor for what may go on in decision-making in higher order processing.

## REFERENCES

1. Hopfield, J. J., Proc. Natl. Acad. Sci. USA 79, 2554-2558 (1982).
2. Perkel, D. H. and Bulloch, T. H. Neurosci. Res. Symp. Summ. 3, 405-527.
3. See, for example, McCarrager, G. and Chase, R., J. Neurobiol. 16, 69-74 (1985).
4. Van Essen, D. C. and Mounsell, J. H. R., Trends Neuro Sci. 6, 370-375 (1983).
5. Hopfield, J. J., Proc. Natl. Acad. Sci. USA 81, 3088-3092 (1984).
6. Denker, J. (private communication); see also Jackson, A. S. "Analog Computation" McGraw Hill, New York (1960) p. 357.
7. Hopfield, J. J. and Tank D. W., Biol. Cybernetics 52 (1985).
8. Garey, M. R. and Johnson, D. S. "Computers and Intractability" W. H. Freeman & Co., New York (1979).
9. Hopfield, J. J., Feinstein, D. I. and Palmer, R. G. Nature 304, 158-159 (1983).
10. Kirkpatrick, S., Gelatt, C. D., and Vecchi, M. P., Science 220, 671-680 (1983).
11. Barahona, F. J. Phys. A 15, 3241 (1982).

# COLLECTIVE PROPERTIES OF NEURAL NETWORKS

P. Peretto[*] and J. J. Niez[**]
[*]DRF/PSC CEN Grenoble BP 85 X - 38041 Grenoble France
[**]LETI/MCS CEN Grenoble BP 85 X - 38041 Grenoble France

## 1. Introduction

A neural network is basically made of interconnected threshold automata i, (i=1,...,N) collecting the signals $S_j$ produced by a set of upstream units j and delivering an output signal $S_i$ to a set of downstream units k. The evolution of the dynamic variables is driven by the following equations :

$$S_i(t) = F(V_i(\{S_j(t-\Delta_{ij})\}) - \theta_i) \qquad (1)$$

where F is a sigmoïd function, $V_i$ the membrane potential of i, $\Delta_{ij}$ the transmission delay between the units j and i and $\theta_i$ the threshold of i.

The first and most simple model of neural dynamics has been proposed by Pitts and Mc Culloch [1]. Their equations read :

$$S_i(n) = \coprod(\sum_j C_{ij}S_j(n-1) - \theta_i) \qquad (2)$$

n labels the steps of a discrete time axis. $\coprod$ is the Heavyside function ($\coprod(x>o)=1$, $\coprod(x<o)=o$) and $C_{ij}$ is the efficacy of the synapse relating the neuron j to the neuron i.

This model relies upon a number of assumptions which are collected below :
- The dynamic variables are binary entities : $S_i \in \{0,1\}$
- The dynamics is synchronous
- It is deterministic
- It is first-order in time
- The model assumes that the efficacies are activity independent
- It also assumes that the somatic integration rule holds i.e. that the impinging activities combine linearly.

To make the model more realistic one or the other of these constraints has been released. For example memory effects due to slowly decaying post-synaptic potentials have been introduced by Caianiello et al. [2]. Most approaches get rid of the first assumption using for $S_i$ the so-called instantaneous frequencies which are continuous quantities [3] [4]. The problem of asynchronism which follows from the existence of incommensurate

NATO ASI Series, Vol. F20
Disordered Systems and Biological Organization
Edited by E. Bienenstock et al.
© Springer-Verlag Berlin Heidelberg 1986

delays has been considered by Feldman et al. [5]. Probabilistic synchronous networks have been introduced by Little [6] [7] who, for the first time, pointed out possible relationships between the theory of neural networks and the statistical mechanics. C. von der Malsburg also has proposed to consider the efficacies $C_{ij}$ as dynamic variables tightly coupled to the neural activities $S_i$ and $S_j$ [8].

All these improvements complicate the original analysis and it seems difficult to release several constraints while keeping tractable models. We suggest that this is possible however, owing to the simplifications which often occur in the theory of large systems by skipping all irrelevant details as in statistical physics for example. This point is hopefully made clear in section 2 which is devoted to the setting of the rules of the stochastic dynamics of neural networks. The model is applied on the one hand to the study of the dynamical retrieval of stored patterns i. e. to long-term memory phenomena (in section 3) and, on the other hand, to short-term memory phenomena by showing how it can be used to recall temporal sequences of stored configurations (in section 4).

## 2. Stochastic dynamics of neural networks

### 2.1 Dynamic variables

The neural activity manifests itself in the form of trains of spikes, the action potentials which are signals of invariant shape. We consider this invariance as a fundamental property of neural networks because it means that the action potentials do not carry other informations than the one associated to their presence or to their absence. It is therefore necessary to consider the internal states of neurons as binary variable as in the model of Pitts and Mc Culloch. The state of a neuron i is "on" $S_i(t) = 1$, if the neuron started to fire an action potential at time $t'$ such that $t-t' < T$ where $T$ is the standard time duration of action potentials ($T \approx 3 \times 10^{-3}$ s). The neuron is "off" $S_i(t) = 0$ otherwise. The overall state $I(t)$ of the network at time t is the set of individual states $S_i(t)$ :

$$I(t) = \{S_i(t)\}$$

## 2.2 Deterministic chaos

Since the states $S_i(t)$ are binary variables the function F in eq. (1) has to be a Heavyside function. It is also assumed that the somatic integration rule holds. This assumption is innocuous as long as the membrane potential V is a monotone increasing function. Equation (1) reads :

$$S_i(t) = \Pi(\sum_j C_{ij}S_j(t-\Delta_{ij})-\Theta_i)$$

(3)

The trajectories generated by eq. (3) are exponentially sensitive to initial conditions for almost all sets $\{C_{ij}\}$ of synaptic efficaces and for almost all sets $\{\Delta_{ij}\}$ of delays. This does not imply that the trajectories are erratic but that the notion of trajectory must be replaced by that of path probability in the phase space. This phenomenon called deterministic chaos occurs in most systems comprising a large number of degrees of freedom driven by non-linear equations of motion.

## 2.3 Time axis discretization

Owing to deterministic chaos, there is no constraint compelling two units to change their states at one and the same time. From now on we assume that the dynamics is perfectly asynchronous and that every neuron is an equally possible candidate for the flipping event at any time. This implies that the time axis can be divided in steps, $\tau$ long, such as no more than one neuron flips its state in one time step. The step duration $\tau$ must satisfy :

$$\tau = T/N$$

(4)

A given neuron is allowed to flip, on average, every N time steps. It stays in a state say $S_i = 1$ for a time $\tau N$. Due to the fixed time length of the action potentials, $\tau N$ has to be of the order of T.

## 2.4 Short time delays

Let $i(n)$ be the label of the candidate neuron at time step n. In practice $i(n)$ is determined by a random process. If one restricts the analysis to systems smaller than $C.T \sim 10$ cm (C is the velocity of action potentials), the delays satisfy :

$$\Delta_{ij} \leqslant T$$

and the joint probability

$$P(S_i(\Delta_{ij}) \mid S_i(o)) \sim 1$$

The equations (3) become :

$$S_i(n) = \Pi(V_i(n-1) - \theta_i) \quad ; \quad i=i(n)$$
$$S_i(n) = S_i(n-1) \qquad\qquad ; \quad i\neq i(n) \tag{5}$$

$$V_i(n-1) = \sum_j C_{ij} S_j(n-1)$$

## 2.5 Introduction of noise

The synapses are sources of noise : the action potential arriving on a synaptic button opens a Gaussian distribution of vesicles and every vesicle releases a Gaussian number of neurotransmitter molecules in the synaptic cleft [9]. The membrane potential is therefore a random variable and the probability for the neuron $i(n)$ to fire is the probability for the potential to be larger than the threshold $\theta_i$. The equations (5) are replaced by probabilistic expressions :

$$P(S_i=1,n)=1-P(S_i=o,n) = \frac{1}{2}\left[1+\mathrm{erf}(\frac{V_i(n-1)-\theta_i}{B})\right] \quad ; \quad i=i(n) \tag{6}$$

where B is the noise parameter. The two equations (6) can be lumped into one equation using the variable $\sigma_i = 2S_i - 1$.

$$P(\sigma_i, n) = \frac{1}{2}\left[1+\mathrm{erf}(\frac{\sigma_i(V_i(n-1)-\theta_i)}{B})\right] \quad ; \quad i=i(n) \tag{7}$$

## 2.6 Monte-Carlo dynamics of neural networks

The various points which have been discussed so far can be used to build two equivalent descriptions of the neuronal dynamics, either algorithmic or probabilistic. The Monte-Carlo algorithm is defined by the following steps :

1. Let N be the number of neurons

   T be the action potential duration

   B the noise parameter

   $C = \{C_{ij}\}$ a set of synaptic efficacies

   $I(n=o) = \{\sigma_i\}$ an initial state

2. The time basis is set to $\tau = T/N$

3. At every time step pick-up a unit $i=i(n)$ at random

4. Compute $V_i = \sum_j C_{ij} \sigma_j$

5. Eventually, change $\sigma_i$ according to the probability law :

$$P(\sigma_i) = \frac{1}{2} (1 + erf(\frac{V_i - \theta_i}{B}))$$

6. Iterate to point 3

It is worth noting that the algorithm is similar to the one proposed by Suzuki to account for the evolution of Ising spin systems (with an erf function replacing the tanh function) [10]. But the present procedure does not assume that the system is driven by a Hamiltonian. In particular the interactions $C_{ij}$ can be asymmetrical.

### 2.7 Probabilistic description of neural dynamics

Due to the probabilistic character of transitions one is led to introduce the quantities $\rho(I, n)$ which are the probabilities for the network to be in state I at time n. The evolution of the probability distribution $\rho$ obeys a Markov equation [11]

$$\rho(I, n) = \sum_J W(I|J) \rho(J, n-1) \tag{8}$$

where the matrix elements are given by

$$W(I|J) = \frac{1}{2N} (1 + erf \frac{\sigma(I) \cdot (V(I) - \theta)}{B}) \tag{9}$$

In eq.(9), the states I and J differ by the reversal of site i(n) and

$$V_i(I) = \sum_j C_{ij}\sigma_j(I)$$

The probabilistic approach allows a natural definition of the concept of instantaneous frequency $\omega$ :

$$\omega_i(n) = \omega_{max} \langle S_i \rangle_n = \omega_{max} \frac{\langle \sigma_i \rangle_n + 1}{2}$$

where $\langle \sigma_i \rangle_n$ is an ensemble average over the distribution $\rho$ :

$$\langle \sigma_i \rangle_n = \sum_I \rho(I,n)\sigma_i(I)$$

and $\omega_{max}$ the maximum firing rate. The evolution of $\langle \sigma_i \rangle_n$ is derived in the next section.

## 3. Application to long-term memory phenomena

### 3.1 Dynamics of instantaneous frequencies

Let $I^+ = (\sigma_1(I), \sigma_2(I), \ldots, \sigma_i(I)=+1, \ldots, \sigma_N(I))$

and $I^- = (\sigma_1(I), \sigma_2(I), \ldots, \sigma_i(I)=-1, \ldots, \sigma_N(I))$

Then $\langle \sigma_i \rangle_n = \sum_{I^+} \rho(I^+,n) - \sum_{I^-} \rho(I^-,n)$

The increase $\langle \sigma_i \rangle_{n+1} - \langle \sigma_i \rangle_n$ in one time step is given by :

$$\langle \sigma_i \rangle_{n+1} - \langle \sigma_i \rangle_n = 2(\sum_{I^-} W(I^+|I^-)\rho(I^-,n) - \sum_{I^+} W(I^-|I^+)\rho(I^+,n))$$

$$= 2(\sum_{I^-} W(I^+|I^-)\rho(I^-,n)\sigma_i(I^-) + \sum_{I^+} W(I^-|I^+)\rho(I^+,n)\sigma_i(I^+))$$

$$= 2\sum_I \sigma_i(I)W(J|I)\rho(I,n)$$

This equation has been first derived by Choi et al. in the context of spin glass theory [12]. Using eq. (9), the equations driving the dynamics of instantaneous activities are obtained :

$$\langle \sigma_i \rangle_{n+1} - \langle \sigma_i \rangle_n = -\frac{1}{N}(\langle \sigma_i \rangle_n - \langle erf\frac{(V_i-\theta_i)}{B} \rangle_n) \tag{10}$$

This is an exact but still untractable result. It simplifies however when the degree of connectivity is high, in large fully connected networks for example. In this limit the fluctuations of the membrane potential $V_i$ are negligible : $\langle V_i^2 \rangle \simeq \langle V_i \rangle^2$

Equation (10) can be replaced by its mean-field expression :

$$\langle \sigma_i \rangle_{n+1} - \langle \sigma_i \rangle_n = -\frac{1}{N}(\langle \sigma_i \rangle_n - erf(\frac{\langle V_i \rangle_n - \theta_i}{B})) \tag{11}$$

with $\langle V_i \rangle_n = \langle \sum_J c_{ij}\sigma_j \rangle_n = \sum_J c_{ij}\langle \sigma_J \rangle_n$

## 3.2 Long-term memory learning rules

Learning is a phenomenon generally associated with a modification of synaptic efficacies driven by the neural activities. Following Hebb [13] the strength of a synapse would be proportional to the correlated activities of the neurons it links :

$$c_{ij} \alpha \overline{\int_{-\infty}^{0} \sigma_i(t)\sigma_j(t-\tau) \, f(\tau)d\tau} \tag{12}$$

where $f(\tau)$ is some monotone decreasing memory function.

In long-term memory, the evolution of the efficacies is irreversible and the bar in eq. (12) is a time average over the learning sessions. Let $I^\alpha = \{\sigma_i^\alpha\}$, $\alpha=1,\ldots,M$ be the patterns experienced by the system during learning. Then eq. (12) reads :

$$c_{ij} = \frac{1}{N} \sum_{\alpha\beta} \Gamma^{\alpha\beta}\sigma_i^\alpha\sigma_j^\beta \tag{13}$$

The matrix elements $\Gamma^{\alpha\beta}$ depend on the details of f and on the learning protocol. A spherical matrix $\Gamma$ yields the usual Hebbian rule :

$$c_{ij} = \frac{1}{N} \sum_{\alpha} \sigma_i^\alpha\sigma_j^\alpha \tag{14}$$

## 3.3 Evolution of order parameters

Equations (11) and (13) determine the time evolution of neuronal activity. One obtains a set of N coupled equations. Using order parameters $Q_\alpha(n)$ defined by :

$$Q_\alpha(n) = \langle \frac{1}{N} \sum_i \sigma_i\sigma_i^\alpha \rangle_n = \frac{1}{N} \sum_i \sigma_i^\alpha\langle\sigma_i\rangle_n \tag{15}$$

reduces the number to M. The equations read :

$$Q_\alpha(n+1)-Q_\alpha(n) = -\frac{1}{N}\left[Q_\alpha(n)-\frac{1}{N}\sum_i \sigma_i^\alpha \, \mathrm{erf}\left[\frac{\sum_{\beta\gamma} \sigma_i^\beta\Gamma^{\beta\gamma}Q_\gamma(n)-\theta_i}{B}\right]\right] \tag{16}$$

The steady solutions are the fixed points of eq. (16). In vector notation the equilibrium equations, quoted in ref. [11], read :

$$\underset{\sim}{Q} = \frac{1}{N} \sum_i \underset{\sim}{\sigma}_i.\mathrm{erf}\left[\frac{\underset{\sim}{\sigma}_i \cdot \underset{\approx}{\Gamma} \cdot \underset{\sim}{Q}-\theta_i}{B}\right] \tag{17}$$

With spherical $\Gamma$ matrices and zero threshold value, they become :

$$Q = \frac{1}{N} \sum_i \sigma_i \cdot \text{erf} \left[ \frac{\sigma_i \cdot Q}{B} \right] \tag{18}$$

These equations have been derived by Provost et al. [14] who used the stationary phase method. The solutions of eq. (18) have been thoroughly studied by Amit et al. [15]. These authors find that the states $Q = Q(1,0,0,....)$, or Mattis states, are the most stable. There also exist metastable states which are mixtures of an odd number of Mattis states, $Q=Q(1,1,1,0...)$ for example.

According to eq. (17) the order parameters vanish above a critical noise $B_C$ which is the largest eigenvalue of the MxM matrix $\gamma$ of patterns correlations

$$\gamma^{\alpha\beta} = \frac{1}{N\sqrt{\pi}} \sum_i \sigma_i^\alpha \Gamma^{\alpha\beta} \sigma_i^\beta$$

As usual in mean field theories, the order parameters behave as $\sqrt{B_C - B}$ for $B \leqslant B_C$. Finally eq. (16) shows that the order parameters evolve exponentially towards their equilibrium values in the vicinity of the critical noise. The time constant $T_R$ is given by :

$$T_R = \frac{T}{1 - \frac{B_C}{B} \exp - \frac{Q(\infty)^2}{4B^2}}$$

The time constant, therefore, does not depend on the size N of the network. It diverges when B tends towards its critical value, a feature related to the known critical slowing down phenomenon of phase transitions.

### 3.4 Computer simulations

Computer simulations have been carried out using networks comprising 50 to 400 neurons. They confirm that, in most cases, the system evolves spontaneously towards a stored pattern, the one which is the most similar to the initial state. The evolution is asymptotically exponential and the time constant diverges as B is closer to $B_C = 1/\sqrt{\pi}$ (fig. 1).

Sometimes the process ends in a spurious state which is a combination of an odd

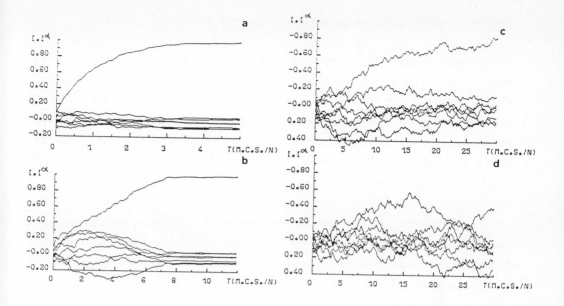

Figure 1 : Computer simulations carried out on a network comprising N = 400 neurons in which M = 8 random patterns have been stored using the Hebbian rule. The various curves correspond to increasing noise parameters. 1a) $B/B_C=0$ ; 1b) $B/B_C=0.4$ ; 1c) $B/B_C=0.75$ and 1d) $B/B_C = 0.95$.

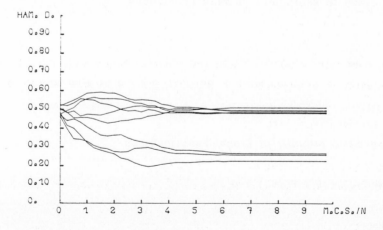

Figure 2 : An example of a simulation process converging to a state made of a mixture of three stored patterns.

number of stored patterns as predicted by Amit et al.

The simulations also shed some light on the topology of the phase space. They suggest for example that the limit of the stability of stored patterns depends linearly upon the size N of the network (fig. 3a). This confirms the analysis of Hopfield [16]. The simulations also show that the basins of attraction of stored patterns occupy the largest portion of the phase space as long as the number of stored patterns does not exceed a limit which varies as LnN (fig. 3b). This result is not explained.

The spectra of the "energy" function $H = \sum_{ij} J_{ij}\sigma_i\sigma_j$ for equilibrium configurations are displayed in fig. 4. They show that the limit for the domination of stored patterns corresponds to the merging of the energy band of memorized configurations into the broad band of spurious states.

One also observes that the size of basins of attraction quickly shrinks when the energy of the corresponding metastable state increases. The number of metastable states seems to follow rather closely the $3^M$ law proposed by Amit.

4. Application to short term memory phenomena

4.1 A general property of the probability matrix
Owing to the peculiar form of eq. (9) the matrix elements of W satisfy :

$$W(I|J) + W(J|I) = 1/N$$

Using the basic property of probabilistic matrices :

$$\sum_I W(I|J) = 1 \; ; \; W(I|J) \geqslant 0$$

one deduces

$$W(I|I) = 1 - \sum_{J \neq I} W(I|J) \quad \text{for } 1/2 < W(I|I) < 1$$

$$\text{and } W(I|I) = \sum_{J \neq I} W(J|I) \quad \text{for } 0 < W(I|I) < 1/2$$

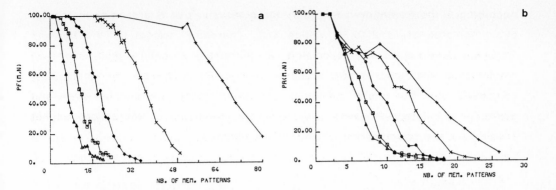

Fig. 3 : Probability for a trajectory to end into one of the stored pattern vs the number of stored patterns. The stored patterns are sets of N random number ±1. Fig. 3a) : the initial state is one of the stored patterns and the final and initial states are identical. Fig. 3b) The initial state is random. From right to left the curves correspond to N = 400, 200, 100, 50 and 25 respectively.

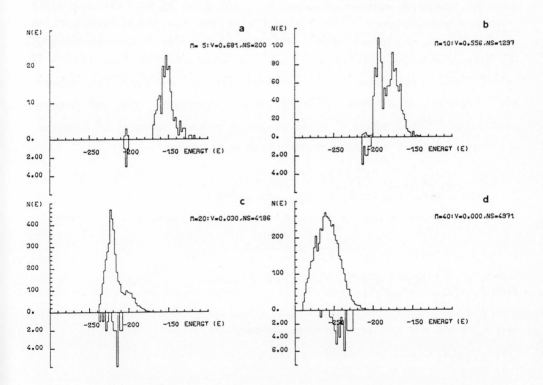

Figure 4. Energy spectra of metastable states in a network comprising 200 neurons and an increasing number of stored patterns (fig. 4a) M=5, (fig. 4b) M= 10, (fig. 4c) M=25, (fig. 4d) M=40. The spectra of stored pattern are drawn under the axis line, the complete spectra are drawn above the axis line. $V_S$ is the volume occupied by the stored patterns, Ns is the number of different metastable states reached after 5000 trials.

According to the Gershgoren theorem, the eigenvalues of W are on or inside the circle centered at (1/2,0) of radius 1/2. Therefore, except for the Perron-Frobenius eigenvalue $\lambda = 1$ associated with the steady probability distribution, all other modes relax towards zero whatever the set {C} of synaptic efficacies. One concludes that the mere existence of asymmetrical interactions does not guarantee oscillatory behaviors : the possible asymmetry of interactions cannot account for the phenomenon of short-term memory.

## 4.2 Short-term time dependence of synaptic efficacies

If short-term memory cannot be explained by any set of fixed interactions, it becomes necessary to assume that the synaptic efficacies are time-dependent [17] [8]. They are presumably made of two parts :

$$C_{ij}(t) = c_{ij}^{L} + c_{ij}^{S}(t)$$

here the long-term synaptic efficacies $c_{ij}^{L}$ are given by eq.(13) and the short-term contributions $c_{ij}^{S}$ are assumed to be reversible and to depend on the current activities of neurons i and j. The short-term memory mechanism perturbs the long-term memory parameters i.e. the elements of the matrix $\Gamma$. It is assumed that, if the system is in a memorized state $I^{\alpha}$, the mechanism weakens the diagonal contribution $\Gamma^{\alpha\alpha}\sigma_i^{\alpha}\sigma_j^{\alpha}$ and strengthens the off-diagonal contribution $\Gamma^{\alpha\beta}\sigma_i^{\alpha}\sigma_j^{\beta}$ where $I^{\beta}$ is the pattern in which $I^{\alpha}$ tends to be changed.

This process is formalized in the following equations

$$c_{ij}^{S}(n+1)-c_{ij}^{S}(n)= -\frac{c_{ij}^{S}(n)}{T_S} +\sum_{\alpha\beta}\Gamma^{\alpha\beta}\sigma_i^{\alpha}\sigma_j^{\beta}\left[\lambda_{\alpha\beta}\sigma_i(n)\sigma_j(n)\sigma_i^{\alpha}\sigma_j^{\alpha}\right]$$

$$= -\frac{c_{ij}^{S}(n)}{T_S} + \frac{\sigma_i(n)\sigma_j(n)}{T_j}$$

with $T_j^{-1} = \sum_{\alpha\beta}\Gamma^{\alpha\beta}\lambda_{\alpha\beta}\sigma_j^{\alpha}\sigma_j^{\beta}$

and $T_S$ is the life-time of short-term perturbations.

## 4.3 Computer simulations

It is assumed that the system experiences a cyclic temporal sequence of patterns $I^{\alpha}$ ; $\alpha=1,2...,M,1,...$ and that the long-term memory mechanism builds a cyclic matrix $\Gamma$. One observes that the patterns $I^{\alpha}$ are stable for $\Gamma^{\alpha\alpha}/\Gamma^{\alpha+1\alpha} \geqslant 3$.

The relaxation times are $T_S$ = 120 Monte Carlo steps per neuron (M.C.S./N), $\lambda^{\alpha\alpha}$

= $- \lambda^{\alpha+1\alpha}$ = 360 M.C.S./N. The two dynamics, the fast one which drives the neuron states and the slow one which modifies the synaptic efficacies, work simultaneously. Figure 5 shows that the system retrieves successfully the temporal sequence. It even seems that initialy faltering behaviors become spontaneously corrected by the algorithm.

The approximate period of retrieval is $\simeq$ 30xT = $10^{-1}$s, which is a realistic value.

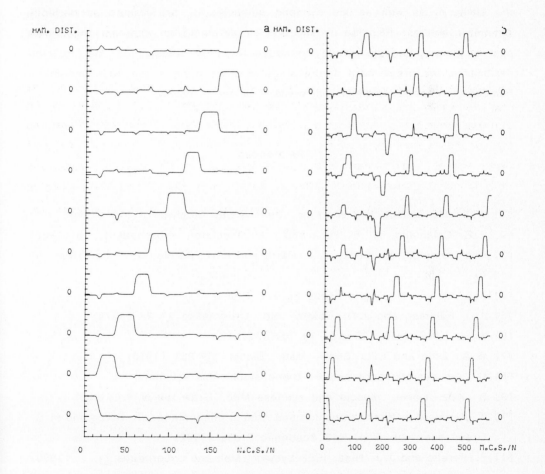

Figure 5. Two examples of retrieval of temporal sequences of stored patterns N = 400 ; M = 10. The two figures have two different sets of stored patterns. In fig. 5b the initial hesitations are spontaneously corrected.

## 5. conclusion

It is claimed in this article that the basic frame of a theory of neural dynamics must be a Monte-Carlo algorithm from which analytical results can eventually be derived. The model we propose gets rid of a number of constraints or shortcomings which are frequently encountered in other approaches. It is probabilistic, asynchronous, it bridges the gap between discrete and continuous models of neural networks, it can be used in mixed dynamics of neuron states and synaptic efficacies. However it still contains over simplifications. In particular the states $\sigma_i$ as well as the synaptic efficacies $C_{ij}$ are treated as perfectly symmetric entities. Also the dynamics is purely Markovian whereas longer-lived post-synaptic potentials would introduce memory effects. In our opinion flexibility is the great merit of the model and we hope it can accommodate a number of refinements which will make it closer to reality.

## References

[1] W.S. Mc Culloch and W. Pitts, Bull. Math. Biophys. $\underline{5}$ 115 (1943)

[2] E.R. Caianiello, A. de Luca and L.M. Ricciardi, Kybernetik $\underline{4}$, 10 (1967)

[3] P.A. Anninos, B. Beck, T.J. Harth and E.H. Pertile, J. Theor. Biol. $\underline{26}$ (1970)

[4] S.I. Amari, Bio. Cybernetics $\underline{26}$ 175 (1977)

[5] J.L. Feldman and J.D. Cowan, Bio. Cybernetics $\underline{17}$ 29 (1975)

[6] W.A. Little, Math. Biosci. $\underline{19}$ 101 (1974)

[7] W.A. Little and C.L. Shaw, Math. Biosci. $\underline{39}$ 281 (1978)

[8] C. Von der Malsburg and E. Bienenstock, this volume

[9] B. Katz. Nerve, muscle and synapse Mac. Graw Hill N.Y. (1966)

[10] K. Kawasaki in Phase transition and critical phenomena. C. Domb and M.S. Green eds. vol. 2 p. 443 Academic Press (1972)

[11] P. Peretto and J.J. Niez, IEEE System, Man and Cybernetics $\underline{14}$, (1986)

[12] M.Y. Choi and B.A. Huberman, Phys. Rev. $\underline{B28}$ 2547 (1983)

[13] D.O. Hebb. The organization of behavior N.Y. Wiley (1949)

[14] J.P. Provost and G. Vallee, Phys. Rev. Lett $\underline{50}$ 598 (1983)

[15] D.J. Amit, H.J. Gutfreund and H. Sompolinsky. Disordered systems and

Biological organization – Les Houches (1985)

[16]J.J. Hopfield. Proc. Natl. Acad. Sci. USA $\underline{79}$ 2554 (1982)

[17]J.P. Changeux. L'homme neuronal. Fayard, Paris (1983)

# DETERMINING THE DYNAMIC LANDSCAPE OF HOPFIELD NETWORKS

**Gérard WEISBUCH** and Dominique D'HUMIERES

Laboratoire de Physique de l'Ecole Normale Supérieure
24 rue Lhomond, 75231 Paris Cedex 5.

The purpose of this paper is to describe in a semi-quantitative manner the basins of attraction of the original Hopfield (1982) networks.

**1 Definitions.** We thus consider a network of n fully connected threshold automata which obey the transition rule:

$$x_i(t) = Y\left[\sum_j a_{ij} \, x_j(t-1) - b_i\right] \tag{1.1}$$

where $x_i$ and $x_j$ are binary variables taking values 0 or 1

$Y(x)$, the Heaviside function, gives 0 if $x < 0$ and 1 otherwise.

The $a_{ij}$ are called the synaptic weights of the i–j link. They verify the following rules:

$$a_{ii} = 0$$
$$a_{ij} = a_{ji}$$

A useful change in notation is to replace the binary vectors **x** by **X** defined as

$$\mathbf{X} = 2x - 1 \qquad \text{(x being 0 or 1, X is -1 or 1)}.$$

In order to retrieve m reference sequences, $\mathbf{X}^1,..,\mathbf{X}^s,..,\mathbf{X}^m$, which are supposed to be random and uncorrelated, the following algorithm, referred to as Hebb's rule, is used to compute the $a_{ij}$

$$a_{ij} = \sum_s X_i^s X_j^s \tag{1.2}$$

A good choice for the thresholds $b_i$ allows to increase the number of retrieved sequences, as discussed in Weisbuch and Fogelman (1985). Expression 1.1 becomes:

$$x_i(t) = Y\left[\sum_j a_{ij} \, X_j(t-1)\right] \tag{1.3}$$

NATO ASI Series, Vol. F20
Disordered Systems and Biological Organization
Edited by E. Bienenstock et al.
© Springer-Verlag Berlin Heidelberg 1986

## 2 Condition for invariance of the reference sequences.

The $n$ automata $X_i^s$, $i=1..n$, of one reference sequence are stable if the $n$ inequalities:

$$0 < \sum_j a_{ij} X_i^s X_j^s \quad ; \quad \forall i \tag{2.1}$$

are satisfied. Using the expression 1.2 for the $a_{ij}$ and taking out from the double sum the terms such that s'=s gives:

$$0 < n-1 + \sum_j \sum_{s' \neq s} X_i^{s'} X_j^{s'} X_i^s X_j^s \tag{2.2}$$

This condition holds for any choice of the reference sequences. If we now use the fact that they are uncorrelated, the double sum is the result of a random walk with $(n-1)(m-1)$ steps of magnitude + or -1. The probability for one inequality not to be satisfied is the probability for the random sum to be smaller than n-1:

$$p = \text{erf} \left( \sqrt{((n-1)/(m-1))} \right) \tag{2.3}$$

Assuming independent samplings of the random sums, the $n$ automata of the $m$ sequences are invariant with a probability

$$P = \text{erf}^{nm} \left( \sqrt{((n-1)/(m-1))} \right) \tag{2.4}$$

This expression was tested by computer simulations. The number of reference sequences such that P=.5 is plotted against n, the number of automata. The solid line corresponds to the theoretical expression (2.4), and the dots were obtained from a computer test of inequalities (2.2) on randomly generated sequences.

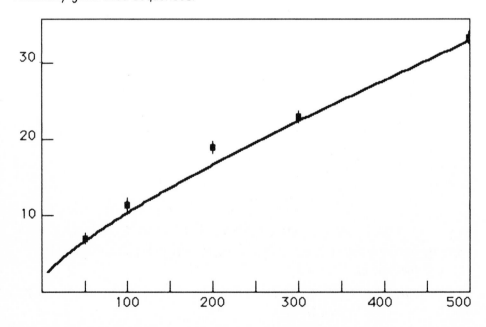

By using the asymptotic expansion of the error function for large arguments, a simple relation allowing to compute m such that P = .5 is obtained:

$$(n-1)/(m-1) + 3 \ln n/m = 4 \ln n .$$ (2.5)

**For large n, m varies then as n/ ln n .**

### 3 Width of the basins of attraction.

With the same method, we are able to answer to the following question: If d automata of a reference sequence $X^s$ are flipped giving an initial state $X^{in}$, what are the probability for the direction of inequalities 2.1 to remain unchanged? A positive answer implies that the network returns to reference $X^s$ in one iteration per automaton.

Inequalities 2.2 now become:

$$0 < n-2d-1 + \sum_j \sum_{s' \ne s} X_i^{s'} X_j^{s'} X_j^{in} X_i^s$$ (3.1)

and their probability to be verified $\mathrm{erf}( (n-2d-1 ) / \sqrt{(n-1)(m-1)} )$

Taking the error function to the n gives a lower bound for the probability of the return to the reference sequence (this is because during the iteration process the flipped automata might be reset, thus increasing the probability for the inequality to be true).

The figure is a plot of the relative distance to the reference, d/(n-1), against the relative number of references m/(n-1),for a return to the reference with a 50 percent probability. Continuous lines correspond to n = 50,100,200,500 from right to left. The dots were obtained by testing directly inequalities 3.1 by computer simulations for n=50.

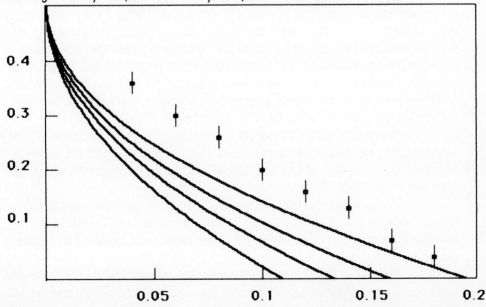

Further approximations allow to obtain the relation between d,m and n corresponding to a return towards the reference with probability 0.5.

$$\frac{(n-2d-1)^2}{(n-1)(m-1)} = 3 \ln n + \ln m - 2 \ln(n-2d) - \ln 2 \tag{3.2}$$

## Conclusions.

The probabilities we have given correspond in fact to convergence in one iteration step per automaton, which implies that the attraction basins are in fact even larger. Thus, provided that the number of references and the distance to a reference are not too large, attraction toward the reference is nearly certain.

Such a strong regularity in the attraction basins and the existence of analytic expressions are exceptional in the theory of networks of automata. They are due to the randomness of the set of sequences. Even though, the results shed some light on more general structures, such as cellular structures discussed by F.Fogelman and by Y. Le Cun in this volume.

We used for the simulations the computer facilities of CNRS GRECO 70, Expérimentation Numérique.

Note added in proof.

During les Houches meeting, D. Amit reported on similar results obtained by solving the mean field equations for magnetization, using a formalism introduced for spin glass ( D. Amit et al.1985). Their approach also allows to describe the spurious states, linear combinations of reference sequences appearing at higher energies. Since the les Houches meeting ,this problem has received some further attention and sometimes apparently different scaling laws were proposed for the maximum number of reference sequences. These apparent discrepancies are due to the fact that different criteria can be used to compute the probabilities.

If one imposes a bound on the probability that the state of the automata of one sequence remain invariant, which implies accepting a finite error rate, one obtains a linear relation between m and n as reported in Hopfield 1982 or Kinzel 1985.

If one is interested in the probability for the whole sequence to remain invariant, one has to compute the error function to the power n, which yields m proportional to n/2ln n for large n as reported by Posner quoted by Amit 1985.

Finally, if one is interested in the probability for all the m sequences to remain invariant, which is the case we considered, one has to compute the error function to the power nm, which yields m proportional to n/4ln n for large n.

References.

1 J.J. Hopfield (1982): Neural networks with emergent collective computational abilities. Proc.Nat.Acad.Sc. USA, **79**,pp. 2554-2558.

2 G.Weisbuch, F.Fogelman-Soulie (1985):Scaling laws for the attractors of Hopfield networks. Lettres au Journal de Physique,**46**,pp. 623-630.

3 D.J. Amit, H. Gutfreund and H. Sompolinsky (1985): Spin glass models of neural networks. Phys. Rev. A, **32**, pp.1007-1018.

D.J. Amit, H. Gutfreund and H. Sompolinsky (1985): Storing infinite number of patterns in a spin glass models of neural networks. Phys. Rev. Lett.,**55**, pp.1530-1533.

4 W. Kinzel (1985) Learning and pattern recognition in spin glass models, Z. Phys. B **60**, pp.205-213.

# HIGH RESOLUTION MICROFABRICATION AND NEURAL NETWORKS

L. D. Jackel, R. E. Howard, H. P. Graf,
J. Denker, and B. Straughn
AT&T Bell Laboratories
Holmdel, N.J. 07733 USA

## ABSTRACT

This paper reviews basic lithographic considerations for current integrated circuit fabrication at the 1 micron level and illustrates methods for making structures with dimensions smaller than 0.1 micron. Many of these methods are well suited to fabricating high-density, resistive neural-networks.

Today, most integrated circuits are patterned using projection optical lithography.[1] An image of the desired pattern is projected onto a photoresist-coated substrate and the developed resist is then used to further pattern the substrate by either etching, ion-implantation, or thin-film liftoff. The minimum feature size is a few times the exposing wavelength, ultimately limited by diffraction of the exposing light. The best projection optics now can produce 1 micron features. With far-ultraviolet exposure, 1/2 micron features are feasible within this decade, and eventually 1/4 micron features may be possible. Smaller features are unlikely using projection optics because there are no transparent optical materials at wavelengths below 1/8 micron. Even attaining 1/2 micron features will be very difficult because of extremely tight focus, alignment, and wafer flatness requirements.

Electron-beam lithography can routinely pattern 1/4 micron features, and has been demonstrated to pattern features as small as 1/100 micron. However, with the exception of customizing one level of wiring, e-beam lithography is not used for direct patterning of commercial circuits. This is because the

NATO ASI Series, Vol. F20
Disordered Systems and Biological Organization
Edited by E. Bienenstock et al.
© Springer-Verlag Berlin Heidelberg 1986

fastest e-beam patterning now takes about 10 times as long as optical patterning due to the serial nature of e-beam exposure compared to the parallel nature of optical exposure. (In e-beam lithography, a focussed beam is scanned across the substrate, individually exposing each pixel.) Faster e-beam machines are being developed, and in some cases the higher resolution and better alignment accuracy of e-beam lithography may be worth the reduced throughput.

It is instructive to compare lithographic requirements of standard transistor circuits with those of neural networks.[2] The smallest devices in an integrated circuit are usually much larger than the minimum lithographic feature. A transistor requires about 5 times the minimum feature lengths on a side, so that with current lithography, a cell with a single transistor requires an area at least 5 microns by 5 microns. Furthermore, 10 levels of lithography are typical in an integrated circuit. The functional element of a neural network is a crossed wire pair, either separated by a dielectric, or joined by a resistor. In either case, the element requires an area only 2 minimum feature lengths on a side. The network also only requires 3 levels of lithography for a pre-programmed memory and 2 levels when using programmable resistors. With modest 2 micron lithography, it is possible to make a matrix with 6 million elements on a cm square chip. Because the network is very defect tolerant, one could expect that yield of such chips would be high. In fact, yield would probably be limited by performance of the conventional circuitry needed for "neuron" amplifiers and the multiplexing circuitry needed to send signals on and off the chip.

Figure 1 shows a 24 x 24 array made with 2 micron optical lithography to test materials for a larger circuit. Refractory metals are used for the wiring and amorphous silicon forms the resistors. (For this test, resistors were only written on the diagonal.) The resistor values are in the Mega-ohm range.

With modest 1/4 micron e-beam lithography, arrays could be made with 400 million elements on a square cm. To demonstrate the feasibility of this element density, we fabricated 12 x 12 matrices in 6 micron square. Part of such an array is shown in Figure 2. This density does not represent any fundamental limit; much higher densities could be achieved.

To obtain better than 1/4 micron resolution in e-beam lithography, some extra effort is required. Obviously, the width of the exposing beam must be smaller than the desired resolution. Even with a very fine beam, resolution is limited by electrons backscattering from the substrate, causing unwanted exposure in regions not addressed by the beam. (This phenomenon is well know in e-beam lithography, and is called the "proximity effect".) We have shown that by using thin resists and high-energy beams the proximity effect can be nearly eliminated, and very high-resolution patterns can be produced. Figure 3 shows 0.01 micron gold alloy lines on a silicon substrate made using a 120 keV electron beam and a thin, high-resolution resist.

Figure 1. A Test neural network array made with 2 micron optical lithography.

Figure 2. A test array made 1/4 micron e-beam lithography

There is no apparent reason why a neural network could not be built with matrix elements occupying an area of 1/10 micron by 1/10 micron. (Such a network would have 10 billion elements/square cm.) There are however, serious practical problems in finding materials that would give adequate performance at these dimensions. This is illustrated by considering a metal for use as a column in a matrix. With this packing density, the strip would be about 0.05 micron wide. It is reasonable that the strip thickness equals the strip width, so the cross section is $0.25 \times 10^{-10} cm^2$. Even if the strip only carries 25 microamps, this implies a current density of $10^6 A/cm^2$, a value near the room-temperature electromigration limit for many materials. The metal must be fine grained or single crystal (or possibly amorphous), so that it can be patterned with high resolution. Furthermore, it should not react or diffuse into the material used as resistors. The materials used in the device shown in Figure 1 satisfy these conditions, and there is no apparent reason why the same materials system could not be used with finer linewidths. Efforts are now underway to make such a high density matrix with e-beam lithography.

REFERENCES
1. For a perpective of the current status of lithography see the proceedings of the 1984 Int. Symposium on Electron, Ion, and Photon Beams, J. Vac. Sci. and Technol. **B3,** (Jan/Feb 1985).
2. J. J. Hopfield, Proc. Natl. Acad. Sci. USA 79, 2554 (1982).

0.1 micron

Figure 3. An example of 10nm AuPd lines on a silicon substrate made by very high resolution e-beam lithography.

# ULTRAMETRICITY, HOPFIELD MODEL AND ALL THAT

M.A. Virasoro

Dipartimento di Fisica - Università di Roma I, La Sapienza

I - 00185 Roma Italy

It is part of our every day experience that when we try to memorize new information we look for all possible relationships with previously stored patterns. If we can classify the new pattern, that is, place it in a hierarchical tree of categories we do it with so much eagerness that sometimes we just censor the data so as to eliminate some exceptional features. Hopfield's model[1] for the brain seems, on the contrary, to be quite happy crunching uncorrelated orthogonal words. This is the more surprising given the fact that otherwise the model simulates quite nicely other features of Human Memory.

The question that then arises is whether there is anything in the architecture of the brain that forces us to categorize.

A related problem concerns the quality of errors. Once granted that a model for the brain should be allowed to make mistakes, it becomes vital for it to recognize a hierarchy among errors because certain mistakes cannot be made more than once in a lifetime.

From the other side, it is known that the Sherrington Kirkpatrick (SK) spin glass, a limiting case of a Hopfield model, has equilibrium states organized ultrametrically[2]. Ultrametricity is a mathematical extension of the metric relation defined on vector spaces. It is equivalent (for our purposes) to the statement that among the state vectors one can define a natural distance such that:

a)  if we define sets of vectors by the property that their

NATO ASI Series, Vol. F20
Disordered Systems and Biological Organization
Edited by E. Bienenstock et al.
© Springer-Verlag Berlin Heidelberg 1986

mutual distance be smaller than d then these sets are disjoint.

b) the corresponding sets for d' < d are proper subsets of the previous ones.

In other words, ultrametricity is equivalent to categorization: patterns can be grouped into disjoint categories, categories can be grouped into metacategoris and so on.

All of these reasons have led N.Parga and myself[3] to try to complement Hopfield's model in such a way that it could process ultrametric patterns rather than orthogonal ones.

Human Memory, as an information processing system, can be conveniently analysed in three subprocesses: Encoding, Storage and Retrieval[4].

In the first stage stimuli are encoded in a form compatible with storage. Therefore we have to construct a hardware that will generate words that are automatically ultrametric.

Imagine for this purpose an encoding stage architectured in succesive layers such that the K-th layer receives the pre processed input from the (K-1)-th layer and simultaneously receives random information from the outside (see Fig.1).

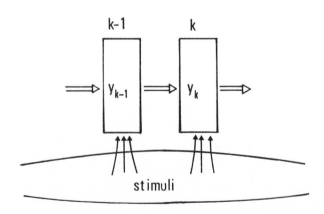

Fig. 1
The encoding stage: layer K receives information from
layer K-1 and random input from outside.

The information from the (K-1)-th layer constrains the K-th layer to a particular set of possible roads while the input data (that can be assumed to be uncorrelated) makes the definite choice.

The net result is an inhomogeneous Markov process. At the K-th layer we obtain a realization of the random variable $y_k$ whose probability distribution is conditioned by the value of $y_{k-1}$

$$P_k ( y_k / y_{k-1} ) \tag{1}$$

In this way the hierarchy of the input (ie the level at which they enter into the encoding unit) results in ultrametricity if every site of the Neural Network proceeds independently.

There is an "embarras de richesse" in possible Markov processes. We have considered in detail just 2 examples chosen because of their amenability to analytical calculation:
A) At level k, $y_k$ takes the values $(-1, 0, 1)$ with the following conditional probabilities:

i) if $y_{k-1} = \pm 1$ then $y_k = \pm 1$ with probability one $\hspace{1cm}$ (2)
ii) if $y_k = 0$ then: $P_k(y_k = -1) = P_k(y_k = 1) = r_k/2$
$$P_k(y_k = 0) = 1 - r_k$$

B) At level k, $y_k$ can take the values $(-r_k, r_k)$ with the following conditional probabilities.

$$P(y_k = \pm r_k) = (1 \pm y_{k-1}/r_k)\%2 \tag{3}$$

There is some definite evidence that the perception system is organized in succesive layers. Furthermore at each layer the processing is parallel and local. There is not such a clear evidence that different pieces of the input enter at different levels but it seems a very natural hypothesis. Therefore there is an exciting possibility that categorization be a consequence of the architecture of the perception system. A bit more of speculation will lead to imagine that Nature copies the same structure for the encoding stage of the memory.

In the "storage" stage the encoded information is stored in the Synaptic connection. If we label $J_{ij}$ (assumed symmetric ie $J_{ij}=J_{ji}$) the synapses between neurons i and j and consider patterns organized as in fig.2 then a careful comparison with the SK spin glass[5] leads to the following proposal.

$$J_{ij} = \sum_{\alpha} \frac{S_i^{\alpha} S_j^{\alpha}}{q} + \sum_{\alpha,\beta} \frac{(S_i^{\alpha\beta}-S_i^{\alpha})(S_j^{\alpha\beta}-S_j^{\alpha})}{1-q} \qquad (4)$$

where $S_i^{\alpha\beta}$ is the spin value at site i in the state $\alpha,\beta$ ; $S_i^{\alpha}$ is defined as:

$$S_i^{\alpha} = \sum_{\beta=1}^{k_{\alpha}} \frac{S_i^{\alpha\beta}}{k_{\alpha}} \qquad (5)$$

q is the average overlap between states belonging to the same category:

$$q = \frac{1}{N} \overline{\sum_{i=1}^{N} S_i^{\alpha\beta} S_i^{\alpha\beta'}} = \frac{1}{N} \overline{\sum_{i=1}^{N} S_i^{\alpha} S_i^{\alpha}} \qquad (6)$$

while in average:

$$\frac{1}{N} \overline{\sum_{i=1}^{N} S_i^{\alpha} S_i^{\alpha'}} = 0 \quad . \qquad (7)$$

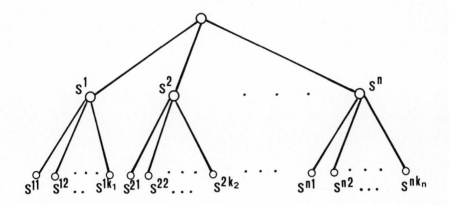

Fig. 2

The ultrametric tree with two generations. The states $\{S^{\alpha 1}, S^{\alpha 2}, \ldots S^{\alpha k_\alpha}\}$ belong to the same category. The average state (the ancestor) is $S^\alpha$.

It is apparent that Ansatz (4) is compatible with a minimally modified Hebb's rule. It does not require complicated feedback from past memories and is fast[6].

Finally we have analysed the "retrieval" stage. We consider it in the thermodynamic limit ($N \rightarrow \infty$, with the number of patterns P such that $P/N \rightarrow 0$). Then as in the Hopfield model, from time to time, we retrieve a spurious solutions[7]. But in our case the spurious solutions are of two different types: either it belongs to a single category (intracluster) ie the spurious pattern has appreciable overlap with patterns belonging to the same category or it mixes patterns of different categories (intercluster).

The retrieval of a spurious solution means that the biosystem is not able to decide about which of the stored patterns is to be associated with the input. The number of spurious patterns grows at least exponentially with the number of stored patterns. Evenmore the total sum of the basins of attraction of the spurious solutions grows with the number of patterns . This leads to a somewhat paradoxical situation: the more the system learns the more it may be easily confused.

The situation becomes clearer when we introduce categories. If can be argued in general (and rigourously proved for
the tree of categories generated by Random Process A) that the
behaviour of spurious solutions intraclusters is similar to
the one occurring in the Hopfield model with the number of
paterns inside the cluster playing the dominant role.

On the other hand spurious solutions interclusters seem
to be sensitive only to the number of clusters. For the
particular case of example B) we have been able to prove that
when the Neural Network instantaneous configuration has large
overlaps with many patterns (and this is the generic
situation) then the relaxation process obeys:

$$m^\alpha(t+\Delta t) = \sum_{S^\alpha=\pm\sqrt{q}} S^\alpha \int \frac{e^{-\frac{x^2}{2r(t)}}}{\sqrt{2\pi r(t)}} \, dx \, \text{sign}\left(\sum_\alpha \frac{s^\alpha m^\alpha(t)}{q} + X\right) \quad (8)$$

$$\sum_{\alpha\beta} \left(m^{\alpha\beta}(t+\Delta t) - m^\alpha(t+\Delta t)\right)\left(m^{\alpha\beta}(t) - m^\alpha(t)\right) \, / \, r(t)(1-q) =$$

$$= \sqrt{\frac{2}{\pi}} \sum_{S^\alpha=\pm\sqrt{q}} \exp\left(-\left(\sum_\alpha \frac{s^\alpha m^\alpha(t)}{q}\right)^2 \, / \, 2r(t)\right) \quad (9)$$

where:

$$m^\alpha(t) = \frac{1}{N} \sum_i S_i(t) S_i^\alpha$$

$$m^{\alpha\beta}(t) = \frac{1}{N} \sum_i S_i(t) S_i^{\alpha\beta}$$

$$r(t) = \sum_{\alpha\beta} \left(m^{\alpha\beta}(t) - m^\alpha(t)\right)^2 \, / \, (1-q) \quad (10)$$

Equation (8) at fixed r(t) is similar to the corresponding equation in the Hopfield's model with temperature. The
fixed point

$$m^\alpha = 0$$

becomes unstable for $r(t) < 2/\pi$. Equation (9) implies that in
general r(t) will decrease. Therefore this feedsback into

equation (8) and induces a final convergence to r(t)=0 and a fixed point of a Hopfield like model whose stored patterns are just the categories.

Three points are worth mentioning about these results:
1) the gradual decrease of r(t) with time is strongly reminiscent of the simulated annealing protocol[7].
2) if we keep the number of categories small the possibility of the system falling into an intercluster spurious state remains small. Therefore we are introducing a hierarchy among errors.
3) Learning lots of patterns intracluster does not lead to confusing categories.

In summary we found that the Hopfield model can be modified as proposed. The problem that one is still facing is that there is too much choice among the possible inhomogeneous Markov Processes that generate the ultrametric tree.

## Acknowledgement

I would like to thank M. Mezard for a very fruitful interaction and G. Toulouse for a crucial discussion at the beginning of this work. Looking backwards I now realize how many of his ideas we have been using freely.

## References

(1) J.J. Hopfield Proc. Natl. Acad. Sci. USA 79 2554 (1982)

(2) M. Mezard, G. Parisi, N. Sourlas, G. Toulouse, M. Virasoro J. Physique 45 (1984) 843.

(3) N. Parga, M. A. Virasoro "The Ultrametric Organization of Memories in a Neural Network", ICTP preprint (Trieste).

(4) See ie R. Klatzky "Human Memory" W. A. Freeman S. Francisco (1975).

(5) M. Mezard, M. A. Virasoro J. Physique (to appear).

(6) For a different compromise in this context see T. Sejno-
    wsky and G. Dreyfus contributions to this Workshop.

(7) D. Amit, H. Gutfreund, H. Sompolinsky, Phys. Rev. A
    32 pp. 1007-1018 (1985).

(8) S. Kirkpatrick, S.D. Gelatt Jr., M. P. Vecchi, Science 220
    671 (1983) and contribution to this Workshop.

# THE EMERGENCE OF HIERARCHICAL DATA STRUCTURES

## IN PARALLEL COMPUTATION

Michel Kerszberg

Institut für Festkörperforschung der Kernforschungsanlage

Jülich GmbH 5170 Jülich/West Germany

We think of the world in hierarchical manner. When dealing with the objects around us, however, we usually concentrate on one level of the hierarchy at a time. The very possibility of such a separation is in fact a precondition for knowledge [1]. When several levels of the hierarchy are simultaneously active, as in economics - where natural conditions, individual, corporate and governmental decisions all interact - we say that the situation is "complex"...

Thus the patterns we recognize are subject to hierarchical treatment: we can interpret the symbols $\mathcal{E}de\mathcal{D}\mathcal{E}\mathcal{L}D$ as the string EDEDEBD, as a series of musical notes, or the beginning of a piece by Beethoven, depending on the context and our own background. It is easy to imagine that this sort of classification will occur quite naturally in a structure where the input data are processed in successive stages, or layers. Thus it may be no coincidence that visual messages - to take the most glaring biological example - travel from multiple retinal to thalamic

NATO ASI Series, Vol. F20
Disordered Systems and Biological Organization
Edited by E. Bienenstock et al.
© Springer-Verlag Berlin Heidelberg 1986

to cortical layers where more and more elaborate treatment is performed.

The layered computers we study are simple arrays of elements, loosely modelled after the mammalian visual system [2]. The following properties are built-in:

- The processors come in layers working in parallel. Information is transmitted from one layer to the next (lower) one.

- All operations are local, i. e. each processor has access only to the output of its neighbours in the previous layer;

- The computing elements are all identical and perform only simple, nonlinear, operations; this processing results in a contraction of phase space, i. e. a given output may be generated by several input configurations (for instance, this is the case if the nonlinearity leads to saturation).

For a given array, we call "input" $S_o$ the set of inputs fed into the top layer; and "output" the group of numbers generated by the last layer.

Arrays of the type defined above have been shown [3] to exhibit "pattern recognizing" activity, in the sense that various inputs lead to the same output ("recognition").

The layered organization, together with the contracting property, lead to a hierarchical structure of the system's state space. Calling $S_i$ the state of layer i, consider two different inputs

$S_o$ and $S_o'$: then in general $S_i \neq S_i'$, i<p, but $S_i = S_i'$, i≥p; the states

thus lie on a <u>tree</u> where the two above branches $S_o$ and $S_o'$ join at

depth p. We say that the system is endowed with an <u>ultrametric</u>

<u>topology</u>, and p is called the ultrametric distance between $S_o$

and $S_o'$. Any $S_o$, $S_o'$, $S_o''$ can always be labelled such that

$$p(S_o S_o') = p(S_o S_o'') \geq p(S_o' S_o'') . \quad (1)$$

We imagine [4,5] for instance an (m lines by n columns) array

consisting of 'majority gates', which accept a string of three

bits and generate an output bit according to the majority rule,

i. e. 001 gives 0 at the output, but 110 gives 1. The gates are

interconnected so that the input string to each gate consists of

the three output bits coming from the neighbours located above it.

At the top, an input string consisting of n bits is fed into the

system; while the 2m additional bits required on the sides of the

array are arbitrarily set to zero. We assume the processing is

instantaneous: each input change is reflected instantly at the

output. We now start from a randomly selected input, and observe

the corresponding output; we then apply noise to the input, which

at each 'time step' may upset any input bit with a certain

probability. The diffusion process in input space generates a

related process in output space, which characterizes the error

resistance of a parallel computer under noisy input. Remarkably,

when the output's autocorrelation (averaged over many initial in-

puts) is plotted, a power law emerges: if output was initially

$S_o$, the probability that is still $S_o$ after t random inversions (per input bit) goes approximately as $t^{-\gamma}$ where $\gamma$ diminishes with m, and does not depend on n ($n \gtrsim 8$).

We conjecture [4] that this dynamical behaviour is related to the ultrametricity of phase space. One way to make this relation clear is to build a distance function for the processor's output space, and see whether triangles in this space satisfy the ultrametricity condition (1). For two outputs $S_m$ and $S_m'$, consider the sets $\{S_o\}$ and $\{S_o'\}$ of all inputs that may lead to them, compute then the average Hamming distance between strings belonging to $\{S_o\}$ and $\{S_o'\}$; the result we call the distance between $S_m$ and $S_m'$: this number is obviously related to the dynamics of the system under the diffusion process described above. We then check (1) on all triangles $S_m$, $S_m'$, $S_m''$: even for small machines, say 10 by 6, (1) is well satisfied by most triangles, as is best seen in a Dalitz-type plot of the triangle sides [5]. Thus the output dynamics itself emerges as hierarchical: analytical evidence [5] of this is also forthcoming.

1. Simon, H.A. (1965), The Sciences of the Artificial, MIT Press.

2. Creutzfeld, O. (1978), p 357, in Architectonics of the Cerebral Cortex, Brazier and Petsche eds., Raven Press, NY.

3. Huberman, B. A. and Hogg, T. (1984), Phys. Rev. Lett 58, 1048.

4. Huberman, B.A. and Kerszberg, M. (1985), J. Phys. A18, L331.

5. Kerszberg, M. and Mukamel, D. (1985) to appear.

# COGNITIVE CAPABILITIES OF A PARALLEL SYSTEM

James A.  Anderson,

Department of Psychology, Brown University,
Providence, RI 02912, U.S.A.

> Any mental process must lead to error.
> Huang Po (9th c.)

There has been recent interest in parallel, distributed, associative models as ways of organizing powerful computing systems and of handling noisy and incomplete data.  There is no doubt such systems are effective at doing some interesting kinds of computations.  Almost certainly they are intrinsically better suited to many kinds of computations than traditional computer architecture.

However such architectures have very pronounced 'psychologies' and though they do some things well, they do many things extremely poorly, and can cause 'errors' as a result of satisfactory operation.  When one talks about psychological systems, the idea of error becomes rather problematical.  One of the tasks of a biological information processing system is to simplify (i.e. distort) the world so complex and highly variable events fall into equivalence classes and can be joined with other events to generate appropriate responses.  Cognitive ideas such as concepts are useful simplifications for deciding what data can be ignored and what is essential.  It is possible that a brain-like computing system would show many of the undesirable features of our own minds such as gross errors, unpredictablity, instability, and even complete failure.  However, such a system might be a formidable complement to a traditional computer because it could then have the ability to make the good guesses, the hunches, and the suitable simplifications of complex system that are lacking in traditional computer systems but at which humans seem to excel.


## Stimulus Coding and Representation.

We have many billion neurons in our cerebral cortex.  The cortex is a layered two dimensional system which is divided up into a moderate number (say 50) of subregions.  The subregions project to other subregions over pathways which are physically parallel, so one group of a large number of neurons projects to another large group of neurons.  It is often not appreciated how much of what we can perceive depends on the details of the way the nervous system converts information from the physical world into discharges of nerve cells.  If it is important for us to be able to see colors, or line segments, or bugs (if we happen to be a frog), then neurons in parts of the nervous system will respond to color, edges, etc.  Many neurons will respond to these properties, and the more important the property, the more neurons potentially will have their discharges modified by that stimulus

NATO ASI Series, Vol. F20
Disordered Systems and Biological Organization
Edited by E. Bienenstock et al.
© Springer-Verlag Berlin Heidelberg 1986

property.

My impression is that much, perhaps most, of the
computational power of the brain is in the details of the neural
codes, i.e. the biological representation of the stimulus
developed by evolution. Perhaps the brain is not very smart. It
does little clever computation but powerful, brute force
operations on information that has been so highly processed that
little needs to be done to it. However the pre-processing is so
good, and the numbers of elements so large that the system
becomes very powerful.

Our fundamental modelling assumption is that information is
carried by the set of activities of many neurons in a group of
neurons. This set of activities carries the meaning of whatever
the nervous system is doing. Percepts, or mental activity of any
kind, are similar if their state vectors are similar. Our basic
approach is to consider the state vectors as the primitive
entities and try to see how state vectors can lawfully interact,
grow and decay. The elements in the state vectors correspond to
the activities of moderately selective neurons. In the language
of pattern recognition we are working with state vectors composed
of great numbers of poor features.

## The Linear Associator.

It is easy to show that a generalized synapse of the kind
first suggested by Donald Hebb in 1949 realizes a powerful
associative system. Given two sets of neurons, one projecting to
the other, and connected by a matrix of synaptic weights A, we
wish to associate two activity patterns (state vectors) f and g.
We assume A is composed of a set of modifiable 'synapses' or
connection strengths. We can view this as a sophisticated
stimulus-response model.

We make two quantit ative assumptions. First, the neuron
acts to a first approximation like a linear summer of its inputs.
That is, the $i\underline{th}$ neuron in the second set of neurons will display
activity g(i) when a pattern f is presented to the first set of
neurons according to the rule,

$$g(i) = \sum_j A(i,j) f(j).$$

where A(i,j) are the connections between the $i\underline{th}$ neuron in the
second set of neurons and the $j\underline{th}$ neuron in the first set. Then
we can write g as the simple matrix multiplication

$$g = A f.$$

Our second fundamental assumption involves the construction
of the matrix A, with elements A(i,j). We assume that these
matrix elements (connectivities) are modifiable according to a
generalized Hebb rule, that is, the change in an element of A,
δA(i,j), is given by

$$\delta A(i,j) \propto f(j)\, g(i).$$

Suppose initially A is all zeros. If we have a column input state vector f, and response vector g, we can write the matrix A as

$$A = \eta\, g\, f^{T}$$

where $\eta$ is a learning constant. Then, if after A is formed, vector f is input to the system, a pattern g' is constructed, which, subject to a multiplicative constant, is in the same direction as g. This model and variants have been discussed in many places. It is powerful, but has some severe limitations. (Anderson, 1970; see especially Kohonen (1977, 1984)).


## Categorization

The model just discussed can function as a simple categorizer by making one assumption. Let us make the coding assumption that the activity patterns representing similar stimuli are themselves similar, that is, their state vectors are correlated. This means the inner product between two similar patterns is large. Now consider the case described above where the model has made the association f → g. Let us restrict our attention to the magnitude of the output vector that results from various input patterns. With an input pattern f' then

$$(\text{output pattern}) = g\, [f,f']$$

This suggests that the perceived similarity of two stimuli should be systematically related to the inner product [f,f'] of the two neural codings. This is a testable prediction in some cases. Knapp and Anderson, (1984) discuss an application of this simple approach to psychological concept formation, specifically the learning of 'concepts' based on patterns of random dots.

There are two classes of simple concept models in psychology. The form a model for concept learning takes depends on an underlying model for memory structure. Two important classes of psychological models exist: 'exemplar' models where details of single presentations of items are stored and 'prototype' models where a new item is classified according to its closeness to the 'prototype' or best example of a category.

Consider a situation where a category contains many similar items. Here, a set of similar activity patterns (representing the category members) becomes associated with the same response, for example, the category name. It is convenient to discuss such a set of vectors with respect to their mean. Let us assume the mean is taken over all potential members of the category.

Specifically consider a set of correlated vectors, {f}, with mean p. Each individual vector in the set can be written as the sum of the mean vector and an additional noise vector, d, representing the deviation from the mean, that is,

$$f_i = p + d_i .$$

If there are n different patterns learned and all are associated with the same response the final connectivity matrix will be

$$A = \sum_{i=1}^{n} g f_i^T$$

$$= n g p^T + \sum_{i=1}^{n} d_i$$

Suppose that the term containing the sum of the noise vector is relatively small, as could happen if the system learned many randomly chosen members of the category (so the d's cancel on the average and their sum is small) and/or if d is not very large. In that case, the connectivity matrix is approximated by

$$A = n g p^T .$$

The system behaves as if it had repeatedly learned only one pattern, p, the mean of the set of vectors it was exposed to. Under these conditions, the simple association model extracts a prototype just like an average response computer. In this respect the distributed memory model behaves like a psychological 'prototype' model, because the most powerful response will be to the pattern p, which may never have been seen. This results is seen experimentally under appropriate conditions.

However if the sum of the d's is not relatively small, as might happen if the system only sees a few patterns from the set and/or if d is large, the response of the model will depend on the similarities between the novel input and each of the learned patterns, that is, the system behaves like an psychological 'exemplar' model. This result can also be demonstrated expermentally. We can also predict when one or the other result can be seen.

We can also see how the system can use partial information to reason 'cooperatively'. Suppose we have a simple memory formed which has associated an input $f_1$ with two outputs, $g_1$ and $g_2$, and an input $f_2$ with two outputs $g_2$ and $g_3$ so that

$$Af_1 = g_1 + g_2 \quad \text{and}$$
$$Af_2 = \qquad g_2 + g_3 .$$

Suppose we then present $f_1$ and $f_2$ together. Then, we have

$$A(f_1 + f_2) = g_1 + 2g_2 + g_3 ,$$

with the largest weight for the common association. This perfectly obvious consequence of superposition has let us pick out the common association of $f_1$ and $f_2$, if we can supress the spurious responses.

The cooperative effects described in several contexts above depend critically on the linearity of the memory. We will suggest below that it is easy to remove the extra responses due to superposition. We want to emphasize that it is the <u>linearity</u> that gives rise to most of the easily testable psychological predictions (many of which can be shown to be present, particularly in relation to simple stimuli) and it is the <u>non-linearity</u> that has the job of cleaning up the output.

## Error Correction.

The simple linear associator works but generates too many errors for many applications: that is, given a learned association $f \rightarrow g$, and many other associations learned in the same matrix, the pattern generated when f is presented to the system may not be close enough to g to be satisfactory. By using an error correcting technique related to the Widrow-Hoff procedure, we can force the system to give us correct associations. Suppose information is represented by vectors associated by $f_1 \rightarrow g_1$, $f_2 \rightarrow g_2$ ... We wish to form a matrix A of connections between elements to accurately reconstruct the association. A vector, f, is selected at random. Then the matrix, A, is incremented according to the rule

$$\Delta A = \eta\,(g - Af)\,f^T$$

where $\Delta A$ is the change in the matrix A and where the learning coefficient, $\eta$, is chosen so as to maintain stability. The learning coefficient can either be 'tapered' so as to approach zero when many vectors are learned, or it can be constant, which builds in a 'short term memory' because recent events will be recalled more accurately than past events. The method is sometimes called the delta method because it is learning the difference between desired and actual responses.

If $f = g$, the association of a vector with itself is referred to as an 'autoassociative' system. One way to view the autoassociative system is that it is forcing the matrix to develop a particular set of eigenvectors. Suppose we are interested in looking at autoassociative systems,

$$A = \eta\,f\,f^T$$

where $\eta$ is some constant.

We can use feedback to reconstruct a missing part of an input state vector. To show this, suppose we have a normalized state vector f, which is composed of two parts, say f' and f'', i.e. $f = f' + f''$. Suppose f' and f'' are subvectors that occupy different sets of elements -- say f' is non-zero only for elements [1..n] and f'' is non-zero only for elements [(n+1)..Dimensionality]. Then, f and f' are orthogonal.

Consider a matrix A storing only the autoassociation of f that is

$$A = (f' + f'')(f' + f'')^T,$$

(Let us take $\eta = 1$).

The matrix is now formed. Suppose at some future time a sadly truncated version of f, say f' is presented at the input to the system.

The output is given by

(output) = A f'

$$= (f'\,f'^T + f'\,f''^T + f''\,f'^T + f''\,f''^T)\,f'.$$

$$= c\,f$$

where c is some constant. The autoassociator can reconstruct the missing part of the state vector.

Let us use this technique. When the matrix, A, is formed, one way information can be retrieved is by the following procedure. It is assumed that we want to get associated information that we currently do not have, or we want to make 'reasonable' generalizations about a new situation based on past experience. We must always have some information to start with. The starting information is represented by a vector constructed according to the rules used to form the original vectors, except missing information is represented by zeros. Intuitively, the memory, that is the other learned information, is represented in the cross connections between vector elements and the initial information is the key to get it out. The retrieval strategy will be to repeatedly pass the information through the matrix A and to reconstruct the missing information using the cross connections. Since the state vector may grow in size without bound, we limit the elements of the vector to some maximum and minimum value.

We will use the following nonlinear algorithm. Let x(i) be the current state vector of the system. f(0) is the initial vector. Then, let x(i+1), the next state vector be given by

$$x(i+1) = \text{LIMIT}\,[\,\alpha A x(i) + \gamma x(i) + \delta f(0)\,].$$

The first term, $\alpha A x(i)$, passes the current state through the matrix and adds more information reconstructed from cross connections. The second term, $\gamma x(i)$, causes the current state to decay slightly. This term has the qualitative effect of causing errors to eventually decay to zero as long as $\gamma$ is less than 1. The third term, $\delta f(0)$, can keep the initial information constantly present if this is needed to drive the system to a correct final state. Sometimes this term $\delta$ is zero and sometimes δ is non-zero depending on the requirements of the task. In Example 1 below, δ was 1, in the other two examples it was zero. In all the simulations to be described α was 0.2 and γ was 0.9. Once the element values for x(i+1) are calculated, the element values are 'limited'. This means that element values cannot be greater than an upper bound or lower than a lower bound. If the element values of x(i+1) have values larger than or smaller than upper and lower bounds they are replaced with the upper and lower bounds repespectively. This process contains the state vector within a set of limits, and we have called this model the 'brain state in a box' or BSB model. In all the examples, the limits were ± 1.3.

Because the system is in a positive feedback loop but is limited, eventually the system will become stable. This may occur when all the elements are saturated or when a few are still not saturated. This final state will be the output of the system. The final state can be interpreted according to the rules used to generate the stimuli. This state will contain the directed conclusions of the information system. The dynamics of this system are closely related to the 'power' method of eigenvector exctraction.

It is at this point that the connection of this model with Boltzmann type models becomes of interest. We have showed in the past (Anderson, Silverstein, Ritz, and Jones, 1977) that in the simple case where the matrix is fully connected (symmetric by the learning rule in the autoassociative system) and has no decay, that the vector will monotonically lengthen. We would like to point out that the dynamics of this system are nearly identical to those used by Hopfield for continuous valued systems. (1984) It is one member of the class of functions he discusses, and can be shown to be minimizing an energy function. In the more general autoassociative case, where the matrix is not symmetric because of limited connectivity and/or there is decay, the system can be shown computationally to be minimizing a quadratic energy function (Golden, 1985). In the simulations to be described the Widrow-Hoff technique is used to 'learn the corners' of the system, thereby ensuring that the local energy 'minima' and the associated responses will coincide.

It is important to emphasize that this is not an information storage system as conventionally implemented. It is poor at handling precise data. It also does not make efficient use of memory in a traditional serial computer. There are several parameters which must be adjusted. Also the output may not be 'correct' in that it may not be a valid inference or it may contain noise. This is the penalty that one must pay for the inferential and 'creative' aspects of the system.

Example One:  A Data Base.

In the specific examples of state vector generation that we will use for the examples, English words and sets of words are coded as concatenations of the bytes representing their ASCII representation. A parity bit is used. Zeros are replaced with minus ones. (I.e. an 's', ASCII 115, is represented by

(-1,1,1,1,-1,-1,1,1) in the state vector.) A 200 dimensional vector would represent 25 alphanumeric characters. This is a 'distributed' coding because a single letter or word is determined by a pattern of many elements. It is an arbitrary coding, but it still gives useful demonstrations of the power of the approach. In the outputs from the simulations the underline, '_', corresponds to all zeros or to an uninterpretable character whose amplitude is below an interpretation threshold. That is, the output strings presented are only those of which the system is 'very sure' because their actual element values were all above a high threshold. The threshold is only for our convenience in interpreting outputs and the full values are used in the computations.

Information in AI systems is often represented as collections of atomic facts, relating pairs or small sets of items together. However, as William James commented in 1890,

> ... the more other facts a fact is associated with in the mind, the better posession of it our memory retains. Each of its associates becomes a hook to which it hangs, a means to fish it up by when sunk beneath the surface. Together, they form a network of attachments by which it is woven into the entire tissue of our thought.
> William James (1890). p. 301.

Information is represented as large state vectors containing correlated information. Each state vector contains a large number of 'atomic facts' together with their mutual co-occurence, so it is hard to specify the exact information capacity of the system.

As a simple example of a distributed data base, a small (200 dimensional autoassociative system) was taught a series of connected facts about antibiotics and diseases. (See the Figures Drugs 1-5). This is a complex, real world data base in that one bacterium causes many diseases, the same disease is caused by many organisms, and a single drug may be used to treat many diseases caused by many organisms. Figure Drugs-1 shows the information used for the data base. The detailed information is taken from Goodman and Gilman (1980). Because only 25 characters are available, the codings are somewhat terse. If each pairwise fact relation in a single state vector is considered an 'atomic fact' there are several hundred facts in this database, though only 31 state vectors.

Figure Drugs-1:

Database Information

| | |
|---|---|
| F[ 1] Staphaur+cocEndocaPenicil | F[ 2] Staphaur+cocMeningPenicil |
| F[ 3] Staphaur+cocPneumoPenicil | F[ 4] Streptop+cocScarFePenicil |
| F[ 5] Streptop+cocPneumoPenicil | F[ 6] Streptop+cocPharynPenicil |
| F[ 7] Neisseri-cocGonorhAmpicil | F[ 8] Neisseri-cocMeningPenicil |
| F[ 9] Coryneba+bacPneumoPenicil | F[10] Clostrid+bacGangrePenicil |
| F[11] Clostrid+bacTetanuPenicil | F[12] E.Coli  -bacUrTrInAmpicil |
| F[13] Enteroba-bacUrTrInCephalo | F[14] Proteus -bacUrTrInGentamy |
| F[15] Salmonel-bacTyphoiChloram | F[16] Yersinap-bacPlagueTetracy |
| F[17] TreponemspirSyphilPenicil | F[18] TreponemspirYaws  Penicil |
| F[19] CandidaafungLesionAmphote | F[20] CryptocofungMeningAmphote |
| F[21] HistoplafungPneumoAmphote | F[22] AspergilfungMeningAmphote |
| F[23] SiEfHypersensOralVPenicil | F[24] SiEfHypersensInjeGPenicil |
| F[25] SiEfHypersensInjeMPenicil | F[26] SiEfHypersensOralOPenicil |
| F[27] SiEfHypersensInje Cephalo | F[28] SiEfOtotoxic Inje Gentamy |
| F[29] SiEfAplasticAInje Chloram | F[30] SiEfKidneys++Inje Amphote |
| F[31] SiEfHypersensOral Ampicil | |

Figure Drugs-2 and -3 show simple retrieval of stored information. The beginning state vector (where '_' corresponds to eight zero's) is followed by the interpretation of the vector after a number of iterations following the BSB algorithm.

Figure Drugs-2: 'Tell about fungal meningitis.'

_____fungMening_____  →  AspergilfungMeningAmphote
     Starting Vector              After 59 Iterations

Partial information combined with the reconstructive properties of the autoassociative system are used to get information out of the system. The usual way we do this is to put in partial information and let the information at a node be reconstructed using feedback and the autoassociator. In the first stimulus (1) above, the '_' indicates zeros in the input state vector. Once the feedback starts working, '_' indicates a byte with one or more elements below interpretation threshold. The appropriate antibiotic for fungal meningitis emerges early, (after 19 iterations) because Amphotericin is used to treat all fungal diseases the system knows about. The specific organism takes longer (59 iterations), but is eventually reconstructed.

Figure Drugs-3: 'What are the side effects of Amphotericin?'

SiEf_____Amphote  →  SiEfKidneys++K__e Amphote
     Starting Vector              After 30 iterations

A prudent therapist checks side effects. Amphotericin has serious ones, involving the kidneys among other organs.

The data base can also 'guess'. When it was asked what drug should be used treat a meningitis caused by a Gram positive bacillus, it responded penicillin even though it never actually learned about a meningitis caused by a Gram positive bacillus. (Figure Drugs-4) It had learned about several other Gram positive bacilli and that the associated diseases could be treated with penicillin. The final state vector contained penicillin as the associated drug. The other partial information cooperated to suggest that this was the appropriate output. This inference may or may not be correct, but it is reasonable given the past of the system. These inferential properties are expected, given the previous discussion.

Figure Drugs-4: 'Meningitis caused by Gram + bacilli.'

_____+bacMening_____ → Co___e_'+bacMeningPenicil
     Starting Vector             After 30 Iterations

This Figure demonstrates the use of the system for generalization. The data base the system learned contains no information about Meningitis caused by Gram positive bacilli. However it does 'know' that other Gram positive bacilli are treated with penicillin. Therefore it 'guesses' that the right drug is penicillin. This may or may not be correct! Notice that the number of iterations to get the answer is fairly long (30), indicating that the system is not totally sure of the answer. There is no internal record of the 'reasoning' used by the system, so errors may be quite hard to correct, unlike rule drive expert systems.

Figure Drugs-5: Use of Converging Information and Consensus

Part I: Urinary Tract Infections

_____UrTrIn_____ → Proteus -bacUrTrInGentamy
     Starting Vector             After 60 Iterations

Part II: Hypersensitivity

____Hypersen_____ → SiEfHypersensInje_Penicil
     Starting Vector             After 51 Iterations

Part III. Hypersensitivity + Urinary Tract Infection

____HypersenUrTrIn_____ → Q__dHypersenUrTrInCephalo
     Starting Vector             After 31 Iterations

Suppose we need to use 'converging' information, that is, find a drug that is a 'second best' choice for two requirements, but the best choice for both requirements together. Figure Drugs-5 demonstrates such a situation. Suppose a nasty medical school pharmacology instructor asked, 'What is a drug causing hypersensitivity and which is used to treat Urinary tract infections.'

If the data base is told 'Urinary Tract Infection', it picks a learned vector, probably the most recent one it saw due to the short term memory effects of the decay term combined with error correction. (This effect is illustrated in Part I. of this Figure.) The drug in this case is Gentamycin, whose side effect is ototoxicity.

Hypersensitivity, used as a probe in Part II, indicates a penicillin family drug. (This is the penicillin 'allergy'.) Since penicillin is the most common drug in the data base, penicillin is the drug most strongly associated with Hypersensitivity. Penicillin is not used (in this data base) to treat urinary tract infections.

One drug that does both is cephalosporin, and given both requirements, as in Part III, this is the choice of the system, which integrated information from both probes and gave a satisfactory answer. Ampicillin would also be a satisfactory answer and has occurred in other tests of this system, with different sequences of learning trials. Notice that the form of the initial vector, where a side effect and a disease occur simultaneously never occurs in the vectors forming the data base.

The number of iterations required to reconstruct the appropriate answer is a measure of certainty: large numbers of iterations either suggest the information is not strongly represented or the inference is weak, small numbers of iterations suggest the information is well represented or the inference is certain.

This system behaves a little like an 'expert system' in that it can be applied to new situations. However it does not have formal codification of sets of rules. It potentially can learn from experience by extracting commonalities from a great deal of learned information, essentially (to emphasize this point again) due to the linear interactions between stored information. The retrieval of information must contain non-linearities to supress spurious responses.

Example Two:  Qualitative Physics.

There is considerable interest among cognitive scientists in the generation of systems capable of 'intuitive' reasoning about physical systems. This is for several reasons. First, much human real world knowledge is of this kind: i.e. information is not stored in 'propositional' form but in a hazy 'intuitive' form is generated by extensive experience with real systems. (It is almost certain that much human reasoning, even about highly structured abstract systems is not of a propositional type, but of an 'a-logical' spatial, visual, or kinesthetic nature. See Davis and Anderson (1979) and, in particular, Hadamard (1949) for examples.) Second, this kind of reasoning is particularly hard to handle with traditional AI systems because of its highly inductive and ill-defined nature. It would be important to be able to model. Third, it is an area where distributed neural systems may be very effective as part

of the system. Riley and Smolensky (1984) have described a
'Boltzmann' type model for reasoning about a simple physical
system, and below we describe another. Fourth, I believe that
the ideal model for reasoning about complicated real systems
will be a hybrid: partly rule driven and partly 'intuitive'.

For an initial test of these ideas we constructed a set of
state vectors representing the functional dependencies found in
Ohm's Law, for example, what happens to E when I increases and R
is held constant. These vectors were in the form of
quasi-analog codings. (Figure Ohms-1) The system was taught
according to our usual techniques. The parameters of the system
were unchanged from the drug data base simulation.


Figure Ohms-1: Autoassociative Stimulus Set

```
F[ 1] E__***__I_____**R**_____    F[12] E___***_I___***_R__***___
F[ 2] E__***__I____***R***_____     F[13] E____***I_____***R__***___
F[ 3] E__***__I__***_R_***_____     F[14] E____**I_____**R__***___
F[ 4] E__***__I_***__R_***_____     F[15] E**_____I_***_R**____
F[ 5] E__***__I_***___R__***___     F[16] E***_____I_***_R***___
F[ 6] E__***__I***____R____***___   F[17] E_***_____I_***_R_***___
F[ 7] E__***__I**_____R_____**_    F[18] E__***_____I_***_R__***___
F[ 8] E**_____I**____R_***___      F[19] E___***_I_***__R__***___
F[ 9] E***_____I***___R_***___      F[20] E___***I_***_R_____***___
F[10] E_***___I_***__R_***___       F[21] E____**I_***_R_____**_
F[11] E__***__I_***__R_***___
```


The three asterisks in these stimuli should be viewed as an
image of a broad meter pointer. The 'E', 'I', and 'R' are for
convenience of the reader. If the 'pointer' deflects to the
left, the value decreases, in the middle, there there is no
change, to the right the value increases. The autoassociative
matrix generated was 45% connected and received about 25
presentations of each stimulus in random order.


Figure Ohms-2: Response to a Learned Pattern

```
E***_____I_***_R_____    →    E***_____I__***__R***_____
     Starting Vector                   After 9 iterations
```


This input pattern simply indicates that the matrix can
respond appropriately to a learned pattern. Noise starts to
appear after 12 or so iterations. These spurious associations
will appear in the blank positions as the system continues to
cycle.

Figure Ohms-3: Response to Consistent Inputs.  Case 1.

```
E_____I***____R***_____   →   E***____I***____R***_____
        Starting Vector                 After 6 Iterations
```

                              Case 2.

```
E____***I***____R_____   →   E____***I***____R____***_
        Starting Vector                 After 7 Iterations.
```

In these two tests, the system sees a pattern it never saw explicitly and it must respond with the 'most appropriate' answer.  Note that although the problem is ill defined, there is a consensus answer.  If we look at Ohm's Law in both the first and second cases, the equation suggests a consistent interpretation.  In the first case, I and R both are decreasing, therefore

$$\downarrow I \downarrow R ==> \downarrow E.$$

In the second case, E is up and I is down, therefore

$$\frac{\uparrow E}{\downarrow I} ==> \uparrow R.$$

Figure Ohms-4: Inconsistent Stimulus Set

```
E***____I***____R_____   →   E***____I***__U_R***_***_
        Starting Vector                 After 14 Iterations
```

There is no such consistency in Figure Ohms-4, and there is no consensus.  Note the answer is 'confused' and shows many possible answers.  Note also that many more iteratons are required to generate an answer, also indicating uncertainty.  In this case, E is down and I is down.  If we look at the equation,

$$\frac{\downarrow E}{\downarrow I} ==> \downarrow\uparrow R$$

the top and bottom of the equation 'fight' each other and there is no agreement.

## Example Three:  Semantic Networks

A useful way of organizing information is as a network of associated information, often called a semantic network.  Next, we show a simple example of a computation of this type.  Information is represented at 200 dimensional state vectors, constructed as strings of alphanumeric characters as before.  By making associations between state vectors, one can realize a simple semantic network, an example of which is presented in Figure Network-1.

Figure Network-1: A Simple 'Semantic' Network

```
Superset           |------------------->  ANML <---------------|
                   |                       |                    |
                   |         (gerbil) <--> animal <--> (elephant) |
                   |            small      ^            large    |
                   |            dart       v            walk     |
                   |            skin    (raccoon)       skin     |
                   |            brown      medium        gray    |
                   |                       climb          ^      |
                   |                       skin           |      |
Subsets           BIRD                     black          |      |
                   |                                       |      |
  (canary) <--> bird <--> (robin)   (Examples)            |      |
     medium       ^          medium    Clyde ---------->|        |
     fly          v          fly       Fahlman          |        |
     seed      (pigeon)      worm                        |        |
     yellow       medium     red                         |        |
       ^          fly                  Jumbo ---------->|         |
       |          junk                    large          |       |
       |          gray                     circus         |      |
       |                                                   |     |
       |--------------------------------Tweetie           |      |
       |                                 small             |      |
                                         cartoon           |      |
                   |--------------------------------------------------
                  FISH
                   |
  (guppy) <--> fish <--> (tuna) <----Charlie
     small       ^          large    StarKist
     swim        v          swim     inadequate
     food     (trout)       fish
  transparent    medium     silver
                 swim
                 bugs
                 silver
```

Each node of the network corresponds to a state vector
which contains related information, i.e. simultaneously present
at one node (the leftmost, under 'Subset') is the information
that a canary is medium sized, flies, is yellow and eats seeds.
This is connected by an upward and a downward link to the BIRD
node, which essentially says that 'canary' is an example of the
BIRD concept, which has the name 'bird'. A strictly upward
connection informs us that birds are ANMLs (with name 'animal').
The network contains three examples each of fish, birds and
animal species and several examples of specific creatures. For
example, Charlie is a tuna, Tweetie is a canary and both Jumbo
and Clyde are elephants. The specific set of associations that
together are held to realize this simple network are given in
Figure Network-2.

Figure Network-2: Stimulus Set

```
F[ 1] BIRD_*_bird___fly_wormred     G[ 1] _____*_robin__fly_wormred
F[ 2] _____*_robin__fly_wormred    G[ 2] BIRD_*_bird___fly_wormred
F[ 3] BIRD_*_bird___fly_junkgry     G[ 3] _____*_pigeon_fly_junkgry
F[ 4] _____*_pigeon_fly_junkgry    G[ 4] BIRD_*_bird___fly_junkgry
F[ 5] BIRD_*_bird___fly_seedylw     G[ 5] _____*_canary_fly_seedylw
F[ 6] _____*_canary_fly_seedylw    G[ 6] BIRD_*_bird___fly_seedylw
F[ 7] ANML*__animal_dartskinbrn     G[ 7] _____*__gerbil_dartskinbrn
F[ 8] _____*__gerbil_dartskinbrn   G[ 8] ANML*__animal_dartskinbrn
F[ 9] ANML_*_animal_clmbskinblk     G[ 9] _____*_raccoonclmbskinblk
F[10] _____*_raccoonclmbskinblk    G[10] ANML_*_animal_clmbskinblk
F[11] ANML__*animal_walkskingry     G[11] _____*elephanwalkskingry
F[12] _____*elephanwalkskingry     G[12] ANML__*animal_walkskingry
F[13] BIRD_____       G[13] ANML_____
F[14] _____Clyde___Fahlman___    G[14] _____*elephanwalkskingry
F[15] ____*__Tweetie_cartoon___     G[15] _____*_canary_fly_seedylw
F[16] _____*Jumbo_____circus___    G[16] _____*elephanwalkskingry
F[17] FISH_____        G[17] ANML_____
F[18] FISH*__fish___swimfoodxpr     G[18] _____*__guppy__swimfoodxpr
F[19] _____*__guppy__swimfoodxpr   G[19] FISH*__fish___swimfoodxpr
F[20] FISH_*_fish___swimbugsslv     G[20] _____*_trout__swimbugsslv
F[21] _____*_trout__swimbugsslv    G[21] FISH_*_fish___swimbugsslv
F[22] FISH__*fish___swimfishslv     G[22] _____*tuna___swimfishslv
F[23] _____*tuna___swimfishslv     G[23] FISH__*fish___swimfishslv
F[24] StarKistCharlieinadequate     G[24] _____*tuna___swimfishslv
```

This is one set of pairs of stimuli that realize the simple 'semantic' network in Figure Network-1. Two matrices were involved in realizing the network, an autoassociative matrix, where every allowable state vector is associated with itself, and a true associator, where f was associated with g. The Widrow-Hoff learning procedure was used. Pairs were presented randomly for about 30 times each. Both matrices were about 50% connected.

When the matrices are formed and learning has ceased, the system can then be interrogated to see if it can traverse the network and fill in missing information in an appropriate way. Figures Network-3 and -4 show simple disambiguation, where the context of a probe input ('gry') will lead to output of elephant or pigeon. Alan Kawamoto (1985) has done extensive studies of disambiguation using networks of this kind, and made some comparisions with relevant psychological data. Kawamoto has generalized the model by adding adaptation as a way of destabilizing the system so it moves to new states as time goes on.

The simulations described in the next few figures display results of successive iterations. The system switches between the autoassociative (Mx 2) and associative (Mx 2) matrix when the system is stable, i.e. there is no change in number of limited elements (Check) for three iterations. The associative matrix (Mx 1) is then applied and the system autoassociates on the output for more iterations. Some iterations are left out in the figures to save space. The next figures are essentially what appears on the terminal when our programs are run.

Figure Network-3: 'Tell me about gray animals'

```
Mx 2         1. ANML___animal_____gry   Check:    0
...
Mx 2.       10. ANML___animal_____gry   Check:  107
...
Mx 2.       20. ANML__*animal___1___i_gry   Check:  133
...
Mx 2.       30. ANML__*animal_walkskingry   Check:  163
...
Mx 2.       38. ANML__*animal_walkskingry   Check:  176
Mx 1.       39. ANML__*_____nwalkskingry   Check:  128
Mx 1.       40. _____*elephanwalkskingry   Check:  136
...
Mx 2.       46. _____*elephanwalkskingry   Check:  152
Mx 1.       47. ANML__*_____nwalkskingry   Check:  128
Mx 1.       48. ANML__*ani_a_nwalkskingry   Check:  160
Mx 1.       49. ANML__*animal_walkskingry   Check:  170
```

Note the simulation will endlessly move back and forth between these two nodes ('animal' and 'elephant') unless jarred loose by some mechanism such as adaptation.

The color, 'gry', appears in several different stimuli, but is disambiguated by the other information. In Figure Network-3 we asked about gray animals, in Figure Network-4 we ask about gray birds.

Figure Network-4: 'Gray birds'

```
Mx 2.        1. BIRD___bird_____gry   Check:    0
...
Mx 2.       10. BIRD___bird___f_____gry   Check:   76
...
Mx 2.       20. BIRD_**bird___f___j__kgry   Check:  127
...
Mx 2.       30. BIRD_**bird___fly_junkgry   Check:  149
...
Mx 2.       32. BIRD_**bird___fly_junkgry   Check:  149
Mx 1.       33. BIRD_**_i__on_fly_junkgry   Check:  112
Mx 1.       34. B____**pi_eon_fly_junkgry   Check:  120
Mx 1.       35. _____**pigeon_fly_junkgry   Check:  122
...
Mx 2.       40. ___L_**pigeon_fly_junkgry   Check:  139
```

The system now tells us about pigeons rather than elephants. Note the confusion where the simulation is not sure whether pigeons are medium sized or large. Note also the intrusion of the 'L', (Iteration 40) probably from ANML, which is the upward association of BIRD.

Another property of a semantic network is sometimes called 'property inheritance'. Figure Network-5 shows such a computation. We ask for the color of a large creature who works in the circus who we find out is Jumbo. Jumbo is an elephant. Elephants are gray.

Figure Network-5: 'Large circus creature.'

```
Mx 2.      1.    _____*_____circus___   Check:    0
...
Mx 2.     10.    _____*J_____circus___   Check:   65
...
Mx 2.     20.    _____*Jumbo_____circus___   Check:   93
...
Mx 2.     25.    _____*Jumbo_____circus___   Check:   97
Mx 1.     26.    _____*_____anw_____   Check:   67
Mx 1.     27.    _____*el_phanwa_ksk_ngr_    Check:  105
...
Mx 1.     30.    _____*elephanwalkskingry    Check:  148
...
Mx 2.     33.    _____*elephanwalkskingry    Check:  149
Mx 1.     34.    ANML__*_____nwalkskingry    Check:  133
Mx 1.     35.    ANML__*ani_a_nwalkskingry    Check:  160
...
Mx 2.     40.    ANML__*ani_al_walkskingry    Check:  173
Mx 2.     41.    ANML__*animal_walkskingry    Check:  174
```

This system uses two separate matrices, to keep the two distinct processes of autoassociation and true association separate. This untidy assumption can be done away with by assuming proper time delays as part of the description of a synapse, but at present it is more convenient to keep it because it separates two distinct operations. Eventually this mechanism will be eliminated.

This work is sponsored primarily by the National Science Foundation under grant BNS-82-14728, administered by the Memory and Cognitive Processes section.

## References

Anderson, J.A. Two models for memory organization using interacting traces. _Mathematical Biosciences._, _8_, 137-160, 1970.

Anderson, J.A. Cognitive and psychological computation with neural models. _IEEE_ Transactions _on Systems, Man, and Cybernetics_, _SMC-13_, 799-815, 1983.

Anderson, J.A. & Hinton, G.E. Models of information processing in the brain. In G.E. Hinton & J.A. Anderson (Eds.), _Parallel_ Models _of_ Associative Memory. Hillsdale, N.J.: Erlbaum Associates, 1981.

Anderson, J.A., Silverstein, J.W., Ritz, S.A. & Jones, R.S. Distinctive features, categorical perception, and probability learning: Some applications of a neural model. _Psychological_ Review. _84_, 413-451, 1977.

Davis, P.J. & Anderson, J.A. Non-analytic aspects of mathematics and their implications for research and education. SIAM Review., 21, 112-127, 1979.

Geman, S. & Geman, D. Stochastic relaxation, Gibbs distributions, and the Bayesian restoration of images. IEEE: Proceedings on Artificial and Machine Intelligence, 6, 721-741, November, 1984.

Goodman, A.G., Goodman, L.S., & Gilman, A. The Pharmacological Basis of Theraputics. Sixth Edition. New York: MacMillan, 1980.

Golden, R. Identification of the BSB neural model as a gradient descent technique that minimizes a quadratic cost function over a set of linear inequalities. Submitted for publication.

Hadamard, J. The Psychology of Invention in the Mathematical Field. Princeton, N.J.: Princeton University Press, 1949.

Hinton, G.E. & Sejnowski, T.J. Optimal pattern inference. IEEE Conference on Computers in Vision and Pattern Recognition.. 1984.

Hopfield, J.J. Neurons with graded response have collective computational properties like those of two-state neurons. Proc. Natl. Acad. Sci. U.S.A., 81, 3088-3092, 1984.

James, W. Briefer Psychology. (Orig. ed. 1890). New York: Collier, 1964.

Kawamoto, A. Dynamic Processes in the (Re)Solution of Lexical Ambiguity. Ph.D. Thesis, Department of Psychology, Brown University. May, 1985.

Knapp, A.G. & Anderson, J.A. Theory of categorization based on distributed memory storage. Journal of Experimental Psychology: Learning, Memory, and Cognition., 10, 616-637, 1984.

Kohonen, T. Associative Memory. Berlin: Springer, 1977.

Kohonen, T. Self Organization and Associative Memory. Berlin: Springer, 1984.

Riley, M. S. & Smolensky, P. A parallel model of sequential problem solving. Proceedings of Sixth Annual Conference of the Cognitive Science Society. Boulder, Colorado: 1984.

# NEURAL NETWORK DESIGN FOR EFFICIENT INFORMATION RETRIEVAL

L. PERSONNAZ, I. GUYON and G. DREYFUS
E.S.P.C.I., Laboratoire d'Electronique
10, rue Vauquelin
75005 PARIS

The ability of neural networks to store and retrieve information has been investigated for many years. A renewed interest has been triggered by the analogy between neural networks and spin glasses which was pointed out by W.A. Little et al.[1] and J. Hopfield[2]. Such systems would be potentially useful autoassociative memories "if any prescribed set of states could be made the stable states of the system"[2] ; however, the storage prescription (derived from Hebb's law) which was used by both authors did not meet this requirement, so that the information retrieval properties of neural networks based on this law were not fully satisfactory. In the present paper, a generalization of Hebb's law is derived so as to guarantee, under fairly general conditions, the retrieval of the stored information (autoassociative memory). Illustrative examples are presented.

DESCRIPTION OF THE NETWORK : we consider a fully connected network of n McCulloch-Pitts formal neurons, with simultaneous, parallel operations. Each neuron is a binary state threshold device having n inputs (the states of all neurons) and one output (its own state). At time t, the state of neuron i is represented by a binary variable $\sigma_i(t)$ which can take the numerical values of +1 or -1. In order to determine its next state $\sigma_i(t+\tau)$, the neuron i performs a weighted sum of its inputs and compares it to a threshold value $\theta_i$ :

$$\sum_{j=1}^{n} C_{ij}\,\sigma_j(t) > \theta_i \Rightarrow \sigma_i(t+\tau) = +1$$
$$< \theta_i \Rightarrow \sigma_i(t+\tau) = -1$$
$$= \theta_i \Rightarrow \sigma_i(t+\tau) = \sigma_i(t)$$

Therefore, the parameters of the network are the (n,n) matrix C of the weights $\{C_{ij}\}$ and the (n) vector $\underline{\theta}$ of the threshold values $\{\theta_i\}$. The state of the network is defined by an

NATO ASI Series, Vol. F20
Disordered Systems and Biological Organization
Edited by E. Bienenstock et al.
© Springer-Verlag Berlin Heidelberg 1986

(n) vector $\underline{\sigma}$ , the components of which are the states $\{\sigma_i\}$ of all the neurons.

NEURAL NETWORKS AS AUTOASSOCIATIVE MEMORIES : for the network to be useful as an autoassociative memory, it should have the following behaviour : if the network is set into a given state $\underline{\sigma} \in \{-1,+1\}^n$ (representing a distorted or incomplete informa- tion), it should evolve until it reaches a stable state (repre- senting the full information to be retrieved). The problem of designing a neural network acting as an autoassociative memory is therefore : given a set of p prototype states to be memori- zed, how should the parameters C and $\underline{\theta}$ be chosen so as to re- trieve these states as faithfully as possible ? Obviously, the minimum requirement is that the prototype states be stable ; very desirable features would be :
(i) the fact that prototype states be stable and act as attrac- tors,
(ii) the absence of cycles,
(iii) the absence of spurious stable states or, at least, their predictability.
In the following, we show how to design networks embodying the- se three features.

THE GENERALIZED HEBB'S LAW : the general stability condition of a network can be expressed as follows[3] if all thre- sholds are taken equal to zero : a given state $\underline{\sigma}$ is stable if and only if there exists a diagonal matrix A, with all ele- ments positive or zero, such that one has :
$$C\underline{\sigma} = A\underline{\sigma} \qquad (1)$$

In order to make the p prototype states $\{\underline{\sigma}^k\}$ stable, relation (1) must hold true for all of them :
$$C\underline{\sigma}^k = A^k \underline{\sigma}^k$$
$A^k$ being a diagonal matrix with all its elements positive or zero. The general problem of finding C has been addressed in Ref. 3. In the present paper we take $A^k = I$ for all k. If the prototype states are linearly independent, the solution is given by : $\qquad C = \Sigma\Sigma^I \qquad (2)$

where $\Sigma$ is the matrix whose columns are the prototype vectors $\{\sigma^k\}$ and $\Sigma^I = (\Sigma^T\Sigma)^{-1}\Sigma^T$ is the pseudoinverse[4] of matrix $\Sigma$. The matrix C is symmetric. It is the orthogonal projection matrix into the subspace spanned by the prototype vectors. It can be computed either directly from relation (2), or recursively[5], without matrix inversion, by introducing each prototype vector once (and only once). This recursive computation is typical of a learning process.

One should notice that relation (2) is a generalized form of the classical Hebb's law $C = (1/n)\Sigma\Sigma^T$ : in the special case where the prototype states are orthogonal , relation (2) reduces exactly to Hebb's law. Therefore, since our storage prescription guarantees the retrieval of the prototype states even if the latter are not orthogonal, it is a generalization of Hebb's law.


PROPERTIES OF THE GENERALIZED HEBB'S LAW : besides the fundamental feature of guaranteeing the stability of the prototype states, we show some additional properties of such networks : the absence of cycles, the characterization of the spurious stable states and the attractivity of the prototype states.

i- Absence of cycles. The "energy" of the network in the state $\sigma$ can be defined by analogy with spin glasses in the absence of external field: $E = -\frac{1}{2}\sigma^T C \sigma$ .

Assume that the system is in an unstable state $\sigma$ and that the next parallel iteration drives it to a state $\sigma'$. It can be shown[6] that : $\sigma^T C \sigma < \sigma'^T C \sigma < \sigma'^T C \sigma'$

Therefore, the energy is an ever decreasing function, thus preventing any cycle to occur even under parallel operation.

It should be noticed that the prototype states and their linear combinations belonging to $\{-1,+1\}^n$ (if any) are the states of lowest possible energy (they are identical to the Mattis states referred to by D. Amit et al. in the present book).

ii- Spurious stable states. Consider a stable state $\sigma$ , which is possibly a non prototype state. Since C is the matrix of the orthogonal projection into the subspace spanned by the prototype vectors, $C\sigma$ is a linear combination of the prototype vectors.

Since $\underline{\sigma}$ is stable, from relation (1), there exists a positive diagonal matrix A such that :

$$C\underline{\sigma} = A\underline{\sigma}$$

Vector $\underline{\sigma}$ is therefore a "normalized" combination of the prototype states. Thus, any stable state is a normalized combination of the prototype states.

iii- <u>Attractivity of orthogonal prototype states</u>. If the prototype states are orthogonal, it has been shown[6] that any state lying within a Hamming distance of n/2p of a given prototype state will converge to that state in one iteration. Similarly, the opposite of any prototype state has a radius of attraction of n/2p.

EXAMPLES : we present an example which illustrates the efficiency of the generalized Hebb's law : a neural network designed after this law is used for error correction purposes. The titles of scientific journals have been chosen as prototype patterns. Each alphabetic character has been coded on six bits. The prototype states are shown in the upper left block of the figure. Each example in the other two blocks has three lines : the first one is the initial state ; the second and third lines are the final states reached by neural networks designed with the generalized Hebb's law and with Hebb's law respectively. These two networks have exactly the same structure; the only difference between them is the analytic expression of the matrix C. As was mentioned above, the numerical computation of that matrix is performed in both cases by an algorithm which yields the exact result (within roundoff errors) after a finite number of steps (equal to the number of prototype vectors). In the upper right block, the retrieval of the prototype states is attempted ; as expected, all the prototype states are retrieved in the first case, whereas several are forgotten in the second case. The lower block shows the error correction properties ; obviously, the generalized Hebb's law is much more efficient than Hebb's law for error correction.

CONCLUSION : we have proposed a generalization of Hebb's law which guarantees a perfect retrieval of the information stored in a neural network. This approach has enabled us to demonstrate the absence of cycles under parallel iteration conditions, to evaluate the attractivity of the prototype states, and to clarify the nature of the spurious stable states. The examples that are presented show that such networks exhibit reliable properties of autoassociative memories.

(1) W.A. LITTLE , G.L. SHAW, Math. Biosciences **39**, 281 (1978).

(2) J. J. HOPFIELD, Proc. Natl. Acad. Sci. USA **79**, 2554 (1982).

(3) L. PERSONNAZ, I. GUYON, G. DREYFUS, J. Physique Lett. **46**, L-359 (1985).

(4) A. ALBERT, Regression and the Moore-Penrose pseudoinverse (Academic Press, 1972).

(5) T.N.E. GREVILLE, SIAM Rev. **2**, 5 (1960).

(6) To be published.

```
   PROTOTYPE STATES          PHYSICAL REVIEW LETTERS    J. PHYSICAL OCEANOGRAPHY
   ****************
                             PHYSICAL REVIEW LETTERS    J. PHYSICAL OCEANOGRAPHY
                             PHYSICAL REVIEW LETTERS    J. PHYSICAL OCEANOGRAPHY
  PHYSICAL REVIEW LETTERS

  JOURNAL OF NEUROBIOLOGY     JOURNAL OF NEUROBIOLOGY    PROG. THEOR. PHYS. KYOTO

                             JOURNAL OF NEUROBIOLOGY    PROG. THEOR. PHYS. KYOTO
  BIOLOGICAL CYBERNETICS      JOURNAL OF NEUROBIOLOGY    PROG. THEOR. PHYS. KYOTO

  BULL. OF MATH BIOPHYSICS    BIOLOGICAL CYBERNETICS     SPECULAT. SCI. TECHNOL.

  J. PHYSICAL OCEANOGRAPHY    BIOLOGICAL CYBERNETICS     SPECULAT. SCI. TECHNOL.
                             CPPKKEAD P RH  QODPLORT    SPECULAT. SCI. TECHNOL.
  PROG. THEOR. PHYS. KYOTO

  SPECULAT. SCI. TECHNOL.     BULL. OF MATH BIOPHYSICS    INDIAN J PURE APPL. MATH

                             BULL. OF MATH BIOPHYSICS    INDIAN J PURE APPL. MATH
  INDIAN J PURE APPL. MATH    COPKK TE O RE BPRIPLOCT    INDIAN J PURE APPL. MATH
```

```
  PHISICAL REFIEX LETTER      BIOL. CYBERN.              MROG. CHEOR. PHYS. TOKYO

  PHYSICAL REVIEW LETTERS     BIOLOGICAL CYBERNETICS     PROG. THEOR. PHYS. KYOTO
  PHYRICAI REVIEW LETTERS     COPKK TE O RE BPRIPLOCT    PROG. THEOR. PHYS. KYOTO

  PHYSICAL REV. LET.          DIOLORICAL CIBERNETICS     SPECULATION ET TECHNOL.

  PHYSICAL REVIEW LETTERS     BIOLOGICAL CYBERNETICS     SPECULAT. SCI. TECHNOL.
  DOQKKFAI O RE BPRIQYPCT     CPPKKEAD P RH  QODPLORT    DOQKK TH O RE BPRLPYPCT

  JØNRNEL AF NOURABIALAGI     J. PHICINAL OCEAN.         INDIEN I PERE EPPL. NATH

  JOURNAL OF NEUROBIOLOGY     J. PHYSICAL OCEANOGRAPHY   INDIAN J PURE APPL. MATH
  COPKK TE O RE BPRIPLOCT     DOQKK TH O RE BPRLPYPCT    INDIAN J PURE APPL. MATH
```

# LEARNING PROCESS IN AN ASYMMETRIC THRESHOLD NETWORK.

YANN LE CUN


Ecole Supérieure d'Ingénieurs en Electrotechnique et Electronique.
89, rue Falguière 75015 PARIS
and
Laboratoire de dynamique des réseaux. 1, rue Descartes 75005 PARIS.

## 1 - INTRODUCTION

Threshold functions and related operators are widely used as basic elements of adaptive and associative networks [Nakano 72, Amari 72, Hopfield 82]. There exist numerous learning rules for finding a set of weights to achieve a particular correspondence between input-output pairs. But early works in the field have shown that the number of threshold functions (or linearly separable functions) in N binary variables is small compared to the number of all possible boolean mappings in N variables, especially if N is large. This problem is one of the main limitations of most neural networks models where the state is fully specified by the environment during learning: they can only learn linearly separable functions of their inputs. Moreover, a learning procedure which requires the outside world to specify the state of every neuron during the learning session can hardly be considered as a general learning rule because in real-world conditions, only a partial information on the "ideal" network state for each task is available from the environment. It is possible to use a set of so-called "hidden units" [Hinton,Sejnowski,Ackley. 84], without direct inter-action with the environment, which can compute intermediate predicates. Unfortunately, the global response depends on the output of a particular hidden unit in a highly non-linear way, moreover the nature of this dependence is influenced by the states of the other cells. Thus, it is difficult to decide whether the output of a hidden unit is wrong for a particular input, and, consequently, how to modify its weights. This last problem has been referred to as the "credit assignment problem" (CAP) in [Hinton & al. 84]. Attempts to find a learning rule taking into account hidden units and generating high order predicates failed until recently, which could explain for the decrease of interest in this field for the past 15 years [Minsky & Papert 68].

In this paper, we consider learning and associative memorization as dynamic processes and show how to describe the evolution of the weights through an "energy" function. This method will be used to solve the CAP and applied to a model of hierarchical associative memory called HLM (Hierarchical Learning Machine).

NATO ASI Series, Vol. F20
Disordered Systems and Biological Organization
Edited by E. Bienenstock et al.
© Springer-Verlag Berlin Heidelberg 1986

## 2- LEARNING AS ITERATIVE MINIMIZATION

In the following, we consider a neural network with state vector $\mathbf{X}$ in $\{-1,+1\}^N$. The time evolution of the network is described by:

$$\mathbf{X}(t+1) = F[\mathbf{W}(t).\mathbf{X}(t)] \tag{1}$$

where $\mathbf{W}(t)$ is the weight matrix at time t and $F$ is the mapping whose $i^{th}$ coordinate has value +1 if the $i^{th}$ coordinate of its argument is positive, −1 otherwise (threshold function).

Learning modifies the weights and can be viewed as minimizing a given cost function or criterion. The simplest way to minimize a function is to use a gradient descent method: given $C(t)$ where t is a discrete time index and whose time average $\langle C \rangle$ is taken as the criterion, the recursive formula:

$$\mathbf{W}(t) = \mathbf{W}(t-1) - \mathbf{K}(t).\,grad_{\mathbf{W}}\,[C(t)]. \tag{2}$$

minimizes $\langle C \rangle$ with respect to $\mathbf{W}$, with a noise level defined by $\mathbf{K}$ (where $\mathbf{W}(t)$ is the weight matrix at time t, and $\mathbf{K}$ is a diagonal positive matrix).

For example, we can choose for C the Hopfield's energy [Hopfield 82]:

$$C(t) = -1/2\,\mathbf{Y}(t)^T.\mathbf{W}(t).\mathbf{Y}(t) \tag{3}$$

where the patterns to be memorized, $\mathbf{Y}(t)$, are presented sequentially and $\mathbf{Y}(t)^T$ is $\mathbf{Y}(t)$ transpose. In this case, (2) gives:

$$W_{ij}(t) = W_{ij}(t-1) + K_i(t)Y_i(t)Y_j(t) \tag{4}$$

which is the classical Hebb's law. Running this procedure will "dig holes" in the energy landscape around states to be memorized. Notice that this process diverges if the patterns are presented indefinitely.

Another possible criterion is:

$$C(t) = [\mathbf{Y}(t) - \mathbf{W}(t).\mathbf{Y}(t)]^T[\mathbf{Y}(t) - \mathbf{W}(t).\mathbf{Y}(t)] \tag{5}$$

in case of auto-association ($\mathbf{Y}(t)$ associated to itself) or

$$C(t) = [\mathbf{Y}(t) - \mathbf{W}(t-1).\mathbf{X}(t-1)]^T[\mathbf{Y}(t) - \mathbf{W}(t-1).\mathbf{X}(t-1)] \tag{6}$$

in case of hetero-association ($\mathbf{Y}(t)$ associated to $\mathbf{X}(t-1)$).

Equation (2) then gives:

$$W_{ij}(t) = W_{ij}(t-1) + K_i(t)[Y_i(t)-A_i(t)]X_j(t-1) \tag{7}$$

where $\mathbf{A}$ is the vector defined by: $\mathbf{A}(t) = \mathbf{W}(t-1).\mathbf{X}(t-1)$ (ie $A_i$ is the total input to cell i).

Equation (7) is known as the Widrow-Hoff rule [Widrow & Hoff. 60] (or least mean square algorithm) and is strongly related to the pseudo-inverse method [Duda & Hart. 73, Kohonen 74, 84].

Under some conditions on $\mathbf{K}$, this method provides an exact solution when the $\mathbf{X}$'s are linearly independent. Otherwise, the process is still convergent, but may find only sub-optimal solutions. Nevertheless, a "good" solution is found in both the separable and the non-separable cases [Duda & Hart. 73].

A similar method can be applied to the Boltzmann Machine [Hinton & al. 84] if we take:

$$C(t) = \mathbf{X}(t)^T.\mathbf{W}(t).\mathbf{X}(t)\rangle - \langle\mathbf{Y}(t)^T.\mathbf{W}(t).\mathbf{Y}(t) \tag{8}$$

where $\mathbf{X}$ is the network state when input cells only are clamped and $\mathbf{Y}$ is the network state when both input and output cells are clamped. (2) now gives:

$$W_{ij}(t) = W_{ij}(t-1) + K_i(t)[Y_i(t)Y_j(t) - X_i(t)X_j(t)] \tag{9}$$

This is a deterministic version of the Boltzmann Machine learning rule. This algorithm solves the "credit assignment problem", because the **Y**'s have a dynamics which is not fully specified by the external environment.


## 3- THE HIERARCHICAL LEARNING MACHINE

In the learning rules described above, the **Y**'s played the role of desired states. In classical neural network models, without hidden units, the desired states are fully specified externally, while in the Boltzmann Machine and the HLM, they are partly computed by the network itself. Solving the CAP requires being able to change the output of each hidden unit so as to satisfy the global criterion. This can be done by computing each local criterion -attached to hidden units- while remaining consistent with all other local criteria.

In the following, we consider "smooth" neural networks, with real valued cell outputs defined by $X_i(t+1) = f[\sum_j W_{ij}(t)X_j(t)]$, where f is an odd function, with strictly positive first derivative. Let us denote $A_i(t)=\sum_j W_{ij}(t)X_j(t)$ and C a global criterion, depending on **A**, and defined by:

$$C(t) = \sum_{j=1}^{N} C_j(t) \tag{10}$$

where $C_j$ is the criterion attached to the j-th cell and depends on $A_j$.

Assume that cell k is a hidden unit and let $C_k$ be its local criterion. An optimal definition for $C_k$ can be chosen such that minimizing $C_k$ with respect to the $W_{k\ell}$ ($\ell=1,...,N$) amounts to minimizing $\sum_{j\neq k} C_j$ (assuming that all $C_j$ , $j\neq k$ are known). This condition can be expressed by:

$$\frac{\partial C_k}{\partial W_{k\ell}} = \sum_{j\neq k} \frac{\partial C_j}{\partial W_{k\ell}} \qquad \ell=1,...,N \tag{11}$$

It is possible to use a weaker constraint:

$$SIGN\left[\frac{\partial C_k}{\partial W_{k\ell}}\right] = SIGN\left[\sum_{j\neq k} \frac{\partial C_j}{\partial W_{k\ell}}\right] \qquad \ell=1,...,N \tag{12}$$

where SIGN is the threshold function . This condition ensures that $C_k$ and $\sum_{j\neq k}C_j$ vary in the same direction. We can compute each term:

$$\frac{\partial C_j}{\partial W_{k\ell}} = \frac{\partial C_j}{\partial A_j}.\frac{\partial A_j}{\partial X_k}.\frac{\partial X_k}{\partial A_k}.\frac{\partial A_k}{\partial W_{k\ell}} = \frac{\partial C_j}{\partial A_j} W_{jk} \frac{\partial A_k}{\partial W_{k\ell}}.\frac{\partial X_k}{\partial A_k} \tag{13}$$

$$\frac{\partial C_k}{\partial W_{k\ell}} = \frac{\partial C_k}{\partial A_k}.\frac{\partial A_k}{\partial W_{k\ell}} \tag{14}$$

Condition (12) implies:

$$\text{SIGN} \left[ \frac{\partial C_k}{\partial A_k} \right] = \text{SIGN} \left[ \sum_{j \neq k} W_{jk} \frac{\partial C_j}{\partial A_j} \frac{\partial X_k}{\partial A_k} \right] \qquad (15)$$

The last term in the bracketed sum is always positive, hence:

$$\text{SIGN} \left[ \frac{\partial C_k}{\partial A_k} \right] = \text{SIGN} \left[ \sum_{j \neq k} W_{jk} \frac{\partial C_j}{\partial A_j} \right] \qquad (16)$$

If we take for C its simplest form: $C_i = - Y_i A_i$ (17)

we obtain: $\text{SIGN}[ Y_k ] = \text{SIGN}[ \sum_{j \neq k} W_{jk} Y_j ]$ (18)

We can choose any absolute value for $Y_k$, in particular:

$$Y_k = \text{SIGN}[ \sum_{j \neq k} W_{jk} Y_j ] \qquad (19)$$

Using vector notation, (19) becomes: $\mathbf{Y} = F(\mathbf{W}^T \mathbf{Y})$ (20)
and we can compute a "desired response" for each hidden cell by back propagating this cost function gradient [Hinton 85].

Minimizing the error between $\mathbf{Y}$ and $\mathbf{X}$ can be done with the Widrow Hoff rule (7). We obtain a system of recursive equations describing the complete behavior of HLM:

$\mathbf{X}(t+1) = F[ \mathbf{W}(t) . \mathbf{X}(t) ]$

$\mathbf{Y}(t+1) = F[ \mathbf{W}(t)^T . \mathbf{Y}(t) ]$

$W_{ij}(t+1) = W_{ij}(t) + K_i(t)[Y_i(t+1) - A_i(t+1)]X_j(t)$ (21)

$A_i(t+1) = \sum_k W_{ik}(t)X_k(t)$

assuming that the X's and Y's of some cells can be clamped externally.

## 4- SIMULATIONS

The learning rule we obtain is totally local in space and time. The global criterion $\langle C \rangle$ is convex provided there is no hidden cell. This implies that several solutions to a particular task can co-exist, and that the weights can be trapped into a local minimum of $\langle C \rangle$. This means too that the weight configuration found out by learning depends on the initial weight configuration. Moreover, it depends on the whole network history: what has been learned before and how.

Two parameters have a great influence on the weight dynamics. The first one is the matrix $\mathbf{K}$ which defines the step size towards the criterion minimum at each iteration. A trade-off must be found between the speed of convergence (proportional to the $K_i$'s) and the accuracy and stability of the stable point (achieved with small $K_i$'s). It must be noticed that large values for $K_i$'s facilitate escaping from local minima because large energy barriers can be crossed over. The second parameter is the nature of the learning pattern sequence. By definition, the $\langle C \rangle$ landscape is statistically defined and can be completely modified by changing the relative occurrence frequency of each pattern. Increasing the occurrence frequency of poorly learned patterns increases the energy of the nearest local minimum, and therefore, lowers the heights of the surrounding barriers. This can define what a "good pedagogy" must be, and can be interpreted as a "smart noise".

A particular structure has been chosen for testing the algorithm. It has some restrictions to facilitate the study of the weight dynamics. First, the cells are threshold automata (i.e. function f is approximated by a threshold function); second, all the interactions are local in a 3-D space; third, there is no loop in the interaction graph, which means that the matrix $W$ is isomorphic to a triangular matrix. The network has a hierarchical architecture with several layers. Each layer is composed of 64 cells (8 by 8 plane) with a toric topology to avoid boundary effects. The first layer is the input plane (retina). Each cell in a layer receives signals only from 9, 25, or 49 cells in the previous layer. There is no interaction between cells in a given layer. The last layer (output plane) has a variable number of cells which "see" the entire previous layer (see fig 1). This structure allows hierarchical information processing, the abstraction level of the representation increases as the information is processed by the successive layers. The locally connected loop-free structure causes the $X$ and $Y$ dynamics to have trivial fixed points, and makes the study of the weight dynamics easier. It must be noticed that in such a struture, since there is no direct path between input and output cells, the information *has* to be processed by the hidden units.

A learning iteration is composed of three phases:
- Present a pattern in the input plane (clamp the $X_i$'s of the first layer cells) and compute the stable $X$ state.
- Present the associated desired response (clamp the $Y_i$'s of the output cells) and compute the stable $Y$ state ( using back-propagation).
- Modify the weights using Widrow-Hoff rule with $X$ and $Y$.

For the simulations, the matrix $K$ is chosen as $K = k.I$ , where $I$ is the unit matrix. k is modified manually during the learning phase.

The most important thing we have to test by simulation is whether the network is able to generalize or not. The generalization is the ability to produce a correct response for a non learned input pattern. Without generalization, learning is only memorizing. We have chosen for that test a set of alphabetic characters presented in several positions on the retina (See fig 2). There are 5 instances of the first 6 characters of the alphabet, each one presented in 4 different positions (total of 120 patterns). There are 6 classes, each of them associated with the activation of one of the 6 output cells. A correct classification (with 100% recognition rate) is obtained with a 6 layer network and connectivity 25. This task is somewhat complex because very different patterns (in the Hamming sense) must be put into the same class, while some close patterns must be separated.

Fig. 3 shows the classification achieved by a 6 layer network with connectivity 49 when 5 to 7% of the pixels have been randomly inverted during learning and recognition phases.

Fig. 4 shows classification made by the previous network on a set of non-learned input patterns. With this kind of data, the observer has a "semantic criterion" (the only one possible) to test the quality of the generalization. Of course, Hamming distance could be used to measure the spatial basins of attraction, but this would have little semantic relevance in this context.

In these examples, the generated boolean function is not linearly separable and requires using hidden units. The learning phase is somewhat long (a few thousands iterations), and needs a sophisticated pedagogy. All the learning set must not be presented at once (new patterns may be added when previous ones are learned). The value for k must be carefully chosen, decreased when learning is successfull, increased when trapped in a local minimum. The patterns must be presented in a random order.

Note that by presenting a pattern until it is recognized instead of changing it at each iteration, the learning time is reduced by a significant amount.

Other simulations have been performed in a more autonomous learning mode that modelizes Pavlovian conditioning [Le Cun 85]. In this last experiment, the network response is taken as the desired output, i.e. the desired response is self-generated.

## 5 - CONCLUSION

Numerous simulations must be made to evaluate the algorithm performances, for instance by using a structure with loops, and real valued outputs.

Nevertheless, this demonstrates the possibility of learning without specifying the states of all neurons.

The behaviour of HLM shows strong analogies with animal learning, especially when considering the effects of a pedagogy on the results.

## REFERENCES

[AMARI S.I.] : "Learning patterns and patterns sequences by self-organizing net of threshold elements". IEEE Trans. Com. Vol C-21, No 11, Nov 72.

[DUDA R.O., HART P.E.] : "Pattern classification and scene analysis". Wiley 73.

[HOPFIELD J.J.] : "Neural networks and physical systems with emergent collective computational abilities". P. Nat. Ac. Sci. USA, Nov 82.

[HINTON G., SEJNOWSKI T., ACKLEY D.] : "Boltzmann Machines, constraint satisfaction networks that learn". CMU Tech. Rep. CS-84-119, May 84.

[HINTON G.] , Private communication 1985.

[KOHONEN T., RUOHONEN M.] : "Representation of associated data by matrix operators". IEEE Trans. Computers, July 1973.

[KOHONEN T.] : "An adaptive associative memory principle". IEEE T. Comp, Apr 74

[KOHONEN T.] : "Self-organization and associative memories". Springer 1984.

[LE CUN Y.] : "A learning scheme for asymmetric threshold network". Proc. of COGNITIVA 85, Paris, June 1985 (in french).

[MINSKY M., PAPERT S.] : "Perceptron". M.I.T. Press, 1968.

[NAKANO K.] : "Associatron, a model of associative memory". IEEE Trans Syst. Man Cyb., Vol SMC -2, No 3, July 1972.

[WIDROW B., HOFF H.E.] : "Adaptive switching circuits". 1960 IRE WESCON Conv. Record, Part 4, 96-104, Aug 1960.

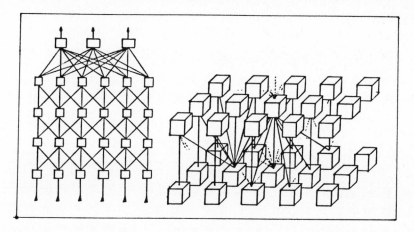

FIG. 1 : The network structure chosen for the simulations is Three-dimensional and composed of several 8 by 8 layers. Each cell in a layer is connected to 9, 25 or 49 cells in the previous layer. The last layer is fully connected to the previous one.

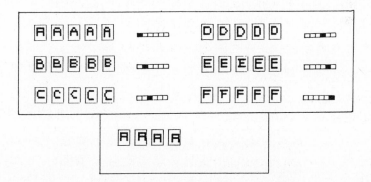

FIG. 2 : Five examples of the first six alphabetic characters are presented in four different positions. Each class is coded by the activation of one of the output cells.

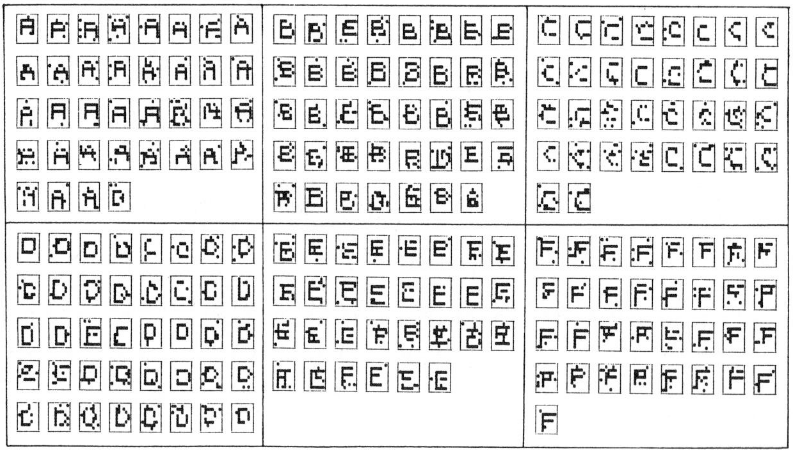

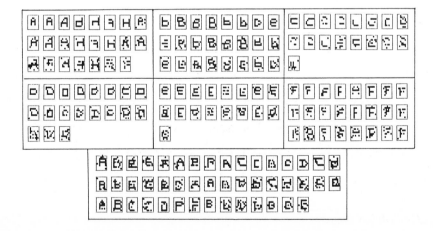

FIG. 3 : The figure shows the classification produced by a six layer network with 7 by 7 receptive fields. 5 to 7% of the pixels were randomly inverted during learning and recognition.

FIG. 4: A generalization test set. Classification produced by the same network as in fig. 3 on distorted noisy patterns. The patterns for which none of the six desired responses has been produced are put together at the bottom of the figure.

# LAYERED NETWORKS FOR UNSUPERVISED LEARNING

D. d'Humières

*Groupe de Physique de Solides de l'Ecole Normale Supérieure*

*24 rue Lhomond, 75231 Paris Cedex 05, FRANCE*

## 1. INTRODUCTION

Among several models of neural networks[1-4], layered structures are particularly appealing as they lead naturally to a hierarchical representation of the input sets, along with a reduced connectivity between individual cells. In Ref. 3 and 4, it was shown that such layered networks are able to memorize complicated input patterns, such as alphabetic characters, during unsupervised learning. On top of that, the filtering properties of the network can be continuously tuned from very sharp discrimination between similar patterns, to broad class aggregation when the selectivity of the cells is decreased. Unfortunately, it was also shown[4] that these properties are obtained with a reduced stability of the learning (the learning process does not converge for some values of the selectivity).

Through the one dimensional case, we present here the most naive rules which underlie the whole family of layered networks. Then we show the limitations of these rules and how they were improved in order to give the learning abilities described in Refs. 3 and 4. As a conclusion, we suggest some modifications of the connection scheme which may avoid most of the complications introduced in Refs 3 and 4, while keeping (if not improving) the learning and filtering properties of the network.

## 2. ELEMENTARY RULES.

### a) Propagation.

For simplicity, in what follows, the study is limited to one dimensional layers, each layer is made of N cells with 2p+1 identical outputs and 2p+1 inputs

NATO ASI Series, Vol. F20
Disordered Systems and Biological Organization
Edited by E. Bienenstock et al.
© Springer-Verlag Berlin Heidelberg 1986

taken from the outputs of 2p+1 cells of the preceding layer. Thus, each layer l can be caracterized by a state vector $S_l = \{s_{i,l}, i=1 \dots N\}$ of length N. The vector $S_1$ of the first layer is given by the external world and, for $l>1$, the vector $S_l$ is given by the outputs of the cells of the preceding layer l–1. The cell (i,l) is connected to the cells (i+j,l–1) with j=–p to p; let $I_{i,l} = \{ s_{i+j,l-1}, j=-p \dots p \}$ be its input vector and $T_{i,l}$ a template vector attached to this cell.

Let $q_{V,W} = V.W/(\|V\|.\|W\|)$ be the overlap between vectors $V$ and $W$, and let the outputs of the cells be related to their input and template vectors by:

$$O_{i,l} = \Phi(Q_0 q_{I_{i,l}, T_{i,l}} - 1) \qquad (1),$$

with $\qquad \Phi(x) = x \qquad$ if $x > 0$

and $\qquad \Phi(x) = 0 \qquad$ if $x \leqslant 0$,

where $Q_0$ is a given constant greater than one, which defines the selectivity of the network. Since $q_{V,W} \leqslant 1$ and $q_{V,W} = 1$ if and only if $V = \lambda W$, the output of each cell reflects how far from one is the overlap between their input and template vectors (with a cut-off below $1/Q_0$), as shown on the following figure.

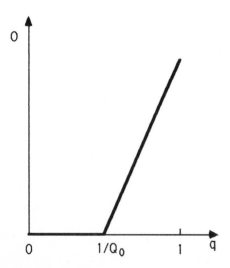

*b) Learning.*

During the learning phase, the network has to build the template vectors of every cell. To do that, each cell within a layer compares its output to the ones of the m cells on its left and the m cells on its right and if its output is a local maximum, its template vector is updated with the following rule:

$$T_{i,l}(t+1)=T_{i,l}(t)+\alpha.I_{i,l}(t) \qquad\qquad (2),$$

where $\alpha$ is a given constant which determines the growth of the template vector.


## 3. REAL IMPLEMENTATION.

In fact, the very simple model described above would have a very low memory capacity. For example, an input pattern made of a single one presented at the first layer propagates to every connected cell and thus, is used as template for $2p+1$ cells at the second layer (all the cells connected to the non-zero cell of the preceding layer) and $2pl+1$ cells at the $l^{th}$ layer. This prevents correct learning of another pattern which will be merged with the preceding one if $Q_0$ is large or ignored if $Q_0$ is close to one.

To prevent this proliferation of spurious templates, the learning rule (2) is replaced by:

$$T_{i,l}(t+1)=T_{i,l}(t)+a.W.I_{i,l}(t),$$

where $W$ is a given weight vector $W=\{w_i, i=-p \ldots p\}$, with a bell shape around $i=0$. With these weights, if a pattern made of a single one is presented at the input of the $i^{th}$ cell of a layer, then the $i^{th}$ cell of the next layer has a locally maximal output and only its template is updated.

In addition, another modification is added to handle the case where the input pattern is made of a long sequence of ones. In this case a large number of cells in the next layer uses a template pattern made only of ones. This effect also reduces the memory capacity and is fixed by splitting each layer in two parts. A first sub-layer has the same adaptive structure as above, but feeds into a second one with given edge detection properties. It is the outputs of this second sub-layer which determine by local comparisons which cells must be updated within the first one, and which feeds into the next first sub-layer. With these new rules, only non-trivial templates are stored, increasing significantly the memory capacity of the network.

# 4. CONCLUSION.

We have shown how the naive algorithm presented in §2 has to be modified in order to get good memory capacities, as it was done in the actual implementation of Ref. 3 and 4. This result was obtained by significantly complicating the very simple rules of §2. These complications hide the basic properties of the network and make very difficult to know what part of the algorithm or what parameters must be changed to improve the network or to better understand the involved phenomena. An alternative method would be to go back to the simple rules of §2, but to modify the connection scheme.

In our opinion, these complications arise from a too large overlap of the input patterns seen by adjacent cells, which differs only by two values (for the one dimension case, 8p+2 for two dimensions). Since each cell can store only one template pattern, when a cell has got a non trivial one, a new template has to be learned by one of the adjacent cells, even if for that relevant informations must be discarded.

A natural way to solve this problem is to modify the connection scheme between layers. For example, each layer can be divided in blocks of length p and the outputs of n blocks (n>1) of a layer can be used as inputs for a block of the next layer. Thus, each block can store up to p templates (one for each cell) which reflect the possible relevant input patterns for the block. The learning rule has also to be modified according this new connection scheme. For example, one can introduce the concept of free cells (a template pattern filled with zeroes). At the beginning all cells are free; for the first input pattern one cell is chosen at random within the block to store the corresponding template pattern. Then the cell producing the maximal non-zero output within the block is updated according to Eqs 2. . But, if all the cells within the block produce a zero output and if there are free cells, one of them updates its template vector. Otherwise, the input vector is merged with the closest template vector or the two closest template vectors are merged to free a cell for the new input vector, depending upon which of the overlaps between the input vector and the template vectors of the block and between these template vectors themselves, is the greatest. In addition, it will be possible at this level to tune the individual selectivity of each cell (for example the selectivity could be decreased, i.e. $Q_0$ increased, when two templates have to be merged).

Of course, these ideas are presently only tentative guesswork and have to be tested by numerical simulations and their validity checked against the results obtained with the previous model.

REFERENCES

1.    Kohonen T. (1978). "Associative Memory" (Springer, New York).

2.    Hopfield J.J. (1982). "Neural Networks with emergent collective computational abilities", Proc. Nat. Ac. Sc. USA, **79**, 2554-2558.

3.    Fukushima K. (1975). "Self-Organizing Multilayered Neural Network", Systems Computers Controls **6,** 15-22.

4.    d'Humières D. and Huberman B.A. (1984) "Dynamics of Self-Organization in Complex Adaptive Networks", J. Stat. Phys. **34**, 361-379 and (1985) in "Dynamical Systems and Cellular Automata" edited by Demongeot J. , Golès E. and Tchuente M. , Academic Press, 187-195.

# Statistical Coding and Short-Term Synaptic Plasticity:
# A Scheme for Knowledge Representation in the Brain

Christoph von der Malsburg
Abteilung für Neurobiologie, Max-Planck-Institut für Biophysikalische Chemie
D-3400 Göttingen W.Germany
and
Elie Bienenstock
Laboratoire de Neurobiologie du Développement
Bâtiment 440, Université de Paris-Sud
F-91405 Orsay Cedex France

**Abstract.** This work is a theoretical investigation of some consequences of the hypothesis that transmission efficacies of synapses in the Central Nervous System (CNS) undergo modification on a short time-scale. Short-term synaptic plasticity appears to be an almost necessary condition for the existence of activity states in the CNS which are stable for about 1 *sec.*, the time-scale of psychological processes. It gives rise to joint "activity-and-connectivity" dynamics. This dynamics selects and stabilizes particular high-order statistical relationships in the timing of neuronal firing; at the same time, it selects and stabilizes particular connectivity patterns. In analogy to statistical mechanics, these stable states, the attractors of the dynamics, can be viewed as the minima of a hamiltonian, or cost function. It is found that these low-cost states, termed synaptic patterns, are topologically organized. Two important properties of synaptic patterns are demonstrated: (i) synaptic patterns can be "memorized" and later "retrieved", and (ii) synaptic patterns have a tendency to assemble into compound patterns according to simple topological rules. A model of position-invariant and size-invariant pattern recognition based on these two properties is briefly described. It is suggested that the scheme of a synaptic pattern may be more adapted than the classical cell-assembly notion for explaining cognitive abilities such as generalization and categorization, which pertain to the notion of invariance.

We are still in almost complete ignorance of how the brain works: there are some very good cues that the Central Nervous System (CNS) is a probabilistic distributed highly non-linear dynamical system—or asynchronous network of "automata"—but we have practically no conceptual tools for studying such knowledge representation devices. The remarkable performances of our cognitive apparatus have received little convincing explanation on the basis of neuronal functioning. One of the most intriguing aspects of cognition is perhaps best described by the notion of an "invariant". Many of the outstanding abilities of the brain, such as categorization and generalization, are directly related to it and could probably be better understood if we had a solid theory of how the brain generates and manipulates invariants. This would require as a first step clarifying the notion of invariance itself. The studies of perception by the Gestalt psychologists at the turn of the century have popularized the idea that

NATO ASI Series, Vol. F20
Disordered Systems and Biological Organization
Edited by E. Bienenstock et al.
© Springer-Verlag Berlin Heidelberg 1986

invariances result from construing objects as structured sets of relationships between their parts, rather than as mere sums or juxtapositions. Modern work in artificial intelligence indeed concentrates on the elaboration of flexible schemes for representing and handling relational information of various types. It should be expected that an understanding of how the brain represents such information using neurons and synapses would constitute a major breakthrough on both the theoretical and the experimental levels.

## 1.The Assembly as Dynamical Object of the Brain

According to the currently dominating view the relevant signal for the function of the brain is contained in the first-order statistics of the electrical activity of neurons: neural signals are independent Poisson processes the rates of which encode signals which can be extracted as a short-term mean. The time-interval over which the mean is to be taken may depend on the location and function of the neuron. This hypothesis of "rate-coding" is strongly rooted in the field of neurophysiology, where it is an everyday experience that the response elicited in a single cortical neuron by peripheral sensory stimulation can hardly if ever be exactly reproduced. There remains an "irreducible" variability in the discharge of the neuron, attributed mainly to fluctuations in the amount of transmitter released in the synaptic cleft upon arrival of a spike in the axon terminal. This variability is a nuisance in most neurophysiological experiments; one eliminates it by averaging over many stimulation trials (technique of "post- or peri- stimulus-time-histogram"). It is also regarded as a nuisance to brain function by most authors. Since the brain cannot afford the time necessary to average over many trials, the hypothesis of rate coding is usually complemented by the notion of using redundancy, i.e., averaging over many neurons carrying out the same function simultaneously[1]. Ensemble averaging, instead of temporal averaging, is thus thought to be the means by which the brain performs "reliable computation in the presence of noise"[2].

The monograph in which Hebb[3] introduced in 1949 the notion of the "cell-assembly" and emphasized its associative-memory properties has inspired much of later brain theory. Roughly, a cell-assembly is a subset of neurons with many mutual excitatory inter-connections. For most authors, including Hebb himself, the functioning of cell-assemblies is to be understood within a strict rate-coding framework. Once a state is initiated in which the mean rate of firing of most or all cells belonging to a given assembly is higher than average, it tends to persist for some time because of the strong excitatory connections between cells within the assembly. Thus, many models use the notion of a preferred or persistent activity state[4,5], which is thought to be more stable than a random activity configuration. The brain would spend about half a second or a second—the "observed duration of a single content in perception"[3]—in one of these states before switching to another persistent state. One may refer to this characteristic range of intervals as to "the psychological time-scale". The internal structure of the space of attractors, or locally stable states, as well as their relationships to external objects, are determined by the connectivity of the system, i.e., the strengths, or efficacies, of the various synapses between the neurons. Connectivity is first globally laid down during early development, and later tuned during various stages of learning.

The description in terms of attractors emphasizes the aspects of brain dynamics linked to discrete events such as single "percepts". It provides a useful model of associative, or content-addressable, also called "error-correcting", memory. The attractors are the memories of the system. Suppose the system is started at a point in the activity space which is near one memory; the system will then with high likelihood settle to this particular memory[5]. One may then say that a content-addressed memory has been retrieved, that appropriate associations have been made, or that errors have been corrected. Alternatively, the information to be processed can be fed into the network by "clamping" part of the activity variables, letting all the others undergo the usual dynamics; the system may be used in this way to solve constraint-satisfaction tasks[6].

Against the notion of the assembly the argument may be raised that the brain seems not at all to be designed to support persistent configurations of firing rates. We know that neuronal integration is highly non-linear: several spikes in excitatory afferents to a neuron which impinge in close neighborhood both in time and in space (location of the nerve endings over the dendritic arborization) are much more effective in triggering an action potential in the postsynaptic cell than the same number of spikes distributed over a somewhat larger spatio-temporal domain. The neuron may consequently be regarded as a "coincidence-detector" in the domain of the millisecond[7]. This time-scale is set by membrane time-constants, by axonal conduction times and by synaptic delays, which are all in the millisecond range. On the other hand, mean firing rates in cerebral cortex are low, typically between 5 and 20 spikes per second, and the excitation level of cells is bound to fluctuate wildly. There is, therefore, a high probability that fluctuations in the actual spike rate would extinguish an assembly which should remain on or would trigger an assembly which should be off. This instability of the assembly is a consequence of the apparent gap between the psychological and the physiological time-scales, half a second and a few milliseconds.

Arguments criticizing the assembly as the fundamental "data format" of the brain have been raised elsewhere[8]. The main point is a lack of flexible representation of relational knowledge by the assembly. Together with the stability arguments above these considerations have lead us to an altogether different approach to brain dynamics. This new approach is based on higher-order statistics as being the essential signal in the brain. As was just mentioned, neurons are extremely sensitive to highly coincident signals, they are strongly activated by such "favorable events". But how can favorable events be reliably produced? An arbitrary network with as dense interconnections as the cortical one is likely to produce flat high-order statistics. Networks of special type are needed to produce favorable events.

It has been suggested that coincidences could be propagated in pathways of particular architecture, called "synfire chains"[7]. Neurons in each link of the chain converge— and also diverge—onto neurons in the next link. Whenever sufficiently many cells at one link become simultaneously active, this synchronized activity elicits synchronized activity in the next link, and so on. In such a network coincidences don't get lost since neuronal activity propagates like a coherent traveling wave. A stable state could be obtained in a synfire chain which would be closed on itself, somewhat reminiscent of the old "reverberating circuits" of Lorente de No[9]. Any cell could be part of several synfire chains, a particular one being selected by the precise set of other neurons with which the cell happens to fire in synchrony. Difficulties with the synfire chain concept arise because (i) long regular chains of this kind are unlikely to develop in the CNS; (ii) the firing of cells in such a chain would be periodic, yet neurons exhibit only little periodicity other than that linked to respiration or blood-flow; and most importantly because (iii) such a model supports only those stable activity modes which are rigidly determined by the structure of the network.

The approach taken here is of a slightly more radical kind: the particular network structure needed to support a particular signal structure is created by the signals themselves on the psychological time scale in a bootstrapping fashion. This is possible with the help of a synaptic mechanism for remembering, at any given synapse, events that occurred a short time ago at this synapse. Such a mechanism would stabilize good timing relationships and thus favor the re-occurrence of events which were successful from the viewpoint of transmission of excitation through the synapse. The ensuing network structures and signals constitute a data format which is very appropriate for typical operations performed in the brain, as will be argued in later sections of this paper.

## 2. Short-Term Plasticity

As was originally proposed elsewhere by one of the authors[8,10], we assume that within the CNS the following rules apply.

Rule A: successful synaptic events enhance the transmission efficacy of the synapse.

Rule B: transmission failures such as presynaptic firing without postsynaptic firing, and possibly also "failures" of the inverse type, i.e., postsynaptic without presynaptic firing, depress synaptic efficacy.

Both types of plastic change become effective within a few milliseconds.

These rules can be modified in an obvious way to include the case of inhibitory synapses. For instance, firing of an inhibitory afferent followed by postsynaptic firing is a transmission failure; such an event depresses the efficacy of the concerned synapse.

Short-term plasticity is restricted to a small range of strengths: the absolute value of the strength saturates after a small number of similar events. In case of successful transmission, it reaches a maximal level. In case of failures, it settles at a minimum.

Short-term plasticity lets connectivity and activity evolve on the same time-scale, leading to the notion of a joint activity-and-connectivity state. It is proposed that the events which underlie brain function are best described by such compound states, rather that by mere activity states. The dynamics of this new type is characterized by positive feedback: a successful event occurring at a given synapse increases the efficacy of this synapse, which in turn increases the likelihood of re-occurrence of the same or a similar event. Failures on the other hand increase the probability for the synapse to fail again. Positive feedback and saturation of synaptic weights ensure stability of the joint activity-connectivity state. We shall also see that positive feedback operates as a selection mechanism which favors and stabilizes a particular type of activity-connectivity configurations.

Short-term plasticity is related to hebbian synaptic modification: "When an axon of cell $A$ is near enough to excite a cell $B$ and repeatedly and persistently takes part in firing it, some growth process or metabolic change takes place in one or both cells so that $A$'s efficiency as one of the cells firing B, is increased"[3]. However, Hebb's rule requiring repeated and persistent coincidences, hebbian modification is generally understood as depending on the correlation between pre- and post-synaptic activities, estimated over a long period of time. It is a slow and long-lasting modification of the network's connectivity, a developmental or learning process which is invoked to explain the "growth of the assembly". In contrast, short-term synaptic modification is in force after a single synaptic event, or a small number of them. It is also reversible, for an unsuccessful event may occur shortly after a successful one and totally undo what the first event did. Long-term plasticity can now be more precisely formulated, as time-integrated short-term modification.

A randomly connected cortex-like network can be described as an entanglement of pathways of different length and delay, so that any fine temporal structure becomes washed out by the dynamics and flat high-order statistics is to be expected. Short-term plasticity as summarized in Rules A and B selects out activity-connectivity states which depart in a systematic way from both independence of activity fluctuations in different neurons and randomness of connectivity. An excitatory synapse can only be successful if there are other active synapses converging on the same target cell and receiving synchronous impulses, thus producing a strong postsynaptic effect; also, few or no activated inhibitory afferents should have fired in the last milliseconds. All synapses participating in a successful event are strengthened, thus making reoccurrence of the event more likely. Clearly then, successful events signify a departure from independent firing. They also signify departure from random connectivity: there must be something in the current connectivity state of the network which makes it likely that all fibres in a particular group fire in synchrony, and makes it unlikely that fibres belonging to another group fire at the same time. Stable activity-connectivity states differ from random states in that there is a good match between activity and connectivity, that is, between successful events and activated synapses, and between failures and deactivated synapses. The positive feedback dynamics achieves such optimal matching under the constraints set by the physical connections and by first-order statistics.

How can stable connectivity states be characterized in distinction to random ones? The following idea may be relevant. In an isolated network, coincidences can be generated by "amplifying" single spikes if there are many parallel alternative pathways of equal length between a pair of neurons. (The length of a multi-synaptic pathway between two neurons in the CNS is obtained by adding up all the axonal conduction times, synaptic delays and somato-dendritic integration times involved.) Whenever the common "source" $i$ emits a spike, this signal diverges along the currently active pathways out of neuron $i$, some of which converge again on the common "target" $j$, with accurate temporal synchrony. The existence of such parallel pathways between pairs of neurons increases the likelihood of coincident events in comparison to a random network. We contend that connectivity states with many such micro-configurations are the most favorable to the occurrence of coincidences. Notice that there should exist an optimal length for the alternative parallel pathways in the favorable micro-configurations. Very short paths don't contribute much because there simply cannot be many of them. Long paths on the other hand are not reliable and introduce scatter in the timing relationships. We leave open the issue of a quantitative estimation of the optimal length.

The number of favorable micro-configurations could be trivially maximized by activating all available excitatory links. This would lead to global synchronization in the network. Inhibition acts to block this route, by limiting the number of cells which are simultaneously active. The number of synapses required to connect any two neurons in cortex—the graph theoretic distance between them—is very small, possibly 5 on the average[11]. If all excitatory synapses were activated, single spikes would be amplified to enormous numbers within a few synaptic delays. Inhibition would counteract this amplification, thereby leading to transmission failure in many synapses and reducing the number of activated synapses. It may therefore be assumed that in a valid network configuration only a small number of synapses to or from each cell is activated. In consequence, stable connectivity states are to be characterized as being a compromise between maximal number of favorable micro-configurations and sparsity. We will later give a more global characterization in terms of topological structure.

From now on, we will use the term *synaptic pattern* for the stable states, or attractors, of the activity-and-connectivity dynamics. The word "pattern" is meant to suggest the existence of internal structure. This stands in contrast with the "assembly" where no internal structure exists apart from stronger-than-average excitatory connectivity. This distinction will become clearer in the next sections.

Synfire chains are a particular instance of synaptic pattern: they clearly contain many of the required micro-configurations, of length between 2 and the length of the chain. They form only a small subclass of the space of synaptic patterns, because their architecture is very regular. The generic synaptic pattern has no global order. It is an irregular high-dimensional mosaic of small multiple-parallel-alternative-paths. The image of a mosaic is, however, slightly misleading, for these preferred micro-configurations are not just laid down on the side of each other like stones in a mosaic. Rather, they are entangled in a complicated, essentially random way: each source-cell or target-cell in one such subgraph is an intermediate cell from the standpoint of others. There are two equivalent ways to look at the stable activity-and-connectivity states: using either connectivity or activity. One could try to characterize the organization in these states by high-order statistical moments, which would tell how the activity propagates in the network. In the case of a synfire chain, there is a well-defined topology which allows the activity to propagate like a traveling wave. In the generic synaptic pattern no such global order exists, yet we shall see that local topological structure is still there. This allows for a type of "propagation" which is much less coherent than in a synfire chain, yet with more coincidences than if all firing processes were statistically independent. In all other respects, the activity in a synaptic pattern looks totally random; it is perfectly aperiodic, has low first-order moments, i.e., average firing rates, and cannot be distinguished from random activity by looking at only one neuron at a time.

No experimental evidence exists at the time in support of short-term synaptic modification. It is in general quite hard to unambiguously demonstrate a change in the permeability of a synapse in the CNS of higher animals. There are several reasons to this, the chief one being the highly complex connectivity which makes it difficult to tell what it is that caused an observed change in the "response properties" of a cell. On the other hand, the hypothesis of short-term plasticity is consistent with current knowledge on molecular processes at the synapse. It has, for instance, been suggested that the dendritic spine undergoes fast geometrical changes[12], which could have significant effects on the strength of the transmitted impulse[13,14,15]. The receptor molecule in the post-synaptic membrane could also be subject to fast reversible conformation changes which would affect its sensitivity to the neurotransmitter; in some cases, such "allosteric" effects are well documented[16,17]. It remains to be proven that such changes may take place on a short time-scale, and obey Rule A and Rule B for enhancement and depression of synaptic efficacy. At any rate, it is a crucial requirement that modification depend on both pre- and post-synaptic activities. The latter is meant in a broad sense: it could be a local signal originating from neighboring synapses on the dendrite, which would not necessarily involve propagated activity, i.e., spiking, in the postsynaptic neuron. Also, it is not required that all synapses in the system be modifiable: short-term plasticity could be restricted to excitatory synapses or to a certain sub-category of these. Whether plasticity of inhibitory synapses would suffice deserves further investigation. The only absolute prerequisite for the theory proposed here is the existence of molecular mechanisms which result in local short-term positive feedback between the occurrence of "successful events" at a synapse and the transmission efficacy of that synapse.

## 3. A Simplified Formulation

It will not be possible for some time to give a precise and realistic formulation of brain dynamics. Most of the detail of interest is not accessible to experiment at present. Even if we had exact knowledge of all relevant neural mechanisms it would not form a convenient basis for mathematical treatment. Those mechanisms are the result of a long history of evolutionary optimization and are bound to be complicated. We shall base all further discussion on a simplified formulation of neural dynamics. This is done in the spirit of statistical physics, which abounds in radical simplifications. The Ising model of magnetism, for instance, is a caricature of real-world physical processes, yet it has proven to be very successful in the study of phase transitions. The elucidation of the fundamental notion of a "universality class" would have been impossible without simplifying steps which capture the essence of the interactions involved.

As a first simplification we restrict explicit treatment to excitatory synaptic weights and will consider neural signals and inhibitory synapses as "hidden variables", the effect of which on the excitatory weights being taken into account implicitly. We denote the *instantaneous efficacy* of the synapse between cell $j$ and cell $i$ by $w_{ij}$. We assume symmetry, $w_{ij} = w_{ji}$, and exclude self-coupling: $w_{ii} = 0$. We furthermore restrict synaptic weights to the values 1 or 0. Dendritic geometry, which probably restricts direct communication between synapses to local pools, is neglected, all synapses converging onto one cell being treated as equivalent. As a consequence of these simplifications, the current state of the system is entirely described by its connectivity which is simply a non-directed *graph*, with no kinks and no parallel edges. This graph has the fixed set of vertices $V = \{1, ..., n\}$, which corresponds to the set of neurons in the network. The set of links, $\{w_{ij} \mid i, j = 1, ..., n\}$, is the dynamic object.

For a connection to be modified within short time it must exist physically. Let us denote by $s_{ij} = 1$ the fact that cells $j$ and $i$ are connected by an axon and a synapse (neglecting, again, the unidirectionality of synapses). $s_{ij} = 0$ then stands for a non-existing physical connection. We correspondingly have the constraint $w_{ij} \leq s_{ij}$. In cortex, a typical neuron has between $10^3$ and $10^4$ synaptic contacts.

The effects of the "hidden variables" are described as rules of cooperation and competition between synaptic weights. The effect of inhibition is to produce competition between the links at each node and keep their number below a certain parameter $p$, where $p$ is much smaller than the number of synaptic contacts of neurons. Neural signals induce cooperation between sets of links which form constellations favorable for the production of synchronous events. As indicated above, we try to capture this effect by introducing cooperativity between alternative paths of equal length between two cells $i$ and $j$. Since we also ignore the orientation of edges, two such paths simply form a *cycle* (a closed circuit) of even length. We further assume that only cycles of a given optimal length $q$, or $q$-cycles, contribute ($q$ would be twice the length of the alternative pathways between a source-neuron and a target-neuron in an oriented-graph version).

To formalize the above dynamic rules we shall adopt the well-known Gibbs-Boltzmann formalism. There, a central role is played by a function of the state $H(w)$ called the energy, or Hamiltonian function. The function $H(w)$ is a sum of local contributions corresponding to all the interactions between the variables in the system. Intuitively, $H(w)$ is a measure of the extent to which these interactions are "satisfied" in the present global state $w$. It may therefore also be considered a "cost function" of an optimization problem. In our case, $H(w)$ is the negative sum of the number of $q$-cycles:

$$H(w) = - \sum_{(i_1, ..., i_q)} w_{i_1 i_2} \cdots w_{i_{q-1} i_q} w_{i_q i_1} \tag{1}$$

with the constraints

$$\sum_j w_{ij} \leq p \qquad i = 1, ..., n \tag{2}$$

$$w_{ij} \leq s_{ij} \qquad i, j = 1, ..., n \tag{3}$$

Constraint (2) limits the number of links at each node, whereas (3) makes sure the link between cells $i$ and $j$ cannot be activated if there is no pre-existing physical connection $s_{ij}$ between $i$ and $j$. The dynamics of the system is now formulated simply by requiring that the probability of finding the system in state $w$ obeys a Gibbs-Boltzmann distribution $P(w) = exp^{(-H(w)/T)}/Z$—where $Z$ is a normalization constant and $T$ is the "temperature" of the system—with the proviso that only states valid under the constraints (2) and (3) are admitted. (The constraints could also be implemented as additive terms in $H(w)$. In our simulations we preferred to impose them kinematically, as formulated here.) According to this formula, low-energy states are more likely than high-energy ones. The shape of the resulting probability landscape also depends in a global way on $T$: lowering $T$ sharpens the peaks and troughs, whereas increasing it flattens the landscape out; at high $T$, all states are nearly equally likely, whereas at low $T$, high-energy states have essentially zero probability. The effect in the present case is to give states with high numbers of $q$-cycles high probability, subject to the constraints.

An algorithm which is often used to study in computer simulations the equilibrium behaviour of such systems at a given finite $T$ is the Metropolis algorithm[18]. It consists of a sequence of tentative moves of the state $w$. The criterion for accepting or rejecting a move is probabilistic and depends on the change in $H(w)/T$: the larger the increase in $H(w)$, the less likely one is to accept that move. In *simulated annealing*, one starts the process with high $T$, thus avoiding trapping in local minima, and slowly reduces $T$ to find one of the absolute energy minima, or states whose energy-level is very close to the absolute minimum[19].

The result of the dynamics described are synaptic patterns in the form of sparse graphs with many $q$-cycles. The formulation given contains non-local interactions between links. This non-locality is the result of suppressing the signal variables. It should be stressed that $H$ and $T$ are by no means a physical energy and a physical temperature. The analogy with physical systems is simply a convenient tool to simulate the dynamics on the computer. In physical systems, $T$ measures the amount of thermal noise, relative to the strength of the local interactions that make up the function $H$. The role of noise in cortex can be played by a non-specific diffuse afferent system to the cortex which injects uncorrelated activity interfering with the mechanism of coincidence detection. The amount of injected noise could even be varied in an appropriate way to achieve an effect analogous to annealing in the brain.

The reader will already have noticed the analogy of the system presented here to some systems studied in statistical physics. There, the term frustration[20] is used to designate large systems of interacting "particles" with the following two properties: (i) there is a certain amount of conflict between local interactions at any point of the system, and (ii) there is quenched disorder, i.e., these local interactions are of random type. The prototypical frustrated system in solid-state physics is the spin glass, but many complex optimization problems such as the traveling salesman problem are also frustrated systems and can be studied within a similar statistical mechanics framework. Frustrated systems exhibit several interesting properties; one of them is their high degeneracy: there exist many almost-minimum energy states which, despite their being very different—they are quite distant from each other in configuration space—have nearly identical energy levels. As solutions to an optimization problem these many different states are virtually as good as the optimal solution(s). The problem of minimizing (1) under constraints (2) can be regarded as a frustrated problem.

## 4. Graph Structures

The low-energy states of (1-2) can be studied under various constraints $s$ in (3). Numerical studies use simulated annealing; an attempted move consists either in a single link activation if it is allowed by constraints (2) and (3), or in the deactivation of a single link. We first consider the "generic" patterns, i.e., the minima of $H$ under constraints (2), the constraints (3) being void: $s_{ij} = 1$ for all $i \neq j$, that is, the graph of permanent links is the complete graph on $V$. We shall see that several simple topological properties of graphs are related to being a low-$H$ state. The argument is intuitive, strongly supported by computer simulations, but by no means a strict mathematical proof. It nevertheless involves some mathematical notions about random graphs.

The topological characterization of patterns involves the comparison with random graphs in the thermodynamic limit, i.e., when $n$, the size of the graph, goes to infinity. Large random graphs contain very few cycles of any given order $k$, when $k$ is small with respect to $n$. Stated accurately[21], the distribution of the number of cycles of a given fixed order $k$ in a random graph obeys, in the thermodynamic limit, a Poisson law, the mean of which depends on $k$ and on the average number of edges per node but not on $n$. In contrast, in a regular lattice in a euclidean space, the number of $k$-cycles for a given $k$ grows linearly with $n$—provided the lattice contains $k$-cycles, e.g., $k$ is an even integer not less that 4 in a rectangular lattice with nearest-neighbor connections. Such a lattice, which contains many short cycles, is an instance of a low-$H$ state, for appropriate values of $q$. As was already mentioned, the graph-theoretic distance between two nodes in any graph is the length of the shortest path between them. In a random graph, the size of the set of nodes within distance $k$ from a given node grows rapidly: it is an exponential function of $k$—as long as $k$ is small with respect to the size of the graph. In contrast, in a regular lattice in a euclidean space, the size of a neighborhood of a given node increases only as a power function of the radius of this neighborhood. For any graph, the exponent in this function is called the fractal dimensionality of the graph. Fractal dimensionality of euclidean lattices is finite—it is identical to the dimension of the space—while it is infinite for random graphs.

It is argued that patterns are in many respects closer to euclidean lattices than to random graphs. Computer simulations show that the links in a pattern obey the following empirical rule: when several short indirect paths connect node $i$ to node $j$, it is likely that the direct connection between $i$ and $j$ exists as well. This is due to the cooperation of edges in $q$-cycles, which tends to make the connectivity "consistent with itself". The neighborhood relationships defined by such a connectivity are quite different from those found in a random graph; in contrast, there is a good match between these relationships and low-dimensional euclidean topology. Parts of a pattern may indeed be embedded in $\mathbf{R}^2$—for instance—in such a way that the graph-theoretic distance and the euclidean distance agree with each other (see ref. 22 for details on this embedding problem). The simulations support the conjecture that the fractal dimensionality of patterns is finite. The chief difference between a pattern and a euclidean lattice is that the former is quite irregular; as a result of this, the embedding in a low-dimensional euclidean space cannot be achieved for the pattern as a whole.

Note that patterns are not just *local* minima of $H$. Given the set of tentative moves used in the annealing, a local minimum of $H$ is a graph such that there are exactly $p$ edges out of each node and each edge belongs to at least one $q$-cycle. In general, this does not ensure low energy; one may achieve lower energy under constraints (2), each edge being part of several $q$-cycles as is the case in a rectangular lattice. Slow annealing is precisely required to avoid the local minima of high energy. Let it be mentioned for the sake of completeness that the *absolute* minima of $H$ are regular graphs of a very particular type. They are disjoint unions of graphs of small size, namely of the order of $p$ or $q$, which in general exhibit strong symmetries. For some values of $p$ and $q$, each of these small graphs is a clique, i.e., a completely connected subgraph. In practice, as soon as $n$ is much larger than $p$ and $q$, these absolute minima are too hard to find, even with very slow annealing. A typical pattern is a connected graph with the topological properties outlined above.

Given $n$, $p$ and $q$, we now define the set of $(n, p, q)$-*patterns* as the set of sub-optimal, i.e., almost best, solutions to the problem of minimizing (1) under constraint (2)—again not a strict mathematical definition. There are two important facts we need to know about $(n, p, q)$-patterns: (i) the edges in an $(n, p, q)$-pattern define a topological structure which is locally that of a low-dimensional euclidean space; and (ii) low-dimensional regular lattices with connections to the $p$-nearest-neighbors are instances of $(n, p, q)$-patterns, for appropriate $p$ and $q$. The notion of a pattern is deliberately ill-defined, yet each occurrence of the word "pattern" may from now on be replaced by the more restrictive "regular $p$-nearest-neighbor lattice" with no appreciable loss.

Remarkable collective properties of our dynamical system arise when constraints $s$ are imposed which already contain patterns. We shall study two such types of constraints. In the first one, the graph $s$ of permanent links is a *random mixture of several patterns*. We shall focus on the case of a mixture of two isomorphic patterns (see Fig. 1), which may be easily generalized. Let $s^1$ be an $(n, p, q)$-pattern. Let $P$ be a *random permutation* of the set of vertices $V = \{1, ..., n\}$, and let $s^2$ be defined as follows: for any $i$ and $j$ in $V$, $s_{ij}^2 = 1$ if and only if $s_{P(i)P(j)}^1 = 1$. Let now $s = s^1 \cup s^2$, i.e., for any $i$ and $j$ in $V$, $s_{ij} = 1$ if and only if $s_{ij}^1 = 1$ or $s_{ij}^2 = 1$. Since the permutation $P$ is random and $p \ll n$, the overlap between the graphs $s^1$ and $s^2$ is small, and there are exactly $2p$ links at almost each $i$ in the graphs $s$. Constraints (2) and (3) therefore imply that at each node $i$, $p$ links have to be selected out of the $2p$ available.

**Proposition 1.** Under the above conditions on $s$, the problem of minimizing $H(w)$ under constraints (2) and (3) has two optimal solutions, namely $w^1 = s^1$, and $w^2 = s^2$.

We do not attempt here to make a mathematical theorem of this proposition. The physical intuition behind Prop. 1 is the following. Since the permutation $P$ is random and $p$ and $q$ are small with respect to $n$, the frequency of "mixed" $q$-cycles using edges belonging to both $s^1$ *and* $s^2$ is vanishingly small. Hence, at each node $i$, $s^1$-edges *compete* with $s^2$-edges for activation. On the other hand, $s^1$-edges belonging to a common $q$-cycle *cooperate* with each other, and so do $s^2$-edges. Cooperation and competition propagate in the system in the following sense. Once a "local decision" is taken at node $i$ to activate $s^1$-edges rather than $s^2$-edges, it is more rewarding to stick to this decision and activate $s^1$-edges at nodes which are neighbors of $i$ in the topology

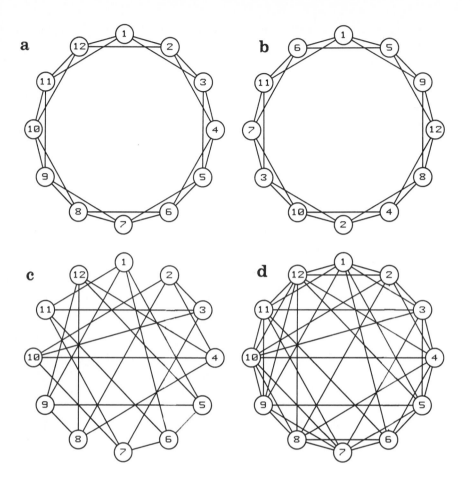

**Figure 1.** A schematic illustration of Prop. 1 in the 1-dimensional case. The patterns stored in the permanent memory (the $s$-variables) are circular graphs. Such a graph is drawn in a). It is obtained by arranging the $n$ vertices on the circle in any given order—in this case the natural order—and by connecting each node to its $p$ first neighbors. Another such graph is drawn in b). The difference between a) and b) is in the ordering of the nodes on the circle: the ordering in b) is a random permutation of the ordering in a). The two graphs are not identical: they are only isomorphic (the isomorphism being achieved by the permutation chosen). In c), the second graph is represented but in the original ordering. The topological structure is not apparent any more, but it still is there, since graphs b) and c) are identical. The superposition of a) and c) is shown in d). It is just the union of the two circular graphs: it contains two *patterns* which are isomorphic yet mixed in a random way. Therefore, only one of the two topological structures is apparent. Graph d) provides the constraint $s$ for Prop. 1. The dynamics is able to sort out the two stored patterns: in a stable state, one of the two sets of edges is selectively activated. In the drawing, the number of nodes $n$ is 12. In reality, $n$ has to be at least of the order of 50 for Prop. 1 to apply.

of $s^1$, since this allows to utilize each edge in a number of cycles as large as can be possibly achieved.

Prop. 1 is stated in the general case where the graph $s^1$ used to define the constraints $s$ is any $(n, p, q)$-pattern. One can make from Prop. 1 a mathematical statement by restricting the definition of $s^1$ to regular lattices. (The statement would have to be probabilistic, for a random permutation enters the definition of $s$, and would be made in the thermodynamic limit.) Prop. 1 can also be stated as follows: with an adequate choice of parameters, *the dynamics of local interactions results in global competition between randomly superimposed patterns.* It is crucial that the stored graphs $s$ be patterns. If the $s$ were random graphs, they could by no means be separated after they had been mixed: in that case, the optimization problem would be highly degenerate, all the "attractors" being random mixtures of the stored random graphs. Prop. 1 generalizes to the superposition of more than two patterns, and to the superposition of patterns which are not isomorphic to each other. It should be mentioned that as soon as more than one pattern is stored in a network, "spurious attractors", i.e., local minima appear. These "solutions" are essentially made of pieces of stored patterns glued together. Simulated annealing is necessary if one wants to avoid these local minima. If more and more patterns are stored in a network of fixed size $n$, "mixed" $q$-cycles become more and more numerous. In a first step, this forces one to use slower and slower annealing schedules. Eventually, when too many patterns are stored, the ability to retrieve them even with very slow annealing schedules breaks down altogether: some patterns which are mixtures of the stored patterns contain more $q$-cycles, hence are of lower energy, than the "pure" ones. Computer simulations have been performed using one-dimensional lattices: the stored patterns $s$ are isomorphic, randomly permuted replicas of the first-$p$-neighbors lattice on the circle. Typical values of the parameters are $n = 150$, $p = 8$, and $q = 4$. Such simulations clearly show the existence of the absolute minima of Prop. 1, and of many local minima.

We now turn to another type of constraint $s$, the last to be studied here. For the sake of clarity, we first state things informally: the system selects out graphs which form continuous maps between stored patterns whenever it is given a chance to do so. Formally, the constraints are the following (see Fig. 2). Let the set of vertices $V$ be divided into two disjoint subsets: $V = V^1 \cup V^2$, with $V^1 \cap V^2 = \emptyset$. $V^1$ and $V^2$ are termed "layers". Let the graph $s$ restricted to $V^1 \times V^1$ be a pattern and let the graph $s$ restricted to $V^2 \times V^2$ be an isomorphic replica of this pattern. Let the graph $s$ restricted to $V^1 \times V^2$ be complete, i.e., $s_{ij} = 1$ for all $(i, j)$ in $V^1 \times V^2$. Finally, let the dynamic graph $w$ restricted to $V^1 \times V^1$ be "clamped" and equal to $s$, and similarly for $V^2$. The dynamics concerns the inter-layer part of $w$, i.e., those links connecting $V^1$ to $V^2$. All existing $q$-cycles contribute to $H$.

**Proposition 2.** Under the above conditions on $s$, the activated links in a minimum $w$ of $H$ under constraints (2) and (3) define a bi-continuous map (both the direct and the inverse maps are continuous) between $V^1$ and $V^2$, where the topology on each of the two layers is induced from the graph $s$.

The intuition in Prop. 2 is the following. Suppose first that only one inter-layer edge is allowed at each $i$ in $V^1$, and similarly at each $j$ in $V^2$ (This is the case illustrated in

Fig. 2). Any graph $w$ with precisely one such edge per node defines a bijection (a map wich is one-to-one and onto) between $V^1$ and $V^2$. If the decision is taken to activate link $w_{ij}$, with $i$ in $V^1$ and $j$ in $V^2$, this decision propagates in both layers: the inter-layer link from a nearest-neighbor of $i$ in $V^1$ should connect it to a nearest-neighbor of $j$ in $V^2$, in order to activate as many "mixed" cycles as possible. Ultimately, the bijection between $V^1$ and $V^2$ in the optimal solution is an isomorphism of graphs. It is therefore an isometry, hence a bi-continuous map. In general, more than one inter-layer edge is allowed at each node: the resulting map is not bijective, yet the same argument shows that it connects neighbors to neighbors, i.e., it is bi-continuous. The "slope" of the projection depends on the parameter $p$, which may be given different values in the two layers. If that is the case, the mapping does not preserve the graph-theoretic distance, i.e., it is not an isometry. This fact is useful when applying Prop. 2 to size-invariant pattern recognition, as is done in the next section. (For some values of the parameters, the optimal map is made of many separate projections between small disconnected patches in the two layers, but we need not be concerned with this case.)

Prop. 2 can be generalized in several ways. In particular, there is no need for two disjoint layers $V^1$ and $V^2$: what is really required is that part of the edges be clamped and contain a pattern. Note that the resulting total graph on $V = V^1 \cup V^2$, which includes both the dynamical inter-layer edges and the clamped edges in each layer, is itself a pattern. Thus, Prop. 2 may be regarded as a generic way to obtain compound patterns out of simple ones. There is also an analogy between Prop. 2 and the development of retinotopic maps. The retina is a 2-dimensional nervous tissue which projects onto central structures in the brain. These structures too are essentially 2-dimensional and it is found that the projection in the mature system preserves the topology: neighboring retinal ganglion cells project to neighboring cells in the Lateral Genicualte Nucleus of the thalamus, and these in turn project to neighboring cells in primary visual cortex. Preserving neighborhood relationships of external objects in higher centers is certainly important for vision, but such bi-continuous maps are also found in many other parts of the brain: they are in general termed topographic projections. A central issue in neurobiology is the development of topographic maps during embryonic and early post-natal life. A large literature exists on this topic, both experimental and theoretical. The reader familiar with it will immediately recognize the similarities with Prop. 2. In both cases, the problem is that of constructing a continuous map between two pieces of nervous tissue. In the case of retinotopy the topology is that of the 2-dimensional euclidean space, and the time-scale is that of ontogenetic development; the map consists of synaptic contacts which are to stay there throughout the life of the organism. In contrast, Prop. 2 deals with "patterns", where the topology is not always as clearly defined. In particular, no strict dimensionality exists, yet if one were to introduce one, it would by no means be restricted to 2. The time-scale is very different, for such maps should be constructed in about $1 sec$. The analogy is still important because the underlying mechanisms are closely related to each other. Short-term modification is something like a very fast version of hebbian modification, which is almost certainly involved in the development of retinotopy[23,24]. This makes short-term plasticity plausible from the point of view of evolution: it was certainly easier to modify an already existing mechanism, than to "invent" short-term plasticity from scratch.

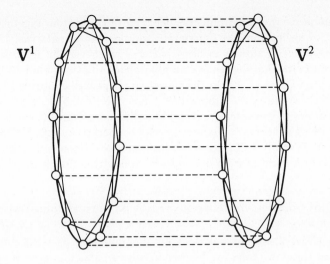

$\mathbf{V}^1$       $\mathbf{V}^2$

**Figure 2.** A schematic illustration of Prop. 2 in the 1-dimensional case. Solid lines are clamped edges: they define, in each of the two "layers", a circular graph as in Fig. 1. Only the inter-layer edges are dynamic, with the constraint that at most one such edge is allowed at each node. The dashed lines represent a stable state, that is, a minimum of $H$ under this constraint: this state evidently realizes an isomorphic projection between the two patterns; it achieves the highest possible number of 4-cycles.

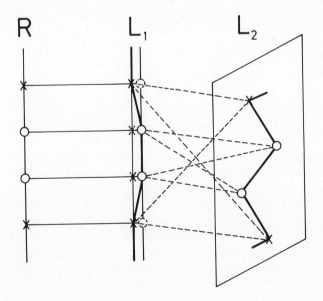

**Figure 3.** A model for size- and position-invariant pattern recognition. For explanation see text. Two types of local feature, or label, are represented by ∘ and ×. The figure illustrates that inter-layer connections are label-preserving.

One may summarize the constraint-free behavior of the system together with Propositions 1 and 2, by saying that the dynamics always generates patterns, and that in doing so it takes advantage of already existing patterns whenever this is possible. It is an interesting question to what degree the simplifying assumptions implicit in the definition of $H$ are essential for establishing the collective properties captured by Propositions 1 and 2. Behavior as stated in Proposition 1 has been demonstrated in a system with full joint dynamics of excitatory and inhibitory signals, transmitted by directed links, and fast-modifying synapses[25]. (A very special cooperative configuration has, however, been used in that study.) Earlier work has validated Proposition 2 in systems with signal-and-synaptic-modification dynamics[23], and with multiple parallel pathways[26].

## 5. A model of position-and-size-invariant pattern recognition

Propositions 1 and 2 are general rules for manipulating patterns. We now give an example of how these two rules can be used to solve a particular type of problem, namely invariant pattern recognition. For this purpose, a *model* of quite specific architecture is proposed. It is one possible solution among many others within the framework given here, and it is meant to illustrate the use of synaptic patterns as a knowledge representation scheme.

As in all models of perception, an essential element is the feature-detector cell. Each such cell is responsible for a specific "elementary" feature—or "primitive"—at a particular position of the sensory surface. Cells of this kind exist in living organisms, and a good deal of sensory physiology is devoted to studying their properties. For instance, some cortical neurons respond selectively to a visual stimulus in the form of an elongated bar or an intensity gradient in a well-defined location of the retina, with a well-defined orientation. We shall not specify here the type of feature-detectors to be employed: an adequate repertory has to be chosen in each particular application, depending on the stimuli one wishes to process and on the desired invariances. Let $r$ be the number of different types of local features. The model contains 3 levels of cells connected in a hierarchical way (see Fig. 3). The lowest level is the retina $R$ where patterns are actually displayed. The intermediate layer $L^1$ contains feature-detector cells arranged in a retinotopic way: each such cell is active if and only if the visual pattern currently displayed on $R$ contains the appropriate local feature at the appropriate location. The connectivity between $R$ and $L^1$ is responsible for these two properties: (i) being a "retinotopic" projection from $R$ to $L^1$, it endows $L^1$ with the 2-dimensional topology of $R$, and we shall assume that $L^1$ is a regular $m \times m$ array; (ii) by extracting local features from the patterns displayed on $R$, it endows each cell in $L^1$ with feature-detector properties. This connectivity is hard-wired and fixed, and not the original part of the model. There are $r$ feature-detectors at each of the $m^2$ positions in $L^1$, and we shall assume that exactly one of them is active at any given time. We also assume that the patterns we have to learn and recognize all have the same first-order statistics in $L^1$: the number of active features of each type is the same for all patterns.

The term pattern is used in the present section in its conventional meaning of a visual stimulus. There is no danger of confusion: patterns on $L^1$ are also patterns according to the graph-theoretic definition of the previous section. We assume that whenever two neighboring feature-detectors in $L^1$ are active, their activities are correlated on a short time-scale, whereas distant units are uncorrelated. Thus, correlation of activity encodes proximity in $L^1$. (This is in good agreement with experimental data on correlations between cells in the retina[27,28]. Proximity-dependent correlations also play a crucial role in the ontogenetic development of retinotopic projections, see above.) Again, the fine temporal scale on which correlations are calculated (a few milliseconds) is not explicit in the model; we imply correlations by assuming that each cell in $L^1$ is connected to all $L^1$-cells in the $p$ first neighboring sites. This results in $r \cdot p$ links per cell since there are $r$ cells at each site in $L^1$. This connectivity is a fixed one: $w_{ij} = s_{ij}$ within $L^1$. In summary, upon presentation of a stimulus, or visual pattern, on the retina, a subset of the cells in $L^1$ is activated, one of the $r$ cells in each location. The links between the activated cells form a 2-dimensional synaptic pattern. Since each of the $m^2$ vertices in $L^1$ is labeled by one of the $r$ features one may call the active object in $L^1$ a *labeled pattern*.

Layer $L^2$ is the third hierarchical level in the model, a non-retinotopic set of $m^2$ feature-detectors. Before any learning has taken place, there is no topological order in $L^2$, and there are no connections at all within $L^2$. Inter-layer connections, i.e., connections between $L^1$ and $L^2$, are only allowed between cells of the same feature-type (see Fig. 3). Save for this constraint, inter-layer permanent connectivity—the set of links denoted by $s$—is complete: one feature-detector of a given type in $L^1$ is connected to all feature-cells of the same type in $L^2$, and vice versa. Thus, any $L^2$-cell is *potentially* activated by an appropriate local feature at any position on $R$. We shall assume that the partition of the $m^2$ $L^2$-cells into the $r$ feature-types is such that there are as many cells of a given type as there are active cells of this type in $L^1$ when a pattern is displayed on the retina; this number was assumed above to be the same for all patterns. These two assumptions will eventually allow us to establish a one-to-one correspondence between cells in $L^2$ and local features in $L^1$. Being by no means essential, they are very convenient for the presentation of the model. In summary, $L^2$-cells are as feature-specific as $L^1$-cells, yet they are totally unselective regarding retinotopic location. In real brains, the transition from a topographic representation to a non-topographic one is done in progressive steps: again the model is a caricature of the CNS.

The dynamics of the model consists in learning and in recognition. Learning affects the "permanent" synaptic weights $s$ within $L^2$. $L^2$ is thus the memory-layer of the system, where patterns are stored. When a pattern is presented on $R$, a labeled pattern results in $L^1$. Storing this labeled pattern in the memory-layer $L^2$ requires the choice of a one-to-one map between $L^1$ and $L^2$. Let $f$ be an *arbitrary* bijection from the set of active feature-cells in $L^1$ onto layer $L^2$. The existence of such bijections is guaranteed by our assumptions. Storage is then done as follows: for any two *active* cells $i$ and $j$ which are neighbors in $L^1$—i.e., the site of $j$ is one of the $p$-first-neighbors of the site of $i$— the link $s_{f(i)f(j)}$ is set equal to 1. This is done for all pairs $i$ and $j$, including cells of different types. In summary, the $s$-graph that is stored in $L^2$ is a 2-D $m \times m$ regular lattice with $p$-nearest-neighbor connections; in addition it is *labeled pattern*-isomorphic to the current labeled pattern on $L^1$, that is, there exists a bijective map between the two which is an isomorphism with respect to the graph-structure and which preserves the labels.

Further learning is done by superimposing patterns on top of each other in the memory-layer $L^2$, that is, by taking the union of the $s$-graphs, as in Prop. 1 of the previous section. We assume for simplicity that the learning sequence consists of $M$ patterns which are totally unrelated to each other; it is then a reasonable policy to take a new arbitrary map $f$ for each new pattern stored. When the $M$ patterns have been stored in $L^2$, permanent connectivity within $L^2$ is a random mixture—in the sense of Prop. 1—of $M$ isomorphic replicas of one 2-D pattern, the array of sites of $L^1$. However, these $M$ graphs are not *labeled pattern*-isomorphic to each other: no mapping exists between two of them which would preserve both the graph-structure and the labels, i.e., the feature types.

The translation of our learning algorithm back to the language of signals and synapses—the activity-and-connectivity variables—is straightforward. We first have to find a map $f$. This is done by using inhibition to reduce the set of active inter-layer links at each node to the required size; in practice, this size may be much larger than 1. Storage in $L^2$ is then achieved through modification of the permanent synaptic weights as follows. Since cells $i$ and $j$ are neighbors in $L^1$, they have correlated activities. As a result of activating inter-layer links which form a *sparse projection* $f$ between $L^1$ and $L^2$, these correlations are carried over to $L^2$: cells $f(i)$ and $f(j)$ now have correlated activities. We then increase the permanent strength $s$ of excitatory synapses between cells with correlated activity. This synaptic modification rule is quite "classical", yet not strictly hebbian: rather than collecting statistics for a long time, the model performs "one-shot learning". The required modification of the $s$-weights need not be much slower than short-term modification of $w$-weights. Yet, whereas modification of $w$ is transitory, immediately reversible and part of current brain function, modification of $s$ is long-lasting. The latter is a process by which a trace of the current (activity-and-) connectivity state $w$ of the brain is stored in its permanent structure $s$. This requires a gating signal or "Now Print" command, for it would be disastrous to alter the brain's permanent structure at all times without discriminating whether the current state is a "favored" one from the standpoint of the fast joint activity-and-connectivity dynamics, and whether it is of biological importance. Chemical or electrical signals which control synaptic modification in a global way are studied in various parts of cortex[29].

The dynamics of recognition concerns $L^1 - L^2$ $w$-links as well as $w$-links within $L^2$—the $w$'s within $L^1$ are always clamped. One of the $M$ learned patterns is presented on $R$, which elicits a labeled pattern in $L^1$. This labeled pattern is generally quite different from the one that was experienced during learning, because the retinotopic position of the stimulus is different. Yet it is isomorphic to it, as a labeled pattern. (For this to be strictly true, "periodic" boundary conditions in $L^1$ should be used: $L^1$ is then a torus.) A minimum of $H(w)$ is then sought under the usual constraints, where only active cells in $L^1$ are taken into consideration. Constraints (2) are as follows: one inter-layer link is allowed at each active $L^1$-cell and at each $L^2$-cell, and $p$ $L^2$-links are allowed at each $L^2$-cell. Constraints (3) are determined in $L^1$ by the fixed $L^1$-connectivity and by the pattern currently presented on $R$—since this pattern determines the set of active cells on $L^1$—and in $L^2$ by the patterns previously stored there. Proposition 1 gives the solution to the problem of minimizing $H$ within the sole layer $L^2$, disregarding $L^1$ and inter-layer connections: since the $s$-graph in $L^2$ is a random mixture of isomorphic

$(m^2, p, q)$-patterns, there are $M$ minima, which are precisely the $M$ stored patterns. Labels are irrelevant to this fact. On the other hand we see from Proposition 2 that a minimum of $H(w)$ in the complete network includes an isomorphism between $L^1$ and $L^2$, if this is made possible by the patterns activated in $L^1$ and in $L^2$. This isomorphism of graphs is necessarily label-preserving too (because the $L^1 - L^2$-connections are). Such a map indeed exists, but only if the activated pattern in $L^2$ is the correct "memory", i.e., the graph which was stored when the current pattern was experienced for the first time. No label-preserving isomorphism exists with the other stored patterns. In summary, the global minimum of $H(w)$ consists of the right pattern in $L^2$ —the stored memory— together with the projection between $L^1$ and $L^2$ which achieves the isomorphism. Since this is true irrespective of the position of the stimulus on $R$, position-invariant pattern recognition has been achieved.

Note that neither the invariant representation nor the map can be found unless the other one is found at the same time. Like some other models[30], the present one has the nice feature that the invariant representation and the map between the two representations are found simultaneously: information flows both top-down and bottom-up. The chief advantage of our solution is that it does not require any "map-units", which specifically code for pieces of maps between the two layers. Such units have a type of specificity which makes them somewhat unlikely to exist in the CNS. In contrast, short-term synaptic modification is a simple local mechanism which ensures that isomorphisms of graphs—actually homeomorphisms—are "automatically" found whenever they exist.

Computer simulations of this model have been performed using the one-dimensional topology of the circle. Feature-extraction being a secondary issue, the model consisted only of the two layers $L^1$ and $L^2$. Features are then mere labels numbered from 1 to $r$, and their only significance is to restrict $L^1 - L^2$-links to cells with identical labels. The following set-up was used for instance. $L^1$ is a circular array of $m = 80$ sites, with connections to the 6 nearest neighbors; there are $r = 8$ label-cells per site; each pattern on $L^1$ consists in activating 16 *blocks* of 5 adjacent cells with the same label. There are two blocks of each label in a pattern, and a pattern is therefore equivalent to a permutation of these 16 blocks. These permutations were chosen at random to obtain different patterns. The block-structure of patterns captures the continuity of objects in the real world: when appropriately discretized, objects can be described as clusters of local features of the same type. Also, the block-structure allows one to contract and dilate the patterns and thereby to address the problem of size-invariant recognition (see below). Contributions to $H$ come from rectangles, i.e., $q = 4$. In this set-up, it was found that one could store and reliably retrieve up to 4 patterns. This storage capacity is small but it must be borne in mind that a pattern stored upon presentation in an arbitrary position can be recognized when presented in any of the 80 positions on the circle. The storage capacity of $L^2$ can be expected to grow with its size as rapidly as the number of statistically independent permutations of its cells.

Size-invariant recognition follows from Proposition 1 and 2 in the same way as position-invariance. In principle, one may infer form Prop. 2 that recognition in our model is invariant under the broader class of all homeomorphic —bijective and bicontinuous— transformations. In practice, there are several rather trivial constraints which restrict the invariance to a class of reasonable deformations. Computer simulations using

1-dimensional block-structured patterns with parameters as above or slightly different, showed that a dilatation or a contraction by a factor of 2 of the pattern used for learning does not affect its recognition. A thorough description of the model will appear elsewhere, including a more detailed account of size-invariance. An application of the model to the recognition of real-world 2-D images after feature-extraction is currently in progress.

The successful recognition of a visual stimulus has to be communicated to other parts of the brain. This is no difficulty if all objects in the brain have the generic form of patterns. Interactions between them may take place according to rules similar to Propositions 1 and 2.

## 6. Discussion

With the available experimental techniques it seems to be difficult or impossible to gain access to the working principle of the brain. Experiment therefore must be complemented by the constructive approach of reverse engineering, i.e., we should try to work backwards from the performance of the brain to its working principles. In contrast to the situation in physics, where simple paradigms can be constructed—and simple laws abstracted from them—the brain confronts us with extremely complicated structures, interposed between basic principles and observable functional properties. Any discussion of working principles must be complemented by hypothetical constructions, e.g., network structures, to establish a connection to the functional level. Such arguments can always be called in question, because 'the brain perhaps does it quite differently'.

The problem is made worse by the difficulties of functional verification of hypothetical structures. By necessity, all interesting and relevant structures are numerically complicated, non-linear and unstable. It is therefore far from trivial to derive the global behavior of a model the local structure of which is hypothetically given. It is very likely that progress with the brain problem is dictated by progress with the functional verification problem. In the past, this step has mostly been based on analogy or even on pure imagination, and much of the discussion has been amateurish in style. New methodological developments are of the utmost importance. The availability of the computer for simulation opened the way for a big step forward, although even computer simulations have their own specific limitations. For one thing the sequential computer still sets very narrow quantitative limits, into which models of the nervous system have to be squeezed with art. Another difficulty is to discriminate numerical artifacts from "real", i.e., robust, phenomena. The observed model behavior may be of measure zero in structure space.

The Boltzmann approach and simulated annealing must in this context be considered a major advance. The systematic introduction of "thermal" noise acts as a filter which passes only essential, generic, behavior of a given model. More importantly, a global probabilistic description of the system can be given directly in terms of local structure—formulated as additive terms of a Hamiltonian—thus providing a firm basis for intuitive reasoning. The method also opens access to the great wealth of experience physics has gathered with self-organizing thermodynamic systems. The method also seems to stress important properties of the nervous system which are not rendered correctly by deterministic models, although, however, it is still possible that fundamental discrepancies will be discovered between the nervous system and systems with Gibbs-Boltzmann statistics.

Computer simulations (and attempted technological applications) have repeatedly hit upon enormous difficulties with seemingly trivial problems with, for instance, low-level vision, figure-ground separation and invariant pattern representation. Many models would work in impoverished, noiseless and unambiguous environments but fail completely under realistic conditions. One of the reasons for failure is local ambiguity, which can only be reduced by imposing systems of mutual constraint, such that solutions have to be found iteratively. Another reason for model failure lies in the fact that realistic environments transcend any pre-established system of classification. Both types of problem are fundamental in nature and stress the necessity of a radical shift in the conceptual framework within which models and theories are constructed. The problems of *brain organization* and of the underlying *data structure* are to be evaluated anew.

Both of these problems are addressed in this paper. The data structure of the brain must form a natural basis for its important operations. One such operation has already been mentioned, the storage and iterative imposition of systems of constraints (this aspect is nicely presented and discussed in ref. 6). Another important operation is the flexible composition of complex objects from pre-existing parts. This is important, for instance, when composing a visually perceived scene from familiar patterns. If the fundamental data structure is the assembly—a set of simultaneously activated cells—then a composite object should be represented by a superassembly, the set-theoretic union of the constituent assemblies. This scheme—although it is the basis for the overwhelming majority of brain theories—has the fundamental flaw that the set-theoretic union cannot be undone: it is impossible to recover the superposed sets from the union. Suppose, for instance, you are imagining a scene involving your grandmother, who happens to frown, and your grandfather, who happens to smile. Suppose, furthermore, your grandmother, your grandfather, the frown and the smile are represented by four assemblies. Then the superassembly representing the whole scene—the union of all four assemblies—is ambiguous and could equally well stand for a situation in which your grandfather had the frown on his face! The cause for this difficulty is the complete lack of internal structure in the assembly (for a thorough discussion see ref. 8).

The absence of internal structure in the assembly makes it difficult or impossible to solve the invariance problem on this basis. Invariance requires that patterns with similar internal structure but on different sets of cells evoke the same response. Now, the optimum of structure in the connections between assemblies can be attained if cells are labeled by feature types, and connections are restricted to pairs of cells with the same feature type. Two assemblies then interact maximally if they are described by identical lists of features. This type of dynamical relationship is too unspecific and generalizes to too large a class of patterns: If pattern $a$ activates assembly $A$ (in a structure corresponding to our $L^1$) which in turn activates an assembly $Z$ (in a structure corresponding to our $L^2$) on the basis of identical composition of feature types, then another pattern $b$ which activates an assembly $B$ with a feature-composition identical to that of $A$ will also activate $Z$. Pattern $b$ could be a transformed (shifted) version with the same internal structure as $a$, but the problem is that $b$ could also be a totally different pattern which happens to be composed of the same types of local features in the same relative proportions. This over-generalization is due to a lack of internal structure in the assembly. In order to show that this problem is not a man of straw, let us cite an example. It has often been proposed to solve the invariance problem by representing visual patterns by the power-spectrum, the absolute value of the spatial Fourier-transform (for references see 31). Fourier components play the role of features, taking the absolute value installs position invariance by destroying information on relative phases. The original image cannot be recovered because very different images may have the same power-spectrum, which illustrates the point made here.

The assembly, hypothesized as universal data format of the brain, with its lack of internal structure, is to be contrasted with virtually all data formats explicitly employed by man. Written and spoken language use the system of neighborhoods inherent in sequential order; mathematical symbols (and in fact all visual images!) use the topology of visual space as their system of relations; and data structures employed by Artificial Intelligence use pointers and adjacency in address-space to establish relations among primitives. It is not to be expected that exactly brain itself should do without a system of relations.

Why should relations in the brain be encoded by temporal coincidences and by dynamic synapses? Aren't there other possibilities, especially ones that can be accommodated within the rate-coding framework? The usual approach represents sets of cells which together form an object by object-specific cells[32], often called cardinal cells or pontifical cells. This 'solution' leads to inflexibility and to enormous administrative problems, and has been discussed elsewhere[8]. A second possibility would be to let activity in neurons encode low-level (binary, ternary,..) relations in a way analogous to the representation by statistical moments discussed in this paper. This must be considered a serious possibility. However, relatively special wiring patterns would be required to implement the scheme. For instance, the interactions of two cells $a$ and $b$ with each other and with third parties have to be controlled by cell $c$ if the activity of $c$ encodes the presence of a binary relation between $a$ and $b$. This machinery is of a kind which cannot be expected to be ubiquitous to the nervous system. It rather would have to be created where needed. The organization of these special circuits is a problem in itself, see below.

Representation of relations within the brain by temporal coincidences of signals and by short-term modifying synapses has the great advantage that the necessary machinery is ubiquitous and that no specialized circuits are required. If two cells can interact at all with each other, or with third parties, the necessary synapses are there to be activated or inactivated, and the corresponding coincidence rates can be enhanced or suppressed. Coincidence rates in turn control nervous interactions in the desired way because neurons are coincidence detectors. Without requirement of any special structures the system forms a natural basis for organization and almost automatically leads to the emergence of connectivity-and-activity patterns which are of a very useful kind, as illustrated by the solution to the invariance problem described in section 5. This is due to the dominance of topological structure in the patterns.

Connectivity dynamics can, with the help of slow plasticity, develop circuit patterns which form a basis for rate-coding of relations as discussed two paragraphs ago. This is best illustrated with the structure of section 5. Each cell in $L^1$ has connections to a number of cells with the same feature specificity in $L^2$. During a particular stimulation, this multiple projection must be reduced, for each active cell in $L^1$, to a single $L^1 - L^2$ connection and a single cell in $L^2$, thus allowing the system a certain freedom of choice. Suppose a new pattern is activated in $L^1$, part of which is homeomorphic to part of a pattern which has already been stored in $L^2$. The system will use its freedom of choice for making the newly activated and stored pattern in $L^2$ part-identical with the old one. This can also be expressed by saying that the system recognizes familiar sub-patterns in new stimuli. There is thus a tendency for cells in $L^2$ to specialize to particular local connectivity structures, and activity in a cell of $L^2$ may be interpreted as encoding relationships between certain sets of other cells. One may, however, argue that even if the mature brain uses a system of rate-coding of relations, high-order statistical coding and short-term plasticity are essential for ontogenesis of the system and for the flexibility needed to handle new situations.

Regarding the organizational aspect mentioned above, the analogy with theories of evolution and of immunology, which has already been proposed in neurobiology[33,34], is very relevant. Selection is indeed a central notion in the present theory: the conflicting excitatory and inhibitory interactions result in the *selection of sparse connectivity graphs w, which are subgraphs of the permanent graph s*. We saw that the states which win in this selection process are those which yield the best possible match between successful synaptic events and activated synapses. In the simplified description which uses only connectivity variables, the favored states are characterized by being as consistent as possible with themselves; this self-consistency can be expressed mathematically by topological properties of graphs. $H(w)$ is an—extremely simplified—intrinsic measure of the self-consistency of the connectivity state $w$, which can be thought of as playing the role of the fitness coefficient in a darwinian evolution. Additional constraints specify this general criterion to a form which is adapted to the problem at hand. For instance, in the situation discussed in section 5, the $M$ patterns stored in $L^2$ stand in mutual competition, and the presentation of a stimulus on $L^1$ results in the selection of one of these $M$ patterns, namely the one which is homeomorphic, as a labeled pattern, to the pattern currently presented on $L^1$. This may be viewed as a "resonance"[34] mechanism, which acts to select one out of many competing patterns. Note that the frustration in the problem of minimizing $H(w)$ in the absence of any constraint on the graph $s$, hence the degeneracy of the solution, allows the dynamics to act, on a slow time-scale, as the "generator of diversity" required in selective theories: a "repertoire" of patterns may be created prior to the interaction with the environment[33,34]. Yet, the main thrust of the present theory is in the faster selection processes. One could say that in the present theory much of what is classically thought to be achieved by ontogenetic, maybe even phylogenetic, evolution processes is passed on to fast connectivity dynamics. The flexibility of such a scheme could perhaps be one of the keys to the remarkable performances of our brain.

Direct experimental verification could concentrate on existence and properties either of high-order statistics in signals, or of short-term modification in synapses. One possible experiment would be the following. In a waking animal, the signals of several sensory cells are recorded which can be activated *a*) by a single stimulus (which is considered by the trained animal as a unit) or *b*) by several stimuli, which are present simultaneously but which are manifestly different and independent of each other. In case *a* the signals should have a higher rate of coincidences in their fine temporal structure than in case *b*. The experiment is a difficult one (especially if the relevant statistics is of rather high order) but would open the window to an entirely new universe of signals in the brain. The experiment may be more feasible on the basis of mass-potentials, created by thousands of cells, if a situation can be found in which the topology of networks and the topology of ordinary space are sufficiently close to each other. Another kind of experiment would concentrate on the detection of short-term synaptic modulation. This may be possible in tissue culture. One would have to create a situation consisting of two cells with a direct synaptic connection. By proper experimental control of the cells' signals, and determination (e.g., by patch-clamp techniques) of the size of the EPSP created in one cell by activity in the other, one could hope to show that coincident activity in the two cells increases the size of the EPSP within a fraction of a second, and asynchronous activity decreases its size. The controlling signal on the post-synaptic side may be the membrane potential, but it may also be the concentration of a second messenger which is influenced by high-order convergence of pre-synaptic signals.

An altogether different "proof" of the feasibility of the system discussed here would consist in the implementation and test in an electronic machine. The machine would have to be massively parallel in order to be able to run networks of realistic size in finite time. Such a machine would differ from the conventional computer in being adapted to exactly one "program", namely the one incorporating the principles of organization implicit in the dynamic laws for individual cells and synapses[35]. It may perhaps be expected that non-Von-Neumann-architecture machines of this or similar kinds will play a significant role in future technology.

Acknowledgements: This work has benefited from many discussions, in particular with S. Geman. Numerical simulations were done on an FPS 164 array processor, a computing facility of GRECO 70 ("Expérimentation Numérique").

# References

1. J. Von Neumann (1956) *Probabilistic logic and the synthesis of reliable organisms from unreliable components.* In: Automata Studies (C. E. Shannon and J. McCarthy, eds.), Princeton University Press, Princeton, NJ.
2. S. Winograd and J. D. Cowan (1963) *Reliable Computation in the Presence of Noise.* The MIT Press, Cambridge, MA.
3. D. O. Hebb (1949) *The Organization of Behavior.* Wiley, New York.
4. W. A. Little (1974) *The existence of persistent states in the brain.* Math. Biosc. **19**, 101–120.
5. J. J. Hopfield (1982) *Neural networks and physical systems with emergent collective computational abilities.* Proc. Natl. Acad. Sci. USA. **79**, 2554–2558.
6. G. E. Hinton, T. J. Sejnowski, and D. H. Ackley (1984) *Boltzmann machines: Constraint satisfaction networks that learn.* Technical Report CMU-CS-84-119, Department of Computer Science, Carnegie-Mellon University, Pittsburgh PA.
7. M. Abeles (1982) *Local Cortical Circuits. An Electrophysiological Study.* (V. Braitenberg, ed.), Springer-Verlag, Berlin.
8. C. von der Malsburg (in press) *Am I thinking assemblies?* In: Proceedings of the 1984 Trieste Meeting on Brain Theory. (G. Palm and A. Aertsen, eds.), Springer Verlag, Heidelberg.
9. R. Lorente de No (1938) *Analysis of the activity of the chains of internuncial neurons.* J. Neurophysiol. **1**, 207–244.
10. C. von der Malsburg (1981) *The correlation theory of brain function.* Internal Report 81-2. Max-Planck Institute for Biophysical Chemistry, Department of Neurobiology, Göttingen, West-Germany.
11. J. Szentagothai and P. Erdi (preprint) *Outline of a general brain theory.* Hungarian Academy of Sciences, Budapest.
12. F. Crick (1982) *Do dendritic spines twitch?* Trends in Neurosci. **5**, 44–46.
13. W. Rall (1978) *Dendritic spines and synaptic potency.* In: Studies in Neurophysiology (R. Porter, ed.), pp. 203-209. Cambridge University Press.
14. C.Koch and T. Poggio (1983) *A theoretical analysis of electrical properties of spines.* Proc. R. Soc. B. **218**, 455–477.
15. J. P. Miller, W. Rall, and J. Rinzel (1985) *Synaptic amplification by active membrane in dendritic spines.* Brain Res. **325**, 325–330.
16. J. P. Changeux, A. Devillers-Thiéry, and P. Chemouilli (1984) *Acetylcholine receptor: an allosteric protein.* Science **225**, 1335–1345.
17. J. P. Changeux, and T. Heidmann (in press) *Allosteric receptors and molecular models of learning.* In: New insights into synaptic function (G. Edelman, W.E. Gall, and W.M. Cowan, eds.), John Wiley Publishers, New York, NY.
18. N. Metropolis, A. W. Rosenbluth, M. N. Rosenbluth, A. H. Teller, and E. Teller (1953) *Equations of state calculations by fast computing machines.* J. Chem. Phys. **21**, 1087–1091.
19. S. Kirkpatrick, C. D. Gelatt Jr, and M. P. Vecchi (1983) *Optimization by simulated annealing.* Science **220**, 671–680.
20. G. Toulouse (1984) *Frustration and disorder, new problems in statistical mechanics: Spin glasses in a historical perspective.* In: Lecture Notes in Physics (J. van Hemmen and I. Morgenstern eds.), Springer Verlag, Heidelberg.

21. P. Erdös and A. Rényi (1960) *On the evolution of random graphs.* Publ. Math. Inst. Hung. Acad. Sci. **5**, 17–61.

22. E. Bienenstock (in press) *Dynamics of central nervous system.* In: Dynamics of Macrosystems. Proc. of a Symposium Held at the I.I.A.S.A., Laxenburg, Austria, Sept. 1984 (J. P. Aubin and K. Sigmund eds.).

23. D. J. Willshaw, and C. von der Malsburg (1976) *How patterned neural connections can be set up by self-organization.* Proc. R. Soc. Lond. B **194**, 431–445.

24. J. T. Schmidt, and D. L. Edwards (1983) *Activity sharpens the map during regeneration of the retinotectal projection in goldfish.* Brain Res.**269**, 29–39.

25. C. von der Malsburg (1985) *Nervous structures with dynamical links.* Ber. Bunsenges. Phys. Chem. **89**, 703–710.

26. A. F. Häussler, and C. von der Malsburg (1983) *Development of retinotopic projections: An analytical treatment.* J. Theoret. Neurobiol. **2**, 47–73.

27. D. N. Mastronarde (1983) *Correlated firing of cat retinal ganglion cells, I and II.— Interactions between ganglion cells in cat retina.* J. Neurophysiol. **49**, 303–365.

28. D. W. Arnett (1978) *Statistical dependence between neighbouring retinal ganglion cells in goldfish.* Exp. Brain Res. **32**, 49–53.

29. Y. Frégnac (in press) *Cellular mechanisms of epigenesis in cat visual cortex.* In: Imprinting and Cortical Plasticity (J. P. Rauscheker, ed.), John Wiley Publishers, New-York, NY.

30. G. E Hinton, and K. Lang (in press) *Shape recognition and illusory conjunctions.* International Joint Conference on Artificial Intelligence.

31. J. Altmann, and H. J. P. Reitböck (1984) *A fast correlation method for scale- and translation-invariant pattern recognition.* IEEE Trans. **PAMI-6**, 46–57.

32. J. A. Feldman, and D. H. Ballard (1982) *Connectionist models and their properties.* Cognitive Science **6**, 205–254.

33. G. M. Edelman (1978) *Group selection and phasic reentrant signalling: A theory of higher brain function.* In: The Mindful Brain (G. M. Edelman and V. B. Mountcastle, eds.), The MIT Press, Cambridge, MA.

34. J. P. Changeux, T. Heidman, and P. Patte (1984) *Learning by selection.* In: The Biology of Learning. Proc. Dahlem Workshop, October 1983 (P. Marler and H. Terrace, eds.), Springer Verlag, Heidelberg.

35. C. von der Malsburg (1985) *Algorithms, brain and organization.* In: Dynamical Systems and Cellular Automata (J. Demongeot, E. Golès, and M. Tchuente, eds.), Academic Press, London.

# A Physiological Neural Network as an Autoassociative Memory

J.Buhmann and K.Schulten

Physik-Department, Technische Universität München

8046 Garching, Fed. Rep. Germany

## 1. Introduction

We consider a neural network model in which the single neurons
are chosen to resemble closely known physiological properties.
The neurons are assumed to be linked by synapses which change
their strength according to Hebbian rules [1] on a short time
scale (100ms) [2]. Each nerve cell receives input from a
primary set of receptors, which offer learning and test
patterns without changing their own properties. The activity of
the neurons is interpreted as the output of the network (see
Fig.1). The backward bended arrows in Fig.1 indicate the
feed-back due to the effect of the neuron activity on the
synaptic strengths $S_{ik}$ between neuron k and i in the neural
network.

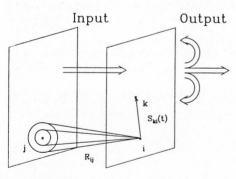

Input     Output

Receptors     Neurons

Figure 1
Schematic presentation of the
model investigated: Receptors
send spikes to a network of
neurons. The connectivity
between the receptors j and
neurons i is given by the
matrix $R_{ij}$, the connectivity
between the neurons is given
by $S_{ik}(t)$. The resulting
activity of the neural network
is affected by an activity-de-
pendent alteration of $S_{ik}(t)$,
i.e. the network experiences a
feed-back as indicated.

Initially the synapses $S_{ik}$ which carry action potentials
from cell k to cell i are chosen at random, i.e. the network is
initially completely uninstructed. The connections between the
receptors and the 'physiological' neurons possess no plasticity
and have a local center-surround-organization. Receptors j
which are lying in the neighbourhood of the receptor i are
connected with the neuron i by excitatory synapses, whereas
receptors arranged in the immediate surrounding of this

NATO ASI Series, Vol. F20
Disordered Systems and Biological Organization
Edited by E. Bienenstock et al.
© Springer-Verlag Berlin Heidelberg 1986

excitatory center have an inhibitory effect on neuron i. The area on the receptor set which affects the neuron i is much smaller than the size of the network. Therefore, the connections between the receptors and the neurons constitute a continuous projection of the input pattern onto the neural net, the projection being locally convoluted with the center-surround function.


## 2. Dynamics of the Cell Potential

The fast dynamics of a neuron involves its cell potential which changes on the time scale of a few milliseconds. In our model two important contributions to the dynamics of the potential are included. A first term describes the relaxation of the cell potential which takes place on the time scale $T_R$ ($T_R$ =2.5ms). The second term accounts for the change of the cell potential due to interactions with other neurons. If the cell k which forms a synapse with neuron i has fired, a postsynaptic potential difference corresponding to the synaptic strengths $S_{ik}$ appears in cell i. The cell i continuously sums up the various excitatory and inhibitory postsynaptic potentials. If a threshold $U_T$ =30mV is exceeded the neuron fires an action potential and excites or inhibits nerve cells connected to it. Sub-threshold potentials relax against the resting potential.

The dynamics of an action potential is simplified by the following rules: If the neuron k fires, a monotonously decreasing function

$$G_i (\Delta t_i /\tau) = \exp(-\Delta t_i /\tau) \quad \text{with } \Delta t_i =t-t_{0i} , \qquad (1)$$

describes the differential change of the postsynaptic potential in the neuron i. In Eq.(1) $t_{0i}$ indicates the time of the latest firing. The effect of the spike of neuron k on the postsynaptic cell i decays with the characteristic time $T_U$ ($T_U$ =1ms).

The kinetic equation which describes the time evolution of the cell potential is

$$dU_i /dt = \begin{cases} -U_i /T_R +\omega\ U_T\ \rho(G_i (\Delta t_i /T_F ))\ A_i (t) & U_F \leqslant U_i (t) \leqslant U_T \\ U_F & \text{else .} \end{cases} \qquad (2)$$

The first term in the upper equation describes the relaxation to the resting potential, the second term the communications of the i-th neuron with the receptors and with other neurons. The key parameter which scales the neuronal communication is the coupling constant $\omega$. This constant $\omega$ can be used to rescale the network dynamics [3] by the equation

$$\omega = \{\langle PSP\rangle\ T_R\ [1-\exp(-T_E /T_R )]\}^{-1} \qquad (3)$$

where $\langle PSP\rangle$ estimates the average postsynaptic potential. The parameter $T_E$ , the effective excitation time of the neuron, determines the time which a neuron requires to reach the threshold $U_T$ if it has rested in the sensitive state and if the connected neurons fire with a average spike rate $(T_E + 2T_F)^{-1}$. Equation (3) furnishes a choice of the coupling coefficient $\omega$ which assures that the neural network avoids the states of epileptic hyperactivity or of abnormal quiescence.

$A_i(t)$ in Eq.(2) is the activity function which sums up all spikes converging on the cell i and weights them with the corresponding synaptic strength $S_{ik}$. External contributions of the receptors presented with an input frequency $T_i^1$ are included in

$$A_i(t) = \Sigma_k S_{ik} G_k (\Delta t_k / T_U) + \Sigma_j R_{ij} G_j^R (\Delta t_j^R / T_U) . \qquad (4)$$

$\rho(G_i)$ in Eq.(2) is a function which accounts for the existence of the total and relative refractory period $T_F = 5ms$. The factor $\rho$ is chosen such that the sensitivity of the neuron i is suppressed or reduced in the total and relative refractory period, respectively. We choose the following functional form

$$\rho[G_i(\Delta t_i / T_F] = \theta(\Delta t_i - T_F) \{1 - G_i [2(\Delta t_i - T_F)/T_F)]\} . \qquad (5)$$

When the threshold potential is reached and the cell fires, the continuous time evolution of the cell potential i is interrupted and the memory function $G_i(\Delta t_i / T_U)$ starts again with the value 1. In addition the cell potential is set to the refractory value of $U_F = -15$ mV:

$$\text{if} \quad U_i(t) \geqslant U_T \quad \text{then} \quad U_i(t) \to U_F \quad \text{and} \quad t_{oi} = t \qquad (6)$$

## 3. Learning through Synaptic Plasticity

In our model of learning information is stored nonlocally in the synaptic connections of the network. The plasticity of the synapse with the strength $S_{ik}$, leading from the neuron k to i, evolves on the time scale $\Omega^{-1} = 300ms$ and is governed by the Eq.

$$dS_{ik}/dt = \begin{cases} -\mathcal{R}(S_{ik}) + \Omega G_k(\Delta t / T_M)\kappa(G_i, G_k) & \text{if} \quad S_u \geqslant |S_{ik}| \geqslant S_1 \\ -\mathcal{R}(S_{ik}) & \text{else} \end{cases} \qquad (7)$$

$$\mathcal{R}(S_{ik}) = [S_{ik}(t) - S_{ik}(0)]/T_S$$

which holds for excitatory and inhibitory synapses. The first term $\mathcal{R}(S_{ik})$ accounts for the relaxation of the synapses to their initial values during the time $T_S \approx 1s$. The second term in (7) causes a growth of the synapses. This term is governed by the function $\kappa(G_i, G_k)$ which distinguishes four different activity states of a pair of neurons i and k as presented below

| $G_i(\Delta t_i / T_M)$ | $G_k(\Delta t_k / T_M)$ | $\kappa(G_i, G_k)$ | $dS_{ik}/dt$ | (8) |
|---|---|---|---|---|
| $> e^{-1}$ | $> e^{-1}$ | $+1$ | $> 0$ | |
| $< e^{-1}$ | $> e^{-1}$ | $-1$ | $< 0$ | |
| $> e^{-1}$ | $< e^{-1}$ | $-1$ | $< 0$ | |
| $< e^{-1}$ | $< e^{-1}$ | $0$ | $= 0$ | |

Figure 2 shows the changes which the strength $S_{ik}$ of an excitatory synapse experiences if the presynaptic neuron k fires at t=0 and the postsynaptic cell i answers with a spike at $t=t_0$. A time delay shorter than $\Delta t_0^* = T_M \ln(2e+1)/(e+2)$ results in an asymptotic synaptic strength above the initial value, otherwise below.

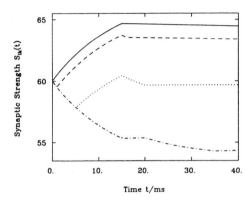

Figure 2
Time-dependence of the synaptic strengths $S_{ik}(t)$ in case of two spikes in the pre- and postsynaptic cells for 4 different spike intervals $\Delta\tau$. For $\Delta\tau=0$ the synapse grows at a maximum rate, an interval $\Delta\tau=20$ms causes a strong decrease of $S_{ik}(t)$ ($T_M=15$ms); (———) $\Delta\tau=0$ms, (-----) $\Delta\tau=1$ms, (·········) $\Delta\tau=5$ms, (-··-··-) $\Delta\tau=20$ms.

## 4. Behavior of the Network with Receptor Input

The first simulation has two different stages. In the first stage the uninstructed network learns the presented figure **brain** (Fig.3a) and changes its synaptic connections. In the second stage the success in learning is tested by the associative task to restore the missing letter i in the test figure **bra n.**

1. stage      0 - 300 ms :   learning of the figure **brain**
             300 - 320 ms :   relaxation of the cell potentials
2. stage    320 - 360 ms :   association of the missing i in **bra n**

The interval of 20ms between stage 1 and 2 in which the network receives no input spikes from the receptors guarantees that only the changed synaptic strengths and not the cell potentials contain information about the learned figure.

The reaction of the network after the presentation of the figure **brain** in stage 1 is presented in Fig.3b which shows the cell potentials after 10 ms. Most of the neurons which receive input from receptors belonging to the figure (figure neurons) have fired and are resting in the refractory phase or sum up postsynaptic potentials in the sensitive phase. A few of the background neurons in the upper half of the network are excited at the beginning of the learning course because they are connected to the figure neurons by enough excitatory synapses. These connections raise their cell potential but not above the threshold.

At t=290ms (Fig.3c) the background cells are strongly inhibited and only the neurons belonging to the figure show a positive cell potential or are in the refractory state.

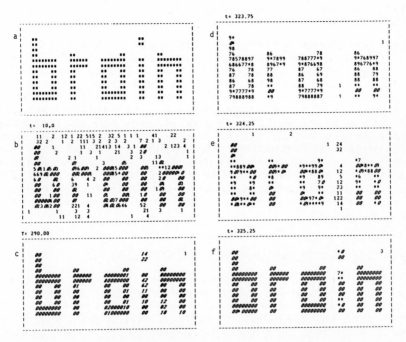

Figure 3
Learning Pattern (a) and Cell Potentials after 10ms (b) and 290ms (c). The values of the cell potentials are divided by the threshold value $U_T$ and represented by the next integer value if positive. The symbol * indicates that the cell potential has reached the threshold. If the memory function $G_i (\Delta t_i / T_v )$ exceeds $1/e$ the integer is italicized. Negative potentials are presented by a blank or by a italic zero $0$ if $G_i (\Delta t_i / T_v )$ exceeds $1/e$. On the right side the cell potentials are presented during the restoration of the missing i.


After the relaxation of the cell potentials the network is excited in stage 2 by the test figure which is identical to the pattern **brain** learned in stage 1 except that the letter i is missing. The time evolution of the cell potentials during the first few milliseconds of this association task is presented in the Figs.3d,e,f. The neurons which obtain input spikes from the receptors react immediately with a raised cell potential. At t=323.75ms, 3.75 ms after the beginning of the association test, several of the neurons belonging to the new figure **bra n** have fired and the potentials of the remaining neurons exceed the value 15mV. At t=325.25ms all except one neurons of this set have fired a spike whereas the potentials of neurons representing the missing i have reached the threshold or are just below the threshold. The Figures 3d,e,f reveal also the mechanism which underlies the associative properties of the network. If a subgroup of the figure neurons, e.g. the neurons of the letters **a** and **n**, fires spikes, strong postsynaptic cell potentials are evoked in the neurons of the missing letter **i**. These postsynaptic potentials compensate the missing receptor inputs and stimulate the postsynaptic cells to fire with a delay of 1-2ms.

In Figure 4 we present both before and after the learning stage the strengths of the synapses which connect the neuron (37,4) representing the point of the i in **brain** with the other neurons. According to the kinetics laid down in Eqs. (7),(8) the synapses between two figure neurons k and i are strengthened. The synaptic strengths $S_{ik}$ of these synapses are saturated either at the value $S_u = 99$ if the corresponding synapse is excitatory, or at the value $-S_l = -1$ if the corresponding synapse is inhibitory. Excitatory and inhibitory synapses connecting the figure neuron (37,4) to background neurons are saturated at the lower boundary values $S_l$ or $-S_u$, respectively. Figure 4 also demonstrates the nonlocal properties of the storage. Each single neuron contains in its synaptic connectivity a blueprint of the pattern stored in the network.

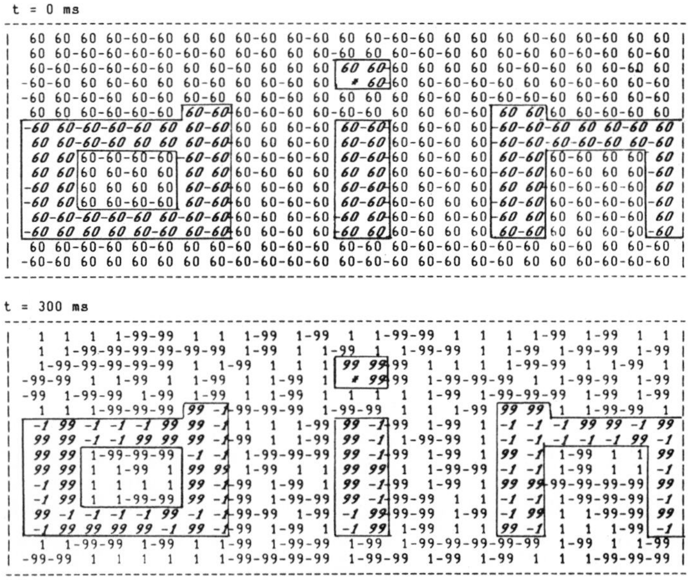

Figure 4
Synaptic changes during the learning of **brain**: The initial synaptic strengths and the strengths at t=300ms of the synapses starting from cell (37,4) are presented. Synapses which end at a figure neuron are indicated by italic, bold numbers.

## 5. Conclusion

In the network model presented above neural units closely related through the dynamics of their cell potential to their physiological counterparts interact by few local rules of synaptic plasticity. These rules induce global cooperation and competition and, thereby, endow the network with the ability for associative storage and recall of patterns. The basis of the rules of synaptic plasticity is a discrimination between states of pairwise synchroneous and asynchroneous neural activity, synchronicity being measured on a time scale of a few ms. Beyond the investigated behavior of the network, also more complex computational properties emerge in the neural net discussed, i.e. the ability to filter strong noise out of a presented pattern, to build up a prototype pattern from a series of varying patterns and to store several patterns [3].

## 6. References

D.O. Hebb: Organization of Behaviour, Wiley 1949

C.v.d. Malsburg: Int.Rep.81/2 Dept. Neurobiol. MPI f. Bio-
    physikalische Chemie, Göttingen (1981)

J. Buhmann, K.Schulten: to be submitted to Biol.Cybern.

# 4 COMBINATORIAL OPTIMIZATION

# CONFIGURATION SPACE ANALYSIS FOR OPTIMIZATION PROBLEMS

Sara A. Solla, Gregory B. Sorkin, and Steve R. White

IBM Thomas J. Watson Research Center

Yorktown Heights, New York 10598, USA.

## 1. Introduction

An interesting analogy between frustrated disordered systems studied in condensed matter physics and combinatorial optimization problems [1] has led to the use of simulated annealing (a stochastic algorithm based on the Monte Carlo method) to find approximate solutions to complex optimization problems. A common feature of these systems is the competition between objectives which favor different and incompatible types of ordering. Such "frustration" leads to the existence of a large number of nearly degenerate solutions which are not related by symmetry.

The dynamical behavior of an algorithm that searches for locally minimal solutions to complex optimization problems is a function of the corresponding configuration space landscape. Analysis of the landscape, in particular the distribution of locally minimal solutions and the structure of the barriers that separate them, thus provides a tool for the characterization of such problems. Recent theoretical developments in the mean field theory of spin glasses [2] have provided a set of concepts: ultrametricity, ergodicity breaking, lack of self-averaging [3], which are useful for the study of configuration space landscapes. These ideas have been recently applied to the analysis of a random-distance version of the traveling salesman problem (TSP) [4], a classic example of NP-complete optimization problems.

NATO ASI Series, Vol. F20
Disordered Systems and Biological Organization
Edited by E. Bienenstock et al.
© Springer-Verlag Berlin Heidelberg 1986

In this paper we look at an optimization problem related to computer wiring and electronic circuit design: a one-dimensional placement problem [5]. The purpose of this work is to sharpen the numerical tools available for the analysis of the configuration space landscape for optimization problems, and to investigate the potential universality of some of the features of such spaces by comparing our results to those available for the TSP [4].

## 2. Combinatorial optimization

Combinatorial optimization refers to the problem of minimizing a function of a large but finite set of variables. The cost function or energy function $E(\vec{x})$ assigns a real number to each possible state $\vec{x}$ of the system. The configuration space X is given by the finite set of all possible system configurations $\{\vec{x}\}$. The problem is then that of finding the global minimum of $E(\vec{x})$ over all $\vec{x}$ in X.

The existence of a large number of local but not global minima of $E(\vec{x})$ in configuration space conspires against the success of heuristic optimization methods. Simulated annealing is an improvement upon downhill algorithms in that it provides a mechanism for getting out of such local minima. It should be noted that while the notion of global minima is an absolute one that depends only on the cost function $E(\vec{x})$, the concept of local minima is relative to the topology since it implies comparing the value of $E(\vec{x})$ with that of $E(\vec{x}')$ for those $\vec{x}'$ which are neighbors of $\vec{x}$.

The topology of configuration space is defined by the neighbor relation. A natural definition of neighborhood for optimization problems is given by the move set: a small rearrangement of the system that produces a new trial configuration from the present one. Distance $d(\vec{x}, \vec{x}')$ between two states is then defined as the minimum number of moves needed to turn one into the other; and the neighborhood of a given state $\vec{x}$ is the set of all states $\vec{x}'$ which can be reached from $\vec{x}$ with one move, i.e. $d(\vec{x}, \vec{x}') = 1$.

## 3. One-dimensional placement

We now proceed to the analysis of configuration space for a specific example of combinatorial optimization: the one-dimensional placement problem [5]. Given N circuits connected by nets of wires, the goal is to find a linear ordering of the circuits which minimizes the total wire length. A net is a set of circuits that are to be interconnected, and the length of a net is the maximum distance between any two circuits in that net. The cost function $E(\vec{x})$ to be minimized is the total wire length, the sum of all the net lengths.

A configuration $\vec{x}$ of the system is an ordering on the N circuits, and is represented by a permutation of the integers from 1 to N. The configuration space is thus the same as for the TSP [4], except for two points. First is a minor difference in the symmetries of the problems: both are symmetric under reversal of the order of the array, but only TSP exhibits cyclic symmetry. Thus this space has N!/2 points while TSP has (N-1)!/2. Second is a difference important because it gives very different natural topologies to the two spaces: Pairwise interchange (exchanging the locations of two circuits within the array) is a good move set for this problem, but not for TSP. The number of neighbors of any configuration is thus N(N-1)/2.

The distance $q = d(\vec{x}, \vec{x}')$ between any two states $\vec{x}, \vec{x}'$ is given by the minimum number of pairwise interchanges needed to turn one permutation into the other one. If the number of cycles of one permutation with respect to the other one is k, then the distance is given by q=N-k. The following example illustrates the distance between two permutations for N=7, k=4, q=3:

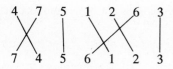

The probability distribution $P(q)$ for distances in this configuration space can be calculated by referring all permutations to the identity, and is

$$P(q) = \frac{S_N^{N-q}}{N!},$$ (1)

where the Stirling number $S_N^k$ counts the total number of permutations of N objects with exactly k cycles [6].

## 4. Low-lying locally-optimal solutions.

We have used simulated annealing [1] to look for good solutions to a one-dimensional placement problem with 105 circuits and 85 nets. The nets are derived from a real logic problem. Those circuits shared among nets are responsible for the frustration in the problem. In this case, 60% of the     105 circuits belong to at most two nets, and the remaining 40% are shared between 3 to 8 nets.

Annealing does work well on one-dimensional placement. Probability distributions for total wire length shown in Figure 1 correspond to three sets of 87 configurations each, obtained by: a) random selection (R), b) iterative improvement (ItIm), and c) simulated annealing (SA). Iterative improvement corresponds to simulated annealing at zero temperature (quenching), and lacks a mechanism for getting out of local minima. Random selection yields an average wire length of 4240 with a standard deviation $\sigma$ of 239. Iterative improvement lowers the average to 1320, but the distribution is still quite broad with a $\sigma$ of 217. Simulated annealing consistently yields good solutions, narrowly distributed around an average of 922 with a $\sigma$ of 46. So SA finds configurations that are better minima than those found by ItIm.

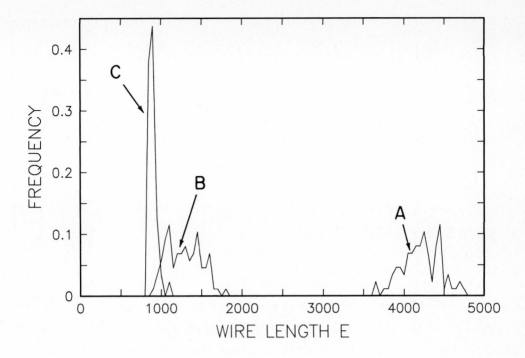

Fig. 1: Probability distribution of total wire length for three sets of configurations obtained by: a) random selection, b) iterative improvement, and c)
simulated annealing.

We now analyze the relative locations of low-lying local minima in configuration space
by calculating the probability distribution P(q) for the distance between any two of them.
The results shown in Figure 2 for $0 \leq q \leq 105$ are based on the same sets of configurations
used to obtain Figure 1. The result for P(q) for random configurations is indistinguishable
in this scale from the analytic result (1) for the complete space. The configurations obtained
by SA yield a distribution with a tail extending into shorter distances, an indication of clustering among the good solutions.

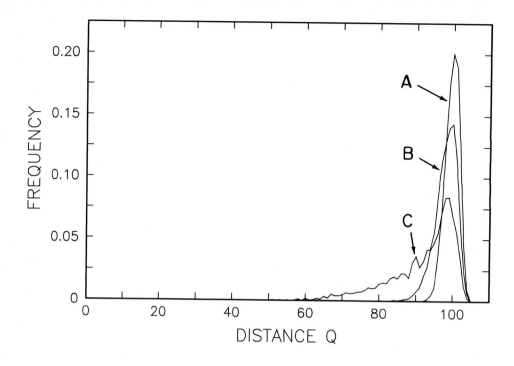

<u>Fig. 2:</u> Probability distribution for distance q between any two configurations within a set obtained by: a) random selection, b) iterative improvement, and c) simulated annealing.

It has been discovered recently that low-lying locally minimal solutions to a class of traveling salesman problems exhibit ultrametricity [4], i.e. that out of any three such states the two of them that are nearer to each other are equidistant from the third one. This property has also been found for the pure states of the spin glass problem [7], and is equivalent to the existence of a tree whose leaves are all at the same level and represent the states. The distance between two leaves is defined as the number of generations to their closest common ancestor [3].

For a tree with constant ramification ratio r (r is the number of descendants of any node in the tree)

$$P(q) = (1 - 1/r) \, r^{\,q-Q} \tag{2}$$

for $1 \leq q \leq Q$. The numerical data on $P(q)$ for the SA configurations shown in Figure 2 is in good agreement with this result in the range $70 \leq q \leq 100$, and yields $r \approx 1.1$. Poor statistics render the data unreliable at low values of q, as $P(q) < 10^{-5}$ for $0 \leq q \leq 70$. The topological organization suggested by the clustering of the SA configurations is thus consistent with that of a uniform tree.

A test of ultrametricity requires triangle statistics. Any three configurations form a triangle with side lengths $q_1$, $q_2$, and $q_3$. If the set of low-lying minima is perfectly ultrametric, all such triangles are isosceles with a small base, and the sides can be labeled such that $q_1 = q_2 \geq q_3$. It is then useful to introduce the correlation function

$$C(q_1, q_2) = \tilde{P}(q_1, q_2) - \tilde{P}(q_1)\tilde{P}(q_2) , \tag{3}$$

where $\tilde{P}(q_1, q_2)$ is the probability that a triangle will have two longest sides of length $q_1$ and $q_2$, and $\tilde{P}(q)$ is the probability that one of the two longest sides will have length q. We have calculated $C(q_1, q_2)$ for the good solutions to the one-dimensional placement problem considered here. The results are shown in Figure 3. A ridge in the distribution along the $q_1 = q_2$ diagonal is evidence for ultrametric structure. Similar features have been observed for the correlations among the good solutions to a TSP with 48 cities and the low energy configurations of a long-range spin glass model with 24 spins [4]. The correlation function considered by the authors of Reference 4 in their numerical analysis differs from the one introduced here in that they consider any two sides of a triangle instead of focusing on the two longest ones. It is the equality of the two longest sides of every triangle that guarantees ultrametricity.

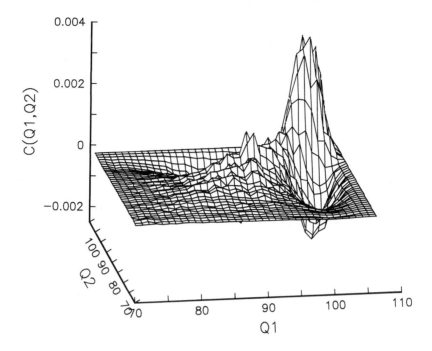

Fig. 3: Statistics of all triangles formed by the good solutions obtained by simulated annealing to the one-dimensional placement problem.

Comparison between degrees of ultrametricity found in different configuration spaces requires a more quantitative measure than the observation of a ridge of variable width in plots of the type shown in Figure 3. Such a measure is provided by the correlation coefficient

$$C = \frac{<(q_1 - <q_1>)\,(q_2 - <q_2>)>}{\sigma_1 \sigma_2}, \qquad (4)$$

where the symbol $<..>$ indicates average with respect to the joint probability distribution $\tilde{P}(q_1, q_2)$ used in Eq.(3), and

$$\sigma_i = <(q_i - <q_i>)^2>^{1/2} \qquad (5)$$

for $i = 1,2$ is the standard deviation. This coefficient measures the degree of correlation between the two longest sides of the triangles, and provides a quantitative parameter for comparing different optimization problems and analyzing deviations from exact ultrametricity due to finite size effects. The results for the one-dimensional placement problem are as follows: $C=0.675$ for the set of low-lying local minima obtained by SA, and $C=0.244$ for the set of local minima obtained by ItIm. For further comparison, $C=0.145$ for the set of random configurations. A large degree of ultrametricity in the set of low-lying local minima is also found in the TSP. We have analyzed the good solutions to a 48-city problem [4], and found $C=0.455$.

These numerical results suggest that ultrametricity might be a general property of the low-lying minima in configuration spaces for complex optimization problems. Such conjecture remains to be proven. The only available analytic result guarantees exact ultrametricity, characterized by $C=1$, for the replica-symmetry-breaking solutions to the infinite range model of a spin glass [7].

## 5. Connection between ultrametricity and annealing

The obvious question that arises at this point is : why do the low-lying local minima exhibit a large degree of ultrametric organization? An intuitive explanation follows from a new definition of distance between local minima, based on the energy function $E(\vec{x})$. If $\vec{x}$ and $\vec{x}'$ are local minima, the "energy distance" $d_E(\vec{x}, \vec{x}')$ is defined as the energy barrier between them, given by the minimax energy of all paths in configuration space connecting $\vec{x}$ to $\vec{x}'$. This distance is ultrametric, since it obeys

$$d_E(\vec{x}, \vec{x}'') \leq \max \{d_E(\vec{x}, \vec{x}'), d_E(\vec{x}', \vec{x}'')\} \tag{6}$$

for all $\vec{x}, \vec{x}', \vec{x}''$.

The numerical experiments discussed here are based on a definition of distance $d(\vec{x}, \vec{x}')$ determined by the length of the shortest path connecting $\vec{x}$ to $\vec{x}'$, or minimum number of moves needed to turn one configuration into the other. (This definition also applies to the spin glass case, since the Hamming distance between two spin configurations is equal to the minimum number of one-spin flips needed to turn one into the other). Since the distance $d_E(\vec{x}, \vec{x}')$ is ultrametric, a large degree of ultrametricity in $d(\vec{x}, \vec{x}')$ is not surprising if the configuration space is such that these two distances are strongly correlated. In such a space, low-lying local minima which are very different from each other are separated by large energy barriers, while small energy barriers separate good solutions with minor differences. We have found such correlations between distances $q = d(\vec{x}, \vec{x}')$ and energy barriers $\Delta E = d_E(\vec{x}, \vec{x}')$ in our analysis of low lying solutions to the one-dimensional placement problem.

Our closing remark refers to the general problem of finding good solutions to combinatorial optimization problems. Given a configuration space $\{\vec{x}\}$ with a topology defined by the distance $q = d(\vec{x}, \vec{x}')$, and a landscape defined by the cost function $E(\vec{x})$, is simulated annealing a useful technique for finding the low-lying local minima? A partial answer is provided by the correlation between q and $\Delta E$, as measured by the degree of ultrametricity of the good solutions under the distance q. A strong correlation between q and $\Delta E$ characterizes configuration spaces with a self-similar structure of energy valleys, in that wide valleys are deep, and narrow valleys are shallow and do not contain good solutions. The simulated annealing algorithm will work well only in such spaces, where gradual cooling will confine the search to those regions of configuration space in the domain of attraction of low-lying local minima [1]. Gross features of the optimal solutions are selected in the early part of the process, while the end part of the search is confined to a fairly small region of configuration space and only allows for minor rearrangements.

These observations lead us to conjecture that if simulated annealing works well in a space, then the low-lying minima are ultrametrically distributed. Further analytic and numerical work is required to establish the validity of this statement. If correct, it provides a useful concept for the categorization of complex optimization problems.

## References

1. S. Kirkpatrick, C.D. Gelatt, Jr., and M.P. Vecchi, Science <u>220</u> , 671 (1983).
2. "Heidelberg Colloquium on Spin Glasses", ed. by J.L. van Hemmen and I. Morgenstern, Lecture Notes in Physics, Vol. 192 (Springer-Verlag, 1983).
3. G. Toulouse, Helv. Phys. Acta <u>57</u> , 459 (1984); M. Mézard, this conference.
4. S. Kirkpatrick and G. Toulouse, to appear in J. Physique.
5. S. Kang in "Proc. of the 20th Design Automation Conference", (IEEE, 1983), p. 457; C. Rowen and J.L. Hennessy in "Proc. of the Custom Integrated Circuits Conference", to be published (IEEE, 1985).
6. M. Abramowitz and I.A. Stegun, "Handbook of Mathematical Functions" (Dover, New York, 1965), p. 824.
7. M. Mézard, G. Parisi, N. Sourlas, G.Toulouse, and M. Virasoro, Phys. Rev. Lett. <u>52</u> , 1156 (1984); N. Parga, G. Parisi, and M. Virasoro, J. Physique Lett. <u>45</u> , L-1063 (1984).

# Statistical Mechanics : a General Approach to Combinatorial Optimization

E. Bonomi, CPT Ecole Polytechnique, F-91128 Palaiseau Cedex

J.L. Lutton, Cnet PAA/ATR/SST, F-92131 Issy les Moulineaux

The aim of this work was to investigate the possibility of using statistical mechanics as an alternative framework to study complex combinatorial optimization problems. This deep connection, suggested in ref. [1] by analogy with the annealing of a solid, allows to design a searching procedure which generates a sequence of admissible solutions. This sequence may be viewed as the random evolution of a physical system in contact with a heat-bath. As the temperature is lowered, the solutions approach the optimal solution in such a way that a well organized structure is brought out of a very large number of unpredictable outcomes. To speed up the convergence of this process, we propose a selecting procedure which favours the choice of neighbouring trial solutions. This improved version of the simulated annealing is illustrated by two examples : the N-city Travelling Salesman Problem [2] and the Minimum Perfect Euclidean Matching Problem [3].

An instance of an optimization problem is a pair $(E,c)$ where $E$ is the set of all possible realizations to achieve some task ; $c$ is the cost-function, a mapping $c: E \rightarrow R$. The problem is to find an element $\pi^* \in E$ for which $c(\pi^*) \leqslant c(\pi)$, for all $\pi \in E$. Such an element $\pi^*$ is called a globally optimal solution to the given instance. Notice that, in combinatorial problems, $E$ is a finite set. In the two following examples, we consider N points inside a bounded domain $A$ of Euclidean space and the distance between every pair of points in the form of an NxN matrix $D=(d_{i,j})$ :

NATO ASI Series, Vol. F20
Disordered Systems and Biological Organization
Edited by E. Bienenstock et al.
© Springer-Verlag Berlin Heidelberg 1986

### 1) Travelling Salesman Problem (TSP)

An instance of the TSP is specified by the distance matrix $D=(d_{i,j})$. A tour is a closed path that visits every point exactly once. The problem is to find the shortest tour : $E=\{$all cyclic permutations $\pi$ on N objects$\}$, $c(\pi)=\sum d_{i,\pi(i)}$ .

### 2) Matching Problem (MP)

An instance of the MP is specified as in the TSP. A perfect matching is a set of N/2 edges such that each point is an end-point of one and only one of the edges. The problem is to find the matching of minimal total length : $E=\{$all perfect matchings $\pi$ on N objects$\}$, $c(\pi)=\sum d_{i,\pi(i)}$ .

From the viewpoint of the theory of computation [4], combinatorial optimization can be divided into easy and hard problems. The MP is a good example of what easy means, namely, it exists a polynomial algorithm which solves it in a number of steps bounded by a function $O(N^3)$. In contrast, the TSP is the classical example of intractable problem belonging to a family of equivalent problems classified as NP-complete, all of which are widely considered unsolvable by polynomial algorithms. It is clear that the MP also requires an important computational effort when the size N increases $(N > 1000)$ ; consequently, in this limit, we pass from an easy to a harder problem. The practical way to approach hard problems is to design approximate-solution algorithms whose running-time is proportional to small powers of N. A simple and natural heuristic method is based on a local search inside the set E of all admissible solutions. Of course, the local search is related to the concept of neighbourhood. This is defined, for each $\pi \in E$, as the set $N(\pi)$ of elements of E close in some sens to $\pi$. For instance, let us consider a transformation $T: E \rightarrow E$, then we may define $N_T(\pi)$ as the set of all elements which are accessible from $\pi$ given $T$ .

Example: In the TSP and the MP, let us consider the 2-OPT transformation rule [4] : "remove 2 edges from $\pi$ and replace them with 2 other edges".

Given an instance $(E,c)$ of an optimization problem and a transformation $T$, we call <u>local</u> <u>minimum</u> an admissible solution $\alpha \neq \pi^*$ such that, for all $\pi \in N_T(\alpha)$, we get $\Delta c = c(\pi) - c(\alpha) > 0$. Obviously, the choice of $T$ selects the set of all local minima.

<u>Example</u>: $E = \{\pi = (x_1, x_2, x_3 x_4) | \; x_i = 0, 1\}$, $c(\pi) =$ random cost. Choose the following transformation rule $T$: "switch from $\pi$ one variable $x_i$". The graph of all possible transitions, fig.1, shows the globally optimal solution and two local minima $\alpha_1$ and $\alpha_2$. Both local minima vanish if we use as transformation rule : "switch from $\pi$ one <u>or</u> two variables $x_i$".

The local search procedure is formally written as follows :

<u>STEP</u> 0. Select an initial solution $t \in E$, $\pi := t$.

<u>STEP</u> 1. Select (in a deterministic way or at random ) one element $s \in N_T(\pi)$.

<u>STEP</u> 2. Compute $\Delta c = c(s) - c(\pi)$, if $\Delta c \leqslant 0$ then $\pi := s$, goto 1.

In this approach, so long as an improved solution exists, we adopt it and repeat the procedure from the new solution until the algorithm is <u>trapped</u> in one local minimum ( or, by chance, in one globally optimal solution ). To release the algorithm from these trapping situations, the idea is to introduce fluctuations towards higher cost solutions. By analogy with statistical mechanics, we modify the local search in the following manner :

<u>STEP</u> 1a. Select at random one element $s \in N_T(\pi)$, with a probability $q_{\pi,s}$ such that $q_{\pi,s} = q_{s,\pi}$.

<u>STEP</u> 2a. Compute $\Delta c$, set $\pi := s$ with a probability $p = \min(1, e^{-\Delta c/\theta})$, goto 1a.

This procedure, controlled by the <u>temperature</u> $\theta$, generates an irreducible Markov chain; hence, eventually, the set $E$ is covered with probability one. The iteration of this procedure leads to a unique stationary distribution [5] : $P_\theta(\pi) \sim e^{-c(\pi)/\theta}$, $\pi \in E$, the <u>Boltzman</u> <u>factor</u>. From this expression, we find:

$$\lim_{\theta \to 0} P_\theta(\pi) = \begin{cases} 1 & \text{if } \pi = \pi^* \\ \\ 0 & \text{otherwise} \end{cases}$$

This implies that decreasing $\theta$ (i.e. exponentially ), the procedure brings out a sequence of solutions corresponding to lower and lower average values of the cost function $<c(\pi)>$, (fig.2: TSP). This operation is called <u>annealing schedule</u> [1]. Obviously, the efficiency of the annealing is related to the random behaviour of $\Delta c/\theta$, depending itself on the transformation rule **T**. In fact, for small $\theta$, an inadequate choice of **T** implies, in average, large positive $\Delta c$ giving rise to small probabilities of accepting a trial solution (STEP 2a) ; globally, the searching procedure will slow down considerably, which is in contrast with the high temperature situation where the ratio of accepted solutions will be in any case large [2]. The natural way to improve the procedure is to select, via **T**, trial solutions belonging to $N_T(\pi)$, such

that $|\Delta c|/\theta < \varepsilon$, where $\varepsilon$ is an arbitrary small constant. For example, in the TSP and the MP, we have divided the domain **A** into a set of disjoint sub-regions, confining the **2-OPT** transformation inside neighbouring regions. As   is shown in fig.3 (average $\Delta c$ versus $\theta$), the effect of this partitioning is to rescale $\Delta c$ and increase substantially the probability of accepting a trial solution, [3]. The efficiency of the method is illustrated in fig.4 (TSP) and in fig.5 (MP), N=10'000. This improved version of the simulated annealing gives a solution within a few percent of the optimal one in better than quadratic times  (N>250).

## REFERENCES

1.  S. Kirkpatrick, C. D. Gelatt and M. P. Vecchi, Science, 220 (1983) 671.
2.  E. Bonomi, J. L. Lutton, SIAM Rev. ,26 (1984) 551.

3. J. L. Lutton, E. Bonomi, "An Efficient Non-deterministic Heuristic for the Matching Problem", submitted to Ann. Op. Res.

4. Ch. H. Papadimitriou, K. Steiglitz, "Combinatorial Optimization : Algorithms and Complexity" , Prentice-Hall, 1982.

5. F. Romeo, A. Sangiovanni-Vincintelli, Memorandum No. UCB/ERL M84/34 University of California, Berkeley.

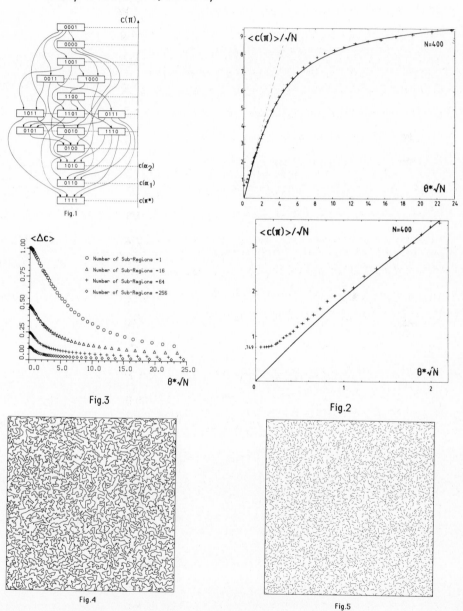

Fig.1

Fig.2

Fig.3

Fig.4

Fig.5

# BAYESIAN IMAGE ANALYSIS

Donald Geman[1]
Department of Mathematics and Statistics
University of Massachusetts
Amherst, MA 01003, USA


Stuart Geman[2]
Division of Applied Mathematics
Brown University
Providence, RI 02912, USA

## I. INTRODUCTION.


In [8] we introduced a class of image models for various tasks in digital image processing. These models are multi-level or "hierarchical" Markov Random Fields (MRFs). Here we pursue this approach to image modelling and analysis along some different lines, involving segmentation, boundary finding, and computer tomography. Similar models and associated optimization algorithms appear regularly in other work involving immense spatial systems; some examples are the studies in these proceedings on statistical mechanical systems (e.g. ferromagnets, spin-glasses and random fields), the work of Hinton and Sejnowski [14], Hopfield [15], and von der Malsburg and Bienenstock [19], in neural modeling and perceptual inference, and other work in image analysis, e.g. Besag [2], Kiiveri and Campbell [17], Cross and Jain [5], Cohen and Cooper [4], Elliott and Derin [7], Deviver [6], Grenander [11], and Marroquin [20]. The use of MRFs and related stochastic processes as models for intensity data has been prevalent in the image processing literature for some time now; we refer the reader to [8] and standard references for a detailed account of the genealogy.

The aforementioned analogy between very large (usually spatial) stochastic systems such as those encountered in digital image processing, computer vision, and neural modelling, and the lattice-based systems of statistical mechanics has been an important theme of our past work. For instance, our computational algorithms are based on a new optimization technique called "simulated annealing", introduced by Černý [3] and

1. Research supported in part by the Office of Naval Research, Contract N000-14-84-K-0531, and The National Science Foundation, DMS-8401927.
2. Research supported in part by the Army Research Office, Contract DAAG-29-83-K-0116, and The National Science Foundation, DMS-8306507 and DMS-8352087.

Kirkpatrik et al [18]. Stochastic relaxation and simulated annealing are briefly discussed in § V and remain the basis of our reconstruction and segmentation algorithms. However, the focus here is on image modelling, statistical inference, and new applications.

Our image models are "hierarchical" and stochastic. First, we regard the "image" as a collection of attribute processes, only one of which is the usual array of intensity or brightness values. The other, mainly geometric, attribute processes are constructs, corresponding to edges, object locations, feature labels, and so forth; they are part of the image model but not of the physical data. We use the term "hierarchical" to reflect the fact that image attributes such as boundaries and texture labels involve increasingly global and contextual information and expectations.

We have chosen the family of MRF priors for "images" for several reasons. First, we believe this formulation provides a solid, theoretical basis for complex image modelling: the class of models is extremely rich and easily accommodates a multi-level framework. Indeed, spatially-invariant, geometric attributes such as edges, curves, and simple polygons (with arbitrary scale and location) can be incorporated in the model in a local fashion. This was illustrated in [8] with the addition of a "line process". Second, the duality between MRFs and Gibbs distributions (see § II) allows the modelling process to be explicit and constructive: we build energy functions to quantify our a priori expectations about imagery. Finally, for many types of degradations (see § III), the conditional independence (= Markov property) of the prior is inherited by the posterior distribution. This is crucial because it guarantees a satisfactory degree of computational feasibility; see §§ IV, V.

For many problems in "low-level" image processing and related fields the current models appear adequate; other needs are more pressing, such as reducing the computational load and developing a rational data-driven method for estimating parameters in the model (see § VI). It remains to be seen whether the hierarchical MRF framework can accomodate the necessary high degree of external knowledge to deal with problems in "high-level" vision (for instance object recognition and texture labeling). Basic concepts such as scale and shape must be merged into the graph and model structures, and in a way that is sufficiently local to avoid unrealistic amounts of computation. Some of our preliminary experiments, and those of others, are encouraging. This paper addresses a "middle-level" of problems in reconstruction and segmentation in which excellent results are possible with some degree of "knowledge engineering", coupled with a careful analysis of the degradation mechanism.

## PRIOR DISTRIBUTIONS ON IMAGES.

Let $\underset{\sim}{X}^P = \{X^P_{ij}\}$, $1 \leqslant i,j \leqslant N$ denote the pixel values associated with an N×N (digitized) picture. Usually, each $X^P_{ij}$ represents the intensity of electromagnetic radiation in some frequency band that is emitted or reflected from a small region in the true "scene" or "object plane". (We regard these as the "ideal" intensities, uncorrupted by the recording system; in § III we will consider the nature of the actual, <u>observed</u> data.) Some examples we have in mind are grey-tone, infrared, and tomographic imagery, but the same analysis applies to other imagery, for example x-rays or a channel of multispectral landsat data.

As already discussed, we view the image as the realization of a compound stochastic process $\underset{\sim}{X} = (\underset{\sim}{X}^P, \underset{\sim}{X}^E, \underset{\sim}{X}^L, ...)$ in which $\underset{\sim}{X}^E$ might denote an array of "edge variables", $\underset{\sim}{X}^L$ certain "label variables", etc. In this paper we shall only consider an edge process $\underset{\sim}{X}^E$ in addition to the "pixel process" or "intensity process" $\underset{\sim}{X}^P$. Since our approach is Bayesian, we are going to impose a <u>prior</u> <u>probability</u> <u>distribution</u> on the set of possible values of $\underset{\sim}{X}$, which we denote by $\Omega$. Thus, for example, if $\underset{\sim}{X} = \underset{\sim}{X}^P$ only, and we are only interested in binary imagery, one would assign probabilities to each of the $2^{N^2}$ elements of $\Omega$.

In order to define our family of priors, we must specify exactly what we mean by $\underset{\sim}{X}^E$. Let s,t denote points in the square lattice. For each pair s,t of adjacent horizontal or adjacent vertical pixels we append an "edge site", denoted <s,t>, to the lattice; it corresponds to the "location" of a putative edge or boundary element between pixels s and t. In the simplest case, the edge variables are binary, with 0 and 1 representing the absence or presence of an edge at <s,t>. Then $\underset{\sim}{X}^E$ consists of the 2N(N-1) variables $X^E_{<s,t>}$.

The totality of pixel and edge sites in denoted by S. Given a <u>neighborhood system</u> $G$ = $\{G_\alpha, \alpha \epsilon S\}$ (see [8]) a stochastic process X on S is a MRF if $P(\underset{\sim}{X}=x)>0$ for all $x \epsilon \Omega$ and

$$P(X_\alpha = x_\alpha \mid X_\beta = x_\beta, \beta \neq \alpha) = P(X_\alpha = x_\alpha \mid X_\beta = x_\beta, \beta \epsilon G_\alpha)$$

for every $\alpha \epsilon S$ and $x = \{x_\gamma\}$, $\gamma \epsilon S$, in $\Omega$. In words, the conditional probability of seeing the value $x_\alpha$ at site $\alpha$ given any other configuration for the remaining sites depends only on the states of the neighbors of $\alpha$. In our case, the $\alpha$'s and $\beta$'s denote pixel or edge sites. The size of the neighborhood determines the range of interactions, and we shall say that $\underset{\sim}{X}$ is "locally-composed" or just "local" if $|G_\alpha|$ is small, say less than ten or twenty. Roughly speaking, these models are computationally feasible to the extent that $\underset{\sim}{X}$ is local. It is now well-known that a process $\underset{\sim}{X}$ on a graph $\{S,G\}$ is a MRF if and only if its joint probability distribution $\Pi(x) = P(\underset{\sim}{X}=x)$, $x \epsilon \Omega$, is a <u>Gibbs</u> <u>distribution</u> <u>on</u> $\{S,G\}$. This means that $\Pi$ has the form

$$\Pi(x) = e^{-U(x)/Z}. \qquad Z = \sum_{x} e^{-U(x)}$$

where the <u>energy</u> <u>function</u> U contains interactions confined to the <u>cliques</u> of the graph. Loosely speaking, this means that $x_\alpha$ and $x_\beta$ may appear together in a term in U only if $\alpha$ and $\beta$ are neighbors. Examples should make this clear and we again refer the interested reader to [8] for a complete discussion. Suffice it to say that the Gibbs formulation is convenient for <u>modelling</u> whereas the Markov property ensures that one can indeed examine samples from such a process.

We restrict ourselves to the following neighborhood system. Each pixel site has eight pixel neighbors, the nearest ones, and four edge neighbors; each edge site <s,t> has six edge neighbors (corresponding to the possible "continuations" of a boundary at <s,t>) and the two pixel neighbors s and t. Sites near the boundary of the lattice have fewer neighbors. Two of these neighborhoods, one for a pixel and one for a "vertical" edge site, are shown in Figure 1, in which the circles and pluses denote pixel and edge sites respectively. (We believe this edge graph originated in [13].)

Figure 1

To illustrate the functional form of the models suppose first that we were only interested in modeling "smoothness" or "regularity" in the intensity array, i.e. the tendency of nearby pixels to have similar intensities. Then a suitable model might be $\underset{\sim}{X} = \underset{\sim}{X}^P$ with

$$P(\underset{\sim}{X} = x) = Z^{-1} \exp\left\{\theta \sum_{(s,t)} \phi(x_s - x_t)\right\} \qquad (1)$$

where the sum extends over all neighbor pairs (s,t) of pixels. (Thus each interior pixel is included in eight terms in the summation.) Here $\phi = \phi(u)$ is even and decreasing for $u > 0$, and $\theta$ is a parameter which corresponds to inverse temperature and controls the degree of regularity. The extreme cases are $\theta = 0$, corresponding to pure noise, and $\theta = \infty$ in which case the distribution is concentrated on images of <u>constant</u> intensity.

We shall consider two examples of these "potentials" $\phi$, depending on the possible intensity values, say $X_\alpha \in \Lambda$, $\alpha \in S$. If $\Lambda$ is discrete, say $\Lambda = \{0,1,2,...,L\}$, and L is small, then one simple choice is

$$\phi_1(u) = \begin{cases} 1, & u = 0 \\ -1, & u \neq 0 \end{cases} \tag{2}$$

In particular, the conditional probability that $X_\alpha = j$ given the eight neighboring values depends only on the number $N_s(j)$ of neighbors which agree with $X_\alpha$. Specifically,

$$P(X_\alpha = j \mid X_\beta, \beta \neq \alpha) = \frac{\exp(2\theta N_\alpha(j))}{\sum_{k=0}^{L} \exp(2\theta N_\alpha(k))}, \quad 0 \leqslant j \leqslant L.$$

For a binary image, this is a weighted majority rule: the log odds of a 1 to a 0 are $2\theta(N_\alpha(1) - N_\alpha(0))$.

If L is large or $\Lambda$ is a <u>continuous</u> interval $[0,L]$, then we have adopted potentials of the form

$$\phi_2(u) = (1 + |\frac{u}{C_1}|^{C_2})^{-1}, \quad u \in [-L,L] \tag{3}$$

where $C_1, C_2$ are parameters; usually $C_2 = 1.5$ or $2.0$ and $C_1$ depends on the dynamic range of the image. One reason for this choice is that if $\phi$ were to decrease too rapidly (e.g. $\phi(u) = -u^2$) we would a priori inhibit (almost <u>pro</u>hibit) adjacent, roughly homogeneous regions of highly separated intensities.

With the inclusion of the edge process $X^E$ we incorporate our expectations about <u>both</u> the interactions between intensities and edges (i.e. where edges "belong") and about clusters of nearby edges. (It should be noted that, at this level of the hierarchy, we are not exactly modelling boundaries but rather <u>segments of boundaries</u>; except in the simplest imagery and with larger neighborhoods, it is essentially impossible to distinguish actual boundary segments from intensity gradients due to lighting, texture, etc.). We conclude this section with an example of an energy function U for $\underset{\sim}{X} = (X^P, X^E)$. The energy $U(x^P, x^E)$ consists of two terms, say $U = U^1(x^P, x^E) + U^2(x^E)$. We want to construct $U^1$ such that the most likely configurations will have $x^E_{<s,t>} = 1$ (resp. $= 0$) when the intensity difference $|x^P_s - x^P_t|$ is large (resp. small). Put differently, we want to break the bond between pixels s and t when their values are "far" apart. Thus we choose

$$U^1(x^P, x^E) = -\sum_{(s,t)} (\theta_1 \phi_2(x^P_s - x^P_t) - \theta_2)(1 - x^E_{<s,t>}) \tag{4}$$

where $\theta_1 > \theta_2 > 0$. The value of u for which $\theta_1 \phi(u) = \theta_2$ represents an intensity difference for which we have no preference in regard to the state of an edge. Finally, the organization of nearby edges is controlled by

$$U^2(x^E) = -\theta_3 \sum_D V_D(x^E) \tag{5}$$

($\theta_3 > 0$), where the sum extends over all subsets $D$ of four neighboring edge sites (the maximal "cliques" in the edge graph), and $V_D$ assigns weights in accordance with our expectations about edge behavior. More specifically, there are six possible clique states (up to rotations):

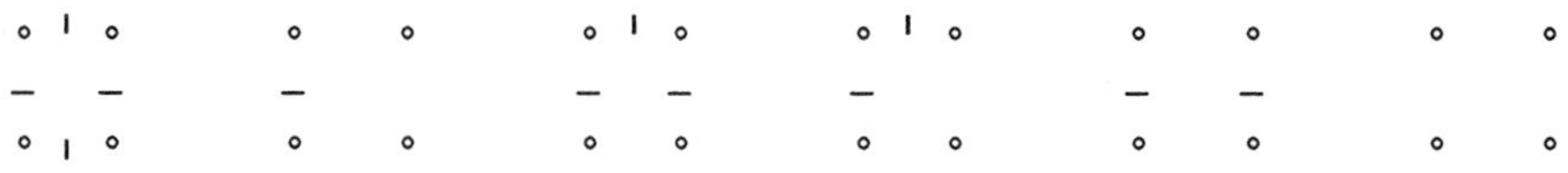

Figure 2

Here the slashes indicate that the edge variable at the indicated site is "on". Let $V_D = \xi_i$, $1 \le i \le 6$, denote the weights assigned to the six above configurations in Figure 2. If we assume that most pixels are not next to boundaries, that edges should continue, and that boundary congestion is unlikely, than we might choose $\xi_1 \le \xi_2 \le \xi_3 \le \xi_4 \le \xi_5 \le \xi_6$. A specific, image-dependent choice is made in the block experiment; see § VII.

A final point: it is useful to rewrite the total energy, up to a constant, as

$$-U(x) = \theta_1 \sum_{(s,t)} \phi(x_s^P - x_t^P)(1 - x_{<s,t>}^E) + \theta_2 \sum_{<s,t>} x_{<s,t>}^E + \theta_3 \sum_D V_D(x^E) \tag{6}$$

For inferential purposes, this shows that our model is an _exponential family_ in $\theta = (\theta_1, \theta_2, \theta_3)$. In addition, the form in (6) is helpful for parameter interpretation; for instance, $\theta_2$ is clearly a "reward" index for edges.


III. DEGRADATIONS.


We actually observe some transformation $\underset{\sim}{Y} = \Gamma(\underset{\sim}{X}^P)$ of the intensity process, three examples being:

(i) _Filtering and Deconvolution._ In many cases the pixel intensities do not represent the radiant energy at the source; rather, this energy is transformed due to the detecting and recording system. This is true both for photochemical (e.g. film) and photoelectronic (e.g. video) systems, and both usually involve blur and noise. A generic model for (space-invariant) degradation is then

$$y_s = g(\sum_{s-r \in D} H_{s-r} \, x_r^P) + \eta_s. \tag{7}$$

Here, D is a symmetric neighborhood of the origin, H is a blurring matrix, g accounts for nonlinearities in the recording device, and $\eta = \{\eta_s\}$ is a random noise due to the sensor, digitization, etc. In some cases the noise may be multiplicative (e.g. in synthetic aperture radar) or the blur may be anisotropic (e.g. in certain infrared scanners). One of the best features of the MRF formulation is that all such degradations are easily handled. We will assume that $\eta$ is white Gaussian noise and statistically independent of $\underset{\sim}{X}$. The general restoration problem is then to recover $\underset{\sim}{X}^P$ from the data $\underset{\sim}{Y} = \{y_s\}$ assuming we are given g , H and the noise statistics.

(ii) <u>Boundary-finding</u>. Another type of information loss occurs in the segmentation of "natural scenes" and other imagery which, for all practical purposes, can be taken as uncorrupted. Since we regard the image as $\underset{\sim}{X} = (\underset{\sim}{X}^P, \underset{\sim}{X}^E)$, what is observed is $\underset{\sim}{X}^P$, whereas $\underset{\sim}{X}^E$ must be inferred. Of course this "transformation", a projection, is to some extent merely an artifact of the model. Nonetheless, from the viewpoint of statistical inference, the information loss is severe.

Whether the primary goal is segmentation or restoration, it can be useful to combine these tasks into a single algorithm. For example, we found in [8] that the inclusion of the edge process $\underset{\sim}{X}^E$ facilitated the restoration of images degraded in accordance with (7), especially when some a priori knowledge about the boundary behavior was available. Conversely, even if the object is segmentation, some de-blurring or noise removal may improve performance. Another advantage of the hierarchical MRF formulation is that these tasks <u>can</u> <u>be</u> <u>combined</u> into a single process. It is well-known that smoothing often degrades boundary behavior, thereby making the segmentation problem more difficult. See Marroquin [20] for a discussion of these issues in the context of surface reconstruction.

(iii) <u>Single Photon Emission Tomography</u>. In this case the pixel lattice corresponds to a discretization of a cross-section of tissue and $X_s = X_s^P$ represents the concentration of some isotope at site s. Particles are emitted in random directions from these sites and follow the usual Poisson laws for radiation counts. In particular, the number of particles emitted from s is a Poisson random variable with rate proportional to $X_s$. (The time interval is fixed, and hence can be ignored.) These particles are received and counted at banks of detectors which are placed around the lattice and in the same plane. However, there is attenuation due to the passage of the photons through the tissue or whatever media is storing the isotope. Thus, the number of received particles at a given detector k (which we denote $y_k$) is Poisson with rate

$$E\left[y_k\right] = \int_{L_\theta} X_t \, A_k(t) dt \tag{8}$$

where $L_\theta$ is the line with direction $\theta$, and A embodies the attenuation factor for the segment from t to the detector k as well as the details of the detector geometry. The object is to recover the isotope density $\{X_s\}$ from the detector counts. See [9] for more information.

## IV. POSTERIOR DISTRIBUTIONS.

Given the data $\underset{\sim}{Y}=y$, the posterior distribution is

$$\pi_y(x) = P(\underset{\sim}{X}=x \mid \underset{\sim}{Y}=y), \ x \epsilon \Omega .$$

This is a powerful tool for image analysis: in principle we can construct the optimal (Bayesian) estimator for $\underset{\sim}{X}$, examine images sampled from $\pi_y$, design near-optimal statistical tests for the presence of special objects, and so forth.

If the data transformation $\Gamma$ is sufficiently "local", then the conditional probability law of $\underset{\sim}{X}$ is also a MRF with a local graph structure. Let $U_y(x)$ denote the energy function in the representation of $\pi_y(x)$ as a Gibbs distribution;

$$\pi_y(x) = e^{-U_y(x)}/Z_y, \quad Z_y = \sum_x e^{-U_y(x)} , \tag{9}$$

Then if $\Gamma$ is local so is $U_y$. The practical import of this observation is that stochastic relaxation methods (such as the "heat-bath" and Metropolis algorithms) are feasible for analyzing $\pi_y$.

The types of degradation we have discussed are mostly "local". For example, the degradation in (7) leads to a locally-composed $U_y$ whenever D is small and $\eta$ is nearly white. In the case of boundary-finding, the posterior distribution is

$$\pi_y(x) = \pi_y(x^E) = P(\underset{\sim}{X}^E = x^E \mid \underset{\sim}{X}^P = x^P)$$

and the posterior energy $U_y$ is then simply the expression in (6) with $y=x^P$ fixed; in particular, the posterior graph is just the subgraph for the edge sites. In contrast, tomogrophy leads to a non-local posterior energy, and potentially severe computational problems. So far, we have largely avoided these by employing more conventional reconstructions as starting points for our Bayesian algorithm (see § VII and Geman and McClure [9] for more details).

The mode(s) of $\pi_y$ is called the maximum a posteriori or MAP estimator of x given y, and much of our previous work has focused on the development of an algorithm to find near-MAP estimates. The computational problem is formidable. We seek to minimize the posterior energy function $U_y(x)$ over $x \epsilon \Omega$. Typically, this function is highly non-linear, has an enormous number of (suitably defined) local minima, and the size of $\Omega$ is at least $2^{1000}$, corresponding to a very small (32×32), binary intensity array and no edge units.

To illustrate the problem, consider the prior in (1) with $\Phi(u) = (1+u^2)^{-1}$ and additive white Gaussian noise with variance $\sigma^2$. Then a simple calculation gives

$$U_y(x) = - \sum_{<s,t>} \frac{\theta}{1+(x_s^P - x_t^P)^2} + \frac{1}{2\sigma^2} \sum_s (x_s^P - y_s)^2. \tag{10}$$

The first term imposes smoothness and the second fidelity to the data, with relative emphasis in accordance with $\theta\sigma^2$.

## V. STOCHASTIC RELAXATION AND SIMULATED ANNEALING.

Stochastic relaxation is an iterative, site-replacement procedure for generating a sample configuration from a Gibbs distribution, $\pi$. In our applications, $\pi = \pi_y$, the posterior distribution given $\underset{\sim}{Y} = y$. We refer the reader to [8] for a complete treatment, including the origins in physics, the precise mathematical formulation, and a comparison with so-called "probabilistic relaxation" [16] or "relaxation labeling". Suffice to say that the algorithm generates a (nonstationary) Markov chain $\underset{\sim}{X}(k)$, $k = 0,1,2,...$, with state space $\Omega$ and asymptotic distribution $\pi$:

$$\lim_{k \to \infty} P(\underset{\sim}{X}(k) = x \mid \underset{\sim}{X}(0) = x^1) = \pi(x) \qquad x, x^1 \epsilon \Omega. \tag{11}$$

To find the mode(s) of $\pi$, we pursue the analogy with statistical mechanics and regard these configurations as the ground states of an (imaginary) physical system with energy $U(x)$. We then simulate the physical process of <u>annealing</u>, in which the slow decrease of <u>temperature</u> T forces the system into its low energy states. Roughly speaking, this occurs because the Boltzmann distribution <u>at temperature</u> T is $\pi_T =$ $\exp(-U(x)/T)/Z_T$ and hence, whereas the ground states are unchanged, their relative weight increases as T decreases. Simulated annealing then refers to the slow decrease of a control parameter T during the generation of the Markov chain. Given a decreasing sequence T(k), k = 1,2,..., the annealing algorithm generates a new Markov chain $\{\underset{\sim}{X}(k)\}$

whose asymptotic distribution as $k \to \infty$ is <u>uniform</u> over the set $\{z \in \Omega : U(z) = \min_X U(x)\}$. The only condition is that $T(k)$ decrease sufficiently slowly, namely that

$$T(k) \geqslant C/\log(1 + k) \tag{12}$$

for a certain constant $C=C(U)$; see [8], [10], and [12].

The algorithm is computationally feasible to the extent that $\pi$ is local because at iteration $k$ a sample must be obtained from the <u>conditional</u> distribution

$$\pi_{T(k)}(X_{s(k)} = \cdot \,| X_r = X_r(k-1), \; r \in G_{s(k)}),$$

where $\{s(k)\}$ is some pre-determined sequence for visiting the sites. (Of course, the computation time also depends heavily on the size of the graph, the intensity range, and other factors.) Finally, the algorithm is highly parallel in the sense that it can be executed by simple and alike processing units acting largely independently. Basically, one can cut the time in half with two processors, in thirds with three, etc. These processors would be assigned to collections of sites, and the exact degree of parallelism would depend on the chromatic index of the graph. For instance, with a nearest neighbor graph, one could update the "red" sites <u>simultaneously</u>, then all the "blacks", etc.

## VI. PARAMETER ESTIMATION.

In our previous work on image restoration, certain parameters which appear in the prior MRF image models, for example $\theta$ in (1), were not estimated from the data, but rather divined, guided by experience and the persistent observation that the quality of the restorations was surprisingly <u>insensitive</u> to the choice of $\theta$, at least over a fair-sized interval. However, in current experiments in tomography, segmentation, and computer vision the models are more complex and are likely to be still more so in envisioned work involving object recognition, texture analysis, etc. These new models (e.g. equation (6)) involve additional parameters whose interpretations are less apparent than, for instance, that of $\theta$ in (1). Moreover there is growing evidence that the algorithms are less robust. Consequently, one needs an accurate, data-driven method of parameter estimation.

Statistical inference is complicated by the high-dimensionality of the data and the severe loss of information in the transformation $\underset{\sim}{X} \to \underset{\sim}{Y}$. Thus it is perhaps not too surprising that the statistics and image processing literature contain very few papers that are relevant for situations akin to ours. The work of Besag ([2], and references therein) on "coding schemes" and "pseudo-likelihood" is an exception, although this work is primarily confined to the case of "complete data", i.e. $\underset{\sim}{Y} = \underset{\sim}{X}$. We of course are basically interested in the case of "incomplete data", as illustrated in § III.

In the statistical terminology, our models are "exponential families", which refers to the fact that the parameters appear _multiplicatively_ in the un-normalized log likelihoods. The major difficulty is that the normalizing constant Z, the so-called _partition function_ in statistical mechanics, is a function of these parameters and entirely intractable.

To illustrate the pitfalls in conventional approaches, we are going to briefly consider the difficulties encountered in maximum likelihood estimation. To simplify matters, however, we make the following assumption:

(i) $\Gamma$ is a projection, i.e. $\underset{\sim}{X} = (\underset{\sim}{W},\underset{\sim}{Y})$ where Y is observed and W is not observed. (Some authors refer to the elements of W as "hidden units".)

(ii) $\underset{\sim}{X}$ is MRF with a local graph structure.

(iii) The parameters $\theta = (\theta_1,\theta_2,...,\theta_J)$ appear _multiplicatively_ in the representation of the distribution of $\underset{\sim}{X}$ as a Gibbs measure:

$$P_\theta(\underset{\sim}{X}=x) = Z^{-1}(\theta) \quad \exp -\sum_{j=1}^{J} \theta_j U_j(x), \qquad x = (w,y).$$

These restrictions are actually less severe than might be expected; indeed, all the examples we have seen so far satisfy (i) - (iii).

Example 1. Consider the case of simple filtering with additive white Gaussian noise with mean 0 and variance $\sigma^2$, no edge process, and a prior on $X^P$ of the form (1). Then with $\underset{\sim}{Y}=\underset{\sim}{X}^P+n$ and $\underset{\sim}{W}=\underset{\sim}{X}^P$, the pair (W,Y) has distribution

$$P_\theta(\underset{\sim}{X}^P=x^P,\underset{\sim}{Y}=y) = Z^{-1}(\theta) \quad \exp -(\theta_1 U_1(x^P) + \theta_2 U_2(x^P,y)) \tag{13}$$

where $\theta_2 = (-2\sigma^2)^{-1}$, $U_1 = \underset{(s,t)}{\Sigma} \Phi(x_s^P - x_t^P)$, $U_2 = \underset{s}{\Sigma} (x_s^P - y)^2$, and $Z(\theta) = Z(\theta_1,\theta_2) = Z(\theta_1)(-\pi/\theta_2)^{N^2/2}$. The joint energy is clearly local. The same reasoning applies to more complex degradations of the family (7).

Example 2. Consider the pixel-edge model $\underset{\sim}{X} = (X^E,X^P)$. Taking $\underset{\sim}{W}=\underset{\sim}{X}^E$, $\underset{\sim}{Y}=\underset{\sim}{X}^P$, we see that the joint law is simply the _prior_ distribution on X. An illustration of the parameter estimation problem is then to estimate $\theta_1,\theta_2$ and $\theta_3$ in (6) based on observations of $\underset{\sim}{X}^P$. This is difficult for several reasons, the main one being that the _marginal_ distributions of $\underset{\sim}{X}^E$ and $\underset{\sim}{X}^P$ have fully-connected graphs! In particular, the "likelihood function" $P_\theta(x^P)$ is intractable.

a) Maximum likelihood estimation. The distribution of the observed variables is

$$P_\theta(y) = \frac{Z(\theta|y)}{Z(\theta)}, \qquad Z(\theta|y) = \underset{w}{\Sigma} e^{\sum_j \theta_j U_j(w,y)}.$$

In the classical case of independent and identically distributed observations $y^{(1)},...,y^{(n)}$, the (normalized) log-likelihood is

$$\frac{1}{n} \sum_{k=1}^{n} \log Z(\theta|y^{(k)}) - \log Z(\theta)$$

and the likelihood equations,

$$\nabla \log \prod_{k=1}^{n} P_{\theta}(y^{(k)}) = 0 \quad \text{reduce to}$$

$$E_{\theta} U_j = \frac{1}{n} \sum_{k=1}^{n} E_{\theta}(U_j|y^{(k)}), \quad j = 1,2,...,J. \tag{14}$$

This system is intractable as it stands: the expected values are impossible to calculate (for the same reasons that mean energies are in spin-glasses and the like) and even when they can be _estimated_ (by sampling) there still remains the problem of _solving_ (14). In particular, the log likelihood is highly _non_-convex.

The "EM" algorithm is an iterative scheme designed for solving systems such as (14), although mainly in more conventional settings involving familiar densities (normal, Poisson, etc.) and much lower dimensional data. The algorithm does not seem suitable in its customary form and we have developed a number of modifications. The basic idea is to generate a sequence $\hat{\theta}^{(k)}$ of estimates intended to converge to a local maximum of the likelihood. We "update" $\hat{\theta}^{(k)}$ by first _sampling_ from the _posterior_ distribution $P_{\theta}(\underset{\sim}{W}=w|\underset{\sim}{Y}=y)$ at $\theta=\hat{\theta}^{(k)}$ and then choosing $\hat{\theta}^{(k+1)}$ to maximize or simply increase the _joint_ likelihood $P_{\theta}(\hat{w}^{(k)},y)$. Results and experiments will be reported elsewhere.

b)  _A Priori Constraints_.  We have started using (a more complex version of) the model in equation (6) for the segmentation of "natural scenes" such as faces and houses, and for the segmentation of infrared imagery.  However in addition to the generic difficulties discussed earlier, statistical inference for $\underset{\sim}{\theta} = (\theta_1,\theta_2,\theta_3)$ from the intensity image $\underset{\sim}{X}^P$ is further complicated by the fact that relatively disparate values of the parameter $\theta$ may induce essentially the _same marginal_ distribution on $\underset{\sim}{X}^P$. Moreover, not all of these values may correspond to "good segmentations"; for example, the likely states of $P_{\theta}(\underset{\sim}{X}^E=\cdot|\underset{\sim}{X}^P)$ may not conform to our prior expectations of where the boundaries "belong" in $\underset{\sim}{X}^P$. Therefore, estimation based on the intensity image is not possible.

Fortunately, we can use these prior expectations to restrict the parameter space.  In fact, we can sometimes identify a small region of the parameter space, $\Lambda \subseteq R^3$, with the following property:   given a class of very simple "training images" $\underset{\sim}{X}_Y$ for which a desired segmentation $\underset{\sim}{X}_Y^E$ is simply and unambiguously defined, the posterior distribution $P_{\theta}(\cdot|\underset{\sim}{X}^P=\underset{\sim}{X}_Y)$ will be maximized at $\underset{\sim}{x}^E=\underset{\sim}{x}_Y^E$ only if $\theta \in \Lambda$. We refer to this as "reparametrization" because the set $\Lambda$ depends on other parameters which _directly_ reflect our characterization of "good segmentations".  For example, one such parameter might be that value of the minimum difference across the boundary between a (candidate) "object" and "background" such that we have no preference whether or not to segment the object. Typically, $\Lambda$ turns out to be a line. Estimation is then reduced to one scale parameter corresponding to temperature, and this is rather easily handled by the variations on EM discussed earlier.


## VIII EXPERIMENTAL RESULTS.


There are three sets of experiments, intended to illustrate a variety of image and degradation models previously described.

a) Blocks. These results appear in [8] and are reproduced here to illustrate the power of the hierarchical approach for image restoration. The original image, Figure 3(a), is "hand-drawn". We added Gaussian noise with mean 0 and variance $\sigma^2 = .49$ to produce Figure 3(b). We then attempted restorations with and without $\underset{\sim}{X}^E$. Figure 3(c) is the restoration with simulated annealing with the prior in (1) with $\theta = 1/3$ and $\Phi$ as in (2). The inclusion of $\underset{\sim}{X}^E$ yields significant improvement - Figure 3(d). The model is essentially (6) with $\Phi$ above, $\theta_1 = 1$, $\theta_2 = 0$, $\theta_3 = .9$, and the following clique weights: $\xi_1 = 0$, $\xi_2 = 1$, $\xi_3 = \xi_4 = 2$, $\xi_5 = \xi_6 = 3$. The reason for favoring straight lines is obvious from Figure 3(a), and nicely illustrates the use of prior knowledge. The second set of block pictures illustrates the flexibility of the model in regard to different degradations g, H, etc. The original was corrupted (Figure 4(a)) according to $y_s = (H(x^P)_s)^{1/2} \cdot \eta_s$, where H puts weight 1/2 on the central pixel and 1/16 on the eight nearest neighbors. The (multiplicative) noise has mean 1 and $\sigma = .1$; the model is the same as before. The restoration, Figure 4(b), is nearly perfect.

b) Infrared. The upper left panel in Figure 5 is an infrared picture. There is one vehicle, with engine running. The intensity data represents corrupted thermal radiation. As with most photoelectronic systems, the imaging system consists of an optical subsystem, arrays of detectors, and a scanner.

There are a number of sources of blur and noise. For example, there is "background noise" due to the fluctuations of black body radiation, noise in the conversion of photons to electric current, and digitization noise. In addition the detectors cause spatial and temporal blurring. Finally, there is attenuation and diffraction at the optical stage.

No effort has been made to model each of these effects, except to note that the model in (7) offers a good first approximation with appropriate choices of g, H and $\eta$. Instead, the picture was segmented and restored under the simple degradation model $y_s = x_s^P + \eta_s$, $\{\eta_s\}$ i.i.d. $N(\theta, \sigma^2)$. The variance, $\sigma^2$, was estimated from the raw grey-level data "by eye", to be 16. The Gibbs prior was on the pixel-edge process $\underset{\sim}{X} = (\underset{\sim}{X}^P, \underset{\sim}{X}^E)$,

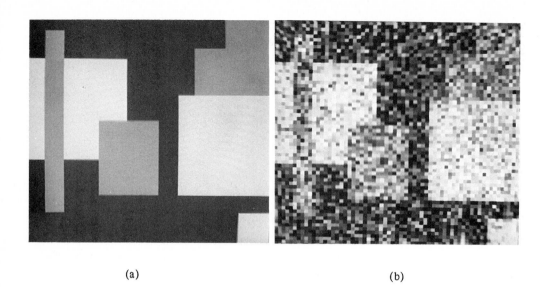

<table>
<tr><td align="center">(a)</td><td align="center">(b)</td></tr>
<tr><td align="center">Original</td><td align="center">Original corrupted by added noise</td></tr>
</table>

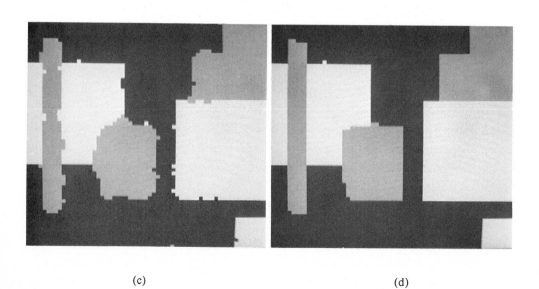

<table>
<tr><td align="center">(c)</td><td align="center">(d)</td></tr>
<tr><td align="center">Restoration without edge process</td><td align="center">Restoration with edge process</td></tr>
</table>

Figure 3

(a)

(b)

Original corrupted by blur, nonlinear transformation, and multiplicative noise

Restoration with edge process

Figure 4

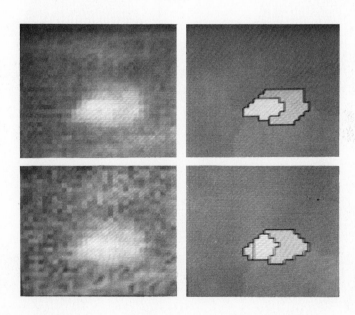

Figure 5

upper left:   infrared image, including one vehicle with hot engine
upper right:  original restored and segmented
lower left:   original corrupted by added noise
lower right:  corrupted restored and segmented

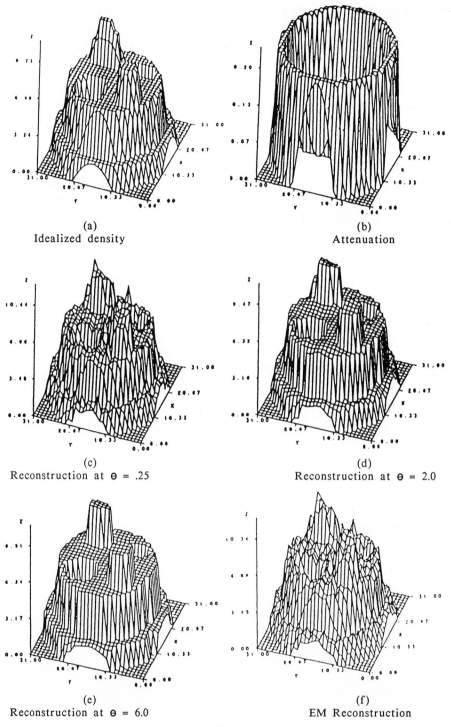

(a)
Idealized density

(b)
Attenuation

(c)
Reconstruction at θ = .25

(d)
Reconstruction at θ = 2.0

(e)
Reconstruction at θ = 6.0

(f)
EM Reconstruction

Figure 6

and the energy was a slight modification of (6), the modification allowing for the breaking of diagonal pixel bonds by the presence of suitable contiguous pairs of edges. The parameter values were chosen by "reparametrization" (see § VI), with the "scale" fixed after experimenting with a range of values.

The upper right panel of Figure 5 is the restored and segmented picture, via simulated annealing. The locations of edges (slightly displaced to coincide with pixel sites) are shown in black. The lower left panel is the original picture corrupted by adding zero mean white Gaussian noise, with variance 16. The same restoration/segmentation was applied to the corrupted picture, except that the assumed noise variance, $\sigma^2$, was adjusted for the additional degradation: $\sigma^2 = 16+16 = 32$. The result is the lower right panel.

c) <u>Tomography.</u> The details of this experiment are in [9]. The object is to reconstruct the idealized isotope concentration shown in Figure 6(a). The observable photon counts $\{y_k\}$ were simulated in accordance with the degradation model given in (8). The detector geometry and assumed attenuation function are incorporated into A. Figure 6(b) shows the attenuation function, which is proportional to the probability of photon absorption per length of travel.

Reconstructions were generated under the prior in (1) with $\Phi$ as in (3), except that diagonal bonds were reduced by a factor of $1/\sqrt{2}$ (again, see [9] for the full story). Figures 6(c), 6(d), and 6(e) are approximate MAP esimators at $\theta=.25$, $\theta=2.0$, and $\theta=6.0$, respectively, and $C_1=.7$, $C_2=2$. Obviously, the value of $\theta$ is important. We have begun to experiment with estimation of $\theta$ using the variations of EM discussed earlier (see § VI), and the preliminary results suggest that satisfactory estimates may be possible from single observations of the Poisson process (photon counts) $\underset{\sim}{Y}=\{y_k\}$.

These reconstructions did not involve annealing. Instead, they were obtained by a simple gradient descent of the posterior energy, starting from the <u>maximum likelihood</u> reconstruction. The latter was achieved by an implementation of EM due to N. Accomando [1]. For comparison, the maximum likelihood reconstruction is shown in Figure 6(f).

REFERENCES

1.  N. Accomando, "Maximum likelihood reconstruction of a two dimensional Poisson intensity function from attenuated projections," Ph.D. thesis, Division of Applied Mathematics, Brown University, 1984.

2.  J. Besag, "On the statistical analysis of dirty pictures," preprint, Department of Statistics, of Durhan, U.K., 1985.

3.  V. Cěrný, "A thermodynamical approach to the travelling salesman problem: an efficient simulation algorithm," preprint, Institue of Physics and Biophysics, Comenius Unive, Bratislava, 1982.

4.  F.S. Cohen and D.B. Cooper, "Simple parallel hierarchical and relaxation algorithms for segmenting noncausal Markovian random fields," preprint, Brown University, 1984.

5.  G.C. Cross and A.K. Jain, "Markov random field texture models," IEEE Transaction Pattern Analysis Machine Intelligence., vol. PAMI-5, pp.25-39, 1983.

6.  P.A. Devijver, "Probabilistic labeling in a hidden second order markov mesh," technical report, Philips Research Laboratory, Brussels, Belgium, 1985.

7.  H. Elliott and H. Derin, "Modelling and segmentation of noisy and textured images using Gibbs random fields," preprint, Department of Electrical and Computer Engineering, University of Massachusetts, 1984.

8.  S. Geman and D. Geman, "Stochastic relaxation, Gibbs distributions, and the Bayesian restoration of images," IEEE Trans. Pattern Anal. Machine Intell., vol. PAMI-6, pp.721-741, 1984.

9.  S. Geman and D.E. McClure, "Bayesian image analysis: An application to single photon emission tomography," 1985 Proceedings of the American Statistical Ass ociation. Statistical Computing Section. (to appear).

10. B. Gidas, "Nonstationary Markov chains and convergence of the annealing algorithm," Journal of Statistical Physics, Vol. 39, pp.73-131, 1985.

11. U. Grenander, "Tutorial in Pattern Theory," Division of Applied Mathematics, Brown University, 1984.

12. B. Hajek, "Cooling schedules for optimal annealing," preprint, Department of Electrical Engineering and the Coordinated Science Laboratory, University of Illinois at Champaign-Urbana, 1985.

13. A.R. Hanson and E.M. Riseman, "Segmentation of natural scenes," Computer Vision Systems. New York: Academic Press, 1978.

14. G.E. Hinton and T.J. Sejnowski, "Optimal perceptual inference," in Proceedings IEEE Conference Computer Vision Pattern Recognition, 1983.

15. J.J. Hopfield, "Neural networks and physical systems with emergent collective computational abilities," Proceedings of the National Academy of Sciences USA, Vol. 79, pp.2554-2558, 1982.

16. R.A. Hummel and S.W. Zucker, "On the foundations of relaxation labeling processes," IEEE Transaction Pattern Analysis Machine Intelligence, vol. PAMI-5, pp.267-287, 1983.

17. H.T. Kiiveri and N.A. Campbell, "Allocation of remotely sensed data using Markov models for spectral variables and pixel labels," preprint, CSIRO Division of Mathematics and Statistics, Sydney, Australia, 1985.

18. J. Kirkpatrick, C.D. Gelatt, Jr. and M.P. Vecchi, "Optimization by simulated annealing," IBM Thomas J. Watson Research Center, Yorktown Heights, N.Y. 1982.

19. C. von der Malsburg, E. Bienenstock, "Statistical coding and short term synaptic plasticity: A scheme for knowledge representation in the brain," (this volume).

20. J.L. Marroquin, "Surface reconstruction preserving discontinuities," Laboratory for Information and Decision Systems, M.I.T., Cambridge, MA, 1984.

# The Langevin Equation as a Global Minimization Algorithm

Basilis Gidas[*]
Division of Applied Mathematics
Brown University

I. Introduction. During the past two years a great deal of attention has been given to simulated annealing as a global minimization algorithm in combinatorial optimization problems [11], image processing problems [2], and other problems [9]. The first rigorous result concerning the convergence of the annealing algorithm was obtained in [2]. In [4], the annealing algorithm was treated as a special case of non-stationary Markov chains, and some optimal convergence estimates and an ergodic theorem were established. Optimal estimates for the annealing algorithm have recently been obtained by nice intuitive arguments in [7].

The annealing algorithm (a variant of the Metropolis algorithm [12]) has been designed mainly for complex systems where the underlying random variables ("spins") have a discrete state space. Motivated by image processing problems with continuous grey-levels, U. Grenander [6], and S. Geman proposed the use of the Langevin equation (with time dependent temperature) as a global minimization algorithm. A convergence theorem for this algorithm was obtained in [3]. Independently, G. Parisi [13] has proposed the Langevin equation as a tool for studying Lattice Gauge Theories.

In [5], we have analyzed the mathematical structure of the Langevin equation (LE), have established the convergence of the algorithm in the entire of $\mathbb{R}^d$, $d \geq 1$, and have obtained (in certain cases) the optimal temperature schedule. Here we summarize some of the results of [5], and indicate their proofs. We have tested the algorithm on the computer for various examples (such as the six-hump camel function [1]) of traditional global optimization [10]. If the function to be minimized has a large number of local minima (we have treated

---

* Partially supported by NSF Grant MCS-8301864.

a function in 1-dimension with nearly 400 local minima, and a
single global minimum), the LE seems to perform better than
other stochastic methods [10].  For low dimensional problems,
a combination of the LE with other stochastic methods might
improve the speed of convergence.  However, for complex systems such as image processing problems, the traditional stochastic methods [10] are impracticable, while the LE is presently the best method.

II.  The Physics of the Langevin Equation.  Here we present a
brief description of the physics which underlies the Langevin
equation (L.E.).  Our mathematical analysis has been guided
by this physics.

The LE at temperature  T  reads

$$dX(t) = \sqrt{2T}\, dw(t) - \nabla f(X(t))dt. \qquad (2.1)$$

Here  $w(t)$  is the standard Brownian (Wiener) process.
The function  $f(x)$, $x \in \mathbb{R}^d$, $d \geq 1$, is bounded below and
grows properly (see next section) as  $|x| \to +\infty$.  Equation
(2.1) defines a d-dimensional diffusion process  $X(t)$.

Equation (2.1) was proposed by Langevin in 1908, as an
equation describing the motion of a particle in a viscuous
fluid.  It generalized Einstein's theory (1905) of the Brownian motion, i.e. the motion of a particle in a fluid without
viscosity.  Langevin's equation was the first mathematical
equation describing non-equilibrium thermodynamics.  The first
term  $\sqrt{2T}\, dw(t)$  corresponds to the microscopic fluctuations
caused by the Brownian force.  While the second term,
$-\nabla f(X(t))dt$, called the "drag" force, is generated by the viscosity of the fluid (it is proportional to the velocity of the
moving particle, and opposite to its direction of motion).
The LE has two time scales: The short time scale where the
Brownian fluctuations dominate, and the long time scale where
the drag force dominates.

Onsager gave a new interpretation to the LE as an irreversible process by a simple correspondence of the two terms:
The velocity of the particle is interpreted as the deviation
of a thermodynamic quantity from its equilibrium value, while

the drag force is interpreted as the "drift" of the thermodynamic systems towards its equilibrium.  In Onsager's interpretation, the thermal fluctuations dominate in the short time,
while the drift dominates in the long time scale.  Thus
Onsager's observation established a mathematical equivalence
between the motion of a particle in a viscous fluid and the
approach to equilibrium of a non-equilibrium state.

When the temperature depends on time, $T = T(t)$, and
$T(t) \rightarrow 0$  (at an appropriate rate) as  $t \rightarrow +\infty$, the LE may be
interpreted as the decay of "metastable" states into the
ground state(s) of a thermodynamic system.  In this case, the
LE has many time scales, corresponding to the length of times
the non-stationary diffusion process takes to get out of local
minima and eventually fall into a global minimum.  Our mathematical analysis isolates all these time scales – the worst of
which determines the best temperature schedule.

From the mathematical point of view, the behavior of
(2.1) as  $T \downarrow 0$  is analogous to the limit  $h \downarrow 0$  of Quantum
Mechanics via the Schrödinger operator [14]  $-h^2\Delta + V(x)$.  The
limit  $T \downarrow 0$  of (2.1) is very similar to Witten's proof [15]
of the Morse inequalities (see [8] for a mathematically rigorous treatment of Witten's result).  Our techniques for studying
equation (2.1) with time-dependent temperature employs the
maximum principle and the concepts of supersolutions and subsolutions for parabolic partial differential equations.

III.  Results.  Let

$$p(t,x) \equiv p(t_0,x_0;t,x) = P\{X(t) = x | X(t_0) = x_0\} \qquad (3.1)$$

be the transition function of the non-stationary diffusion
defined by (2.1) with  $T = T(t)$.  Also, let

$$\pi(t,x) \equiv \pi_{T(t)}(x) = e^{-\frac{1}{T(t)} f(x)} \Big/ \int e^{-\frac{1}{T(t)} f(x)} dx. \qquad (3.2)$$

For  $T$  independent of  $t$, $\pi_T(x)$  is the equilibrium (invariant)
measure of the stationary process defined by (2.1).

Here is our first result.

Theorem 3.1.  Let  $f(x)$, $x \in \mathbb{R}^d$, $d \geq 1$  be a  $C^\infty$  function such
that  $f(x)$  and  $|\nabla f(x)|$  go to  $+\infty$  as  $|x| \rightarrow +\infty$.  Furthermore,

assume that $-\Delta f/(1+|\nabla f|^2)$ is bounded below. Then there exists a constant $\Delta > 0$ such that if $T(t) \to 0$ as $t \to +\infty$, and

$$\int_{t_0}^{+\infty} e^{-\frac{\Delta}{T(t)}} \, dt = +\infty \tag{3.3}$$

then in the weak sense

$$|p(t_0,x_0;t,\cdot) - \pi(t,\cdot)| \to 0 \quad \text{as} \quad t \to +\infty. \tag{3.4}$$

The smoothness assumption on $f(x)$ can be weakened [5]. Before we indicate the proof of Theorem 3.1, we give the optimal value of $\Delta$ in a special case: Assume that the critical set of $f(x)$ consists of isolated points, and let $a_1, a_2, \ldots, a_N$ be the minima of $f(x)$. For simplicity, we assume that $f(x)$ has only one global minimum $a_N$. Furthermore, we assume that $a_1, \ldots, a_N$ are nondegenerate, i.e. the Hessian of $f(x)$ at $a_1, \ldots, a_N$ is positive definite. For each $a_i$, $i = 1, \ldots, N-1$, we consider curves $\gamma(s)$, $s \in [0,1]$ such that $\gamma(0) = a_i$, $\gamma(1) = a_N$ and define

$$\Delta_i = \min_{\gamma} \max_{s \in [0,1]} f(\gamma(s)) - f(a_i) \tag{3.5$_a$}$$

$$\Delta = \max_{i=1,\ldots,N-1} \Delta_i . \tag{3.5$_b$}$$

In one dimension ($d = 1$), the best value of the constant $\Delta$ in (3.3) is given by (3.5$_b$)(of course in $d = 1$, there is only one curve $\gamma$ that connects $a_i$ and $a_N$ — the straight line segment between $a_i$ and $a_N$). In higher dimensions, $d \geq 1$, we have obtained [5] the same result only for $N = 2$. Our proof of Theorem 3.1 is based on the "forward" parabolic equation

$$\frac{\partial p(t,x)}{\partial t} = T(t)\Delta p + \nabla(\nabla f p) \tag{3.6$_a$}$$

$$p(t,x)|_{t \downarrow t_0} = \delta(x - x_0) . \tag{3.6$_b$}$$

We transform this equation by defining $q(t,x)$, $r(t,x)$ via

$$p(t,x) = \pi(t,x)q(t,x), \quad q(t,x) = (\pi(t,x))^{-\frac{1}{2}}r(t,x). \tag{3.7}$$

A straightforward computation gives

$$\frac{\partial q}{T(t)\partial t} = \Delta q - \frac{1}{T(t)}\nabla f \cdot \nabla q - \frac{T'}{T^3}(f(x) - <f>^{(t)})q, \tag{3.8$_a$}$$

$$q(t,x)\big|_{t\downarrow t_0} = \frac{\delta(x - x_0)}{\pi(t_0,x_0)} \tag{3.8$_b$}$$

and

$$\frac{\partial r}{T(t)\partial t} = \Delta r - \frac{1}{(2T(t))^2}\{|\nabla f|^2 - 2T(t)\Delta f\}r - \frac{1}{2}\frac{T'}{T^3}(f(x)-<f>^{(t)})r, \tag{3.9$_a$}$$

$$r(t,x)\big|_{t\downarrow t_0} = \frac{\delta(x - x_0)}{\pi^{\frac{1}{2}}(t_0,x_0)} \tag{3.9$_b$}$$

where $T' = \frac{dT}{dt}$, $<f>^{(t)} = \int f(x)\pi(t,x)dx$. The transition from (3.8) to (3.9) is equivalent to the transformation which connects the Cameron-Martin-Girasov formula and the Feynman-Kac formula. The proof of (3.4) is reduced to proving that $q(t,x) \to 1$, as $t \to +\infty$. This is achieved by using the Maximum Principle and constructing appropriate barrier functions (i.e. supersolutions and subsolutions) which bound $q(t,x)$ from above and below and converge to 1 as $t \to +\infty$. The details of the construction are lengthy and will appear in [5]. The first step consists in studying the eigenvalues and eigenfunctions of the operator

$$L = -\Delta + \frac{1}{T}\nabla f \cdot \nabla \tag{3.10}$$

for time independent but arbitrarily small temperature $T$. The operator $L$ is unitarily equivalent to

$$H = -\Delta + \frac{1}{(2T)^2}\{|\nabla f|^2 - 2T\Delta f\}. \tag{3.11}$$

In fact, $L = (\pi_T(x))^{-\frac{1}{2}}H(\pi_T(x))^{\frac{1}{2}}$. $L$ is self-adjoint on $L_2(\mathbb{R}^d, \pi_T(x)dx)$, while $H$ is self-adjoint on $L_2(\mathbb{R}^d, dx)$. Under the assumptions of Theorem 3.1 for $f(x)$, $L$ and $H$ have a discrete spectrum. Let $\lambda_1(T) < \lambda_2(T) \le \lambda_3(T) \le \ldots$, be the eigenvalues of $L$ and $H$. We note that $\lambda_1(T) = 0$, and the corresponding eigenvector of $H$ is $(\pi_T(x))^{\frac{1}{2}}$ (the corresponding eigenvector of $L$ is 1). Let $\varphi_n(x)$, $n = 1,2, \ldots$ be the eigenfunction of $H(\varphi_1(x) = \pi_T^{\frac{1}{2}})$. For time independent $T$ we have

$$q(t,x) = 1 + e^{-\lambda_2(T)Tt}\frac{\varphi_2(x_0)}{\varphi_1(x_0)}\frac{\varphi_2(x)}{\varphi_1(x)} +$$

$$\sum_{n=3}^{+\infty} e^{-\lambda_n(T)Tt}\frac{\varphi_n(x_0)}{\varphi_1(x_0)}\frac{\varphi_n(x)}{\varphi_1(x)}. \tag{3.12}$$

Under the assumptions of Theorem 3.1 for $f(x)$, $\lambda_2(T)$ converges to zero exponential as $T \downarrow 0$. Under the conditions on $f(x)$ we mentioned before and after (3.5)(i.e. $d = 1$, or $N = 2$), we have proven

Theorem 3.2. With $\Delta$ given by $(3.5_b)$, we have

$$-\lim_{T \downarrow 0} T \log \lambda_2(T) = \Delta \qquad (3.13)$$

Remark. In the proof of Theorem 3.1, condition (3.3) appears as $\int_{t_0}^{+\infty} \lambda_2(T(s))ds = +\infty$. As we mentioned before, the proof uses barrier functions for (3.8). These barrier functions are constructed in terms of some approximants of the eigenfunctions of H. Although the technical details are complicated, the intuitive origin of (3.3) is apparent from (3.12) and (3.13).

References

1. Dixon, L.C.W., and G.P. Szegö (eds.): Towards Global Optimization 2, North-Holland, (1978).
2. Geman, S. and D. Geman: "Stochastic Relaxation, Gibbs Distributions, and the Bayesian Restoration of Images", IEEE transactions, PAMI 6(1984), 721-741.
3. Geman, S. and C.R. Huang: "Diffusions for Global Optimization", preprint, 1984.
4. Gidas, B.: "Non-Stationary Markov Chains and Convergence of the Annealing Algorithm", J. Stat. Phys. 39 (1985), 73-131.
5. Gidas, B.: "Global Minimization via the Langevin Equation" in preparation.
6. Grenander, U.: Tutorial in Pattern Theory, Brown University, (1983).
7. Hajek, B.: "Cooling Schedules for Optimal Annealing", preprint, 1985.
8. Helffer, B. and I. Sjöstrand: "Puits Multiples en Mecanique Semi-Classique IV, Etude du Complexe de Witten", preprint, 1984.
9. Hinton, G., T. Sejnowski, and D. Ackley: "Boltzmann Machine: Constraint Satisfaction Networks that Learn", preprint, 1984.
10. Kan, A., C. Boender, and G. Timmer: "A Stochastic Approach to Global Optimization", preprint, 1984.
11. Kirkpatrick, S., C.D. Gelatt, and M. Vecchi: "Optimization by Simulated Annealing", Science 220, 13 May (1983) 621-680.
12. Metropolis, N., et. al.: "Equations of State Calculations by Fast Computing Machines", J. Chem. Phys. 21 (1953), 1087-1091.
13. Parisi, G.: "Prolegomena To any Further Computer Evaluation of the QCD Mass Spectrum", in Progress in Gauge Field Theory Cargese (1983).
14. Simon, B.: "Semiclassical Analysis of Low Lying Eigenvalues I. Non-degenerate Minima: Asymptotic Expansions", Ann. Inst. Henri Poincaré 38 (1983), 295-307.
15. Witten, E.: "Supersymmetry and Morse Theory", J. Diff. Geometry 17 (1982), 661-692.

# SPIN GLASS AND PSEUDO-BOOLEAN OPTIMIZATION

I. G. Rosenberg
Mathematics and Statistics
Universite de Montreal
C.P.6128, Succ."A"
Montreal, Quebec H3C 3J7 Canada

1. The minimization of pseudo-boolean functions started in the early sixties and since then has been studied as a discrete optimization problem in operations research, mathematical programming and combinatorics. The author was quite surprised to learn that more recently the problem emerged as the spin glass problem in statistical mechanics. The probabilistic approach developed there has been applied to various combinatorial problems and even to biology or cellular automata.

In the spin glass problem the variables $c_i$ range over $\{-1,1\}$ while those below take values $\{0,1\}$. Using the well-known transformation $x' = \frac{1}{2} + \frac{1}{2}x$ we can translate the former into the latter and so there is no loss of generality in the 0-1 approach.

2. Let $\underline{2} := \{0,1\}$ and $n > 0$ integer. The set $\underline{2}^n$ consists of n-tuples over $\underline{2}$ which are also called the vertices of the n-dimensional hypercube. An n-*ary pseudo-boolean function* $f$ associates to each $x \in \underline{2}^n$ a real value $f(x)$. The name was introduced by Hammer and Rudeanu in the early sixties and a survey of the early developments is in [2]. If all the values of $f$ are in $\underline{2}$ we say that $f$ is a *boolean* (or logical or switching) *function*. In the last 137 years, boolean functions became important for logic, switching theory, 0-1 optimization and other fields. We start with the following ([2], II Thm. 6):

*To every n-ary pseudo-boolean function f there exists a unique family $\Phi$ of subsets of $\{1,\ldots,n\}$ and unique reals $c_F$ $(F \in \Phi)$ such that for all $x_1,\ldots,x_n \in \underline{2}$*

$$f(x_1,\ldots,x_n) = \sum_{F \in \Phi} c_F \prod_{i \in F} x_i . \qquad (1)$$

Note that the n-variable real polynomial (1) is the sum of the products of

NATO ASI Series, Vol. F20
Disordered Systems and Biological Organization
Edited by E. Bienenstock et al.
© Springer-Verlag Berlin Heidelberg 1986

distinct variables, i.e. contains no square or higher power and thus is linear in each variable.

3. The minimization of $f$ is the problem (P):

(P)     minimize $f(x)$  subject to  $x \in \underline{2}^n$     (2)

requiring to find a *minimizing point* $a \in \underline{2}^n$ such that $f(a)$ is the absolute minimum of $f$ on $\underline{2}^n$ (i.e. $f(a) \leq f(x)$ for all $x \in \underline{2}^n$).

At first glance (P) seems to be easy but this apparent simplicity is deceiving. It is shown in [2,3] that many NP-complete problems (cf. [1]) may be translated into (P). This is often done through a "Lagrange multiplier" approach (cf. [2] VI.2 or [3]) together with a lower bound on it (which is important for computations to avoid overflowing). This simple trick has been reinvented many times.

4. The problem (P) is trivial if $f$ is linear. The hamiltonian in the spin glass problem is quadratic and so (P) for a quadratic $f$ is the first interesting problem. It is essentially the only problem in the following sense. It is shown in [4] that (P) for more than quadratic $f$ can be reduced to (P) with a quadratic $g$ at the price of introducing additional (slack) variables. The idea is to replace a fixed product $x_i x_j$ everywhere in $f$ by a new variable $x_{n+1}$ and add a penalty term
$\lambda(x_i x_j + (3-2x_i-2x_j)x_{n+1})$ to $f$ (a bound for $\lambda$ is given in [4]). Repeating this many times in the end we get a quadratic $g$. From the computational point of view the addition of many new variables is a serious matter, but, at least in principle, the quadratic case is the general one. Anyway it is certainly sensible to start with the quadratic case.

5. In (P) we minimize $f$ over the discrete set $\underline{2}^n$. If we relax $x_i \in \{0,1\}$ to $0 \leq x_i \leq 1$ we obtain the following continuous problem

(C)     minimize $f(x)$  subject to  $x \in H := [0,1]^n$.

It is somewhat surprising [5] that (P) and (C) are identical (in the sense that the minimizing points of (C) make up the faces of the hypercube $H$

spanned by the minimizing points of (P)) but it is a simple consequence
of (1).

6. Using 5, we may try to find the local minimizing points of $H := [0,1]^n$
in the following way. First define the H-*gradient*. For $y \in H$ let
$\nabla(y) = (g_1,...,g_n)$ be the gradient of $f$ at $y$ (i.e. $g_i \equiv \partial f/\partial x_i|y$).
Put $\nabla_H(y) = (g_1^*,...,g_n^*)$ where (i) $g_i^* = \max(0,g_i)$ if $y_i = 0$ (ii) $g_i^* =$
$\min(0,g_i)$ if $y_i = 1$, and (iii) $g_i^* = g_i$ if $0 < y_i < 1$ (i=1,...,n).
We say that $y \in H$ is a *trap* if $\nabla_H(y)$ is the zero vector $\underline{0} = (0,...,0)$,
a strong trap if $y$ is a local minimizing point and a weak trap other-
wise. Due to (1) a trap $y \in \underline{2}^n$ is always strong. An oriented curve in
H is a *line of steepest H-descent* if at each point $x$ the directed tangent vector is $-\nabla_H(x)$. Such a line agrees with the standard line of
steepest descent on the interior of H and clearly ends up if it hits a
trap.

A point $x \in H$ *belongs* to a trap $y$ if either $x = y$ or the line of
steepest H-descent ends up in $y$. The set $A_y$ of all $x \in H$ belonging to
$y$ is the *region drained* by $y$. For $n = 2$ we can visualize the situation
as follows. The surface $z = f(x,y)$ may be viewed as terrain (with
valleys, mountains, crests, passes, etc.) over the unit square H which is
surrounded by walls on all four sides. The set $A_y$ is the part of the
map (the region) draining rainwater into y. Here $A_y = \{y\}$ of $y$
corresponds to a pass (= saddle point). A line of steepest H-descent is
also the path (on the map) of a trace of a skier who chooses always the
most precipitous direction and $A(y)$ consists of all points from which
the skier would reach y. The regions $A_y$ clearly partition H.

We have the following procedure to get to a locally minimizing point.
Start from a point $x \in H$ (say the center $(\frac{1}{2},...,\frac{1}{2})$) and go to the trap
$y$ to which $x$ belongs. If $y$ is a weak trap, perturb slightly the
coordinates of $y$ to obtain $x'$ and restart with $x'$ instead of $x$.
When a strong trap $y$ we record both $y$ and the value $f(a)$ and we compare it with the current best local minimal value.

7.  We could be satisfied with the value found.  If we want to do better
or to be sure that we cannot improve any further we have to visit other
local minimizing points.  (It is a common experience that in similar problems such a checking may take over 90p.c. of computing time).  Thus
suppose we are at a local minimizing point  y  and wish to restart at a
new point  x  preferably outside the already visited regions.  We can
imagine the following strategies:

    1. Select x randomly

    2. Choose  $x = (1-y_1,\ldots,1-y_n)$  (the opposite vertex of H)

    3. Put  $\alpha = \nabla_H(y)$  and consider  $p(t) := f(y+\alpha t)$  for positive
       real  t.  Let  $t_1$  be the greatest value so that  $p(t_1) \in H$.
       We can choose  $x := y+\alpha t_1$.  If  p  is quadratic, concave and
       with peak at  $t_2 < t_1$  we can also choose  $x = y+\alpha t$  for some
       $t_2 < t < t_1$.

The main problem is that we do not know the regions  $A_y$.  There are several
interesting questions (e.g. what are the quadratic  f  with few  $A_y$, say
1,2, or 3;  the situation is clear for  n = 2).

Assuming a reasonably smooth behaviour it does not make sense to choose a
new point  x  close to a part of a line of steepest H-descent already visited.  This could help for strategy 1 if we can manage to organize the
storage and retrieval of the necessary information.

8.  The sections 6-7 refer to any pseudo-boolean function.  For a quadratic
one the line of steepest descent may be found as follows.  Let  M  be the
matrix of the quadratic part of f.  There is an orthogonal matrix  U  such
that  $U^T MU$  is a diagonal matrix with diagonal  $(\lambda_1,\ldots,\lambda_n)$.  The lines of
steepest descent are obtained from  $x_i = y_i \exp(-2\lambda_i t)$  $(i=1,\ldots,n)$  through
the rotation given by the diagonal  $(\lambda_1,\ldots,\lambda_n)$. This may be too expensive but i
gives a good idea about the shape of the lines of steepest descent.  In
general the lines of steepest H-descent will be approximated by curves
consisting of short line segments.

9.  Paragraphs 6-8 were essentially presented in [6] but have not been
published.  Due to space limitation we cannot mention all the more recent
important work done by P.L. Hammer, P. Hansen, L.Lovász, M. Minoux,
B. Simeone and many others.

REFERENCES

[1]  Garey, M.R.,  Johnson, D.S.,  Computers and intractability,  A guide
     to the theory of NP-completeness,  W.H. Freeman and Co. 1979.
[2]  Hammer, P.L.,  Rudeanu  S., Boolean Methods in Operations Research
     and Related Areas,  Springer Verlag 1968, French translation, Dunod
     1969.
[3]  Hammer, P.L.,  Rosenberg, I.G., Equivalent forms of zero-one pro-
     grams.  In Applications of number theory to numerical analysis,
     Academic Press 1972 pp. 453-463.
[4]  Rosenberg, I.G.,  Reduction of bivalent maximisation to the quadratic
     case,  Cahiers Centre Etudes Rech. Oper. 17 (1975) 71-74.
[5]  Rosenberg, I.G.,  0-1 optimization and non-linear programming,
     R.A.I.R.O. 6 V-2 (1972) 95-97.
[6]  Rosenberg, I.G.,  0-1 optimization and non-linear programming.  Talk
     presented at TIMS/ORSA Meeting Miami.Fla,Nov.3-5 1976, # FP 21.1 p 266

# LOCAL VERSUS GLOBAL MINIMA, HYSTERESIS, MULTIPLE MEANINGS.

Y. L. KERGOSIEN

Département de Mathématique, Batiment 425,
Université Paris-Sud, 91405 ORSAY Cédex, France.

## INTRODUCTION.

Although the macroscopic behaviour of a system is quite different according to the fact that the states are described by
either the global or the local minima of a potential, the actual simulation of the evolution of physical systems shows the
unity of mecanisms. The absence of exterior information to compare the different states (what would not be true of e.g. some
economic agents) and the existence of fluctuations, make the
choice a matter of time scale and temperature. We shall use
some concepts of catastrophe theory (although this theory has
not been concerned so far with microscopic mecanisms) like the
loop of hysteresis, to study some deterministic ways of finding
global minima by acting on a parameter of the system in the
context of pattern matching (such as multiresolution) and shall
then compare for both alternatives (global or local minima) the
functional possibilities relative to A.I.

## GLOBAL MINIMA.

Global minima are generally (generically in the absence of
constraints like symmetry) unique, a very desirable quality for
a decision purpose, but the solution is not a continuous function of the potential, what is a drawback for control. Let us
consider a problem of pattern recognition, e.g. in medical imaging : we want to identify an image, i.e. find the parameters
to give to a model (a family of images) in order to make it
fit best our given image. The distance to our data defines a
function on the space of parameters of our model, the global
minimum of which we will look for.
This search for a minimum should itself be optimal in terms
of expected computation time. This second optimum is complex
to study but we can expect to find some of the usual features
of non linear optimization. For instance, independant travel-
lers who want to minimise their costs will balance the advantage of grouping themselves together with the related drawback
of a detour; they will probably reach a solution comparable
to the vascular or bronchial tree in higher animals, or the
ramification of some fracture phenomena, sparks, etc... A related scheme is used in the well known problem of finding the
false coin among n coins using a minimum expected number of
weightings : one must maximise the entropy of each step and
make the successive weightings explore independant attributes
of the coins, what is achieved by performing the comparisons
on some groups of coins that can serve to code the coins opti-
mally. Moreover, with proper order of operation, the size of
the compared groups can be progressively reduced.

NATO ASI Series, Vol. F20
Disordered Systems and Biological Organization
Edited by E. Bienenstock et al.
© Springer-Verlag Berlin Heidelberg 1986

It is this principle that is used in the multiresolution methods of image analysis, where one scans a frequencies-like space to extract the information contained in the different bands : One usually partitions the image into connected subsets of pixels, e.g. squares, to form a first reduced image (a quotient), each pixel of it corresponding to a subset of the original image, with value the average value of the subset pixels. This image is itself used to form a second reduced image, and so on. Each pixel of the original image is thus coded by the positions of its representatives within their respective subsets.

Consider the problem of matching two identical maps (HORN, BACHMAN 1978) or functions of two variables, one being fixed, the other controlled by displacement parameters so as to minimise a functionnal of their difference (a distance). The functionnal induces a potential on the space of the displacement parameters, the minimum of which we look for. The matching is first performed on the most degraded functions, obtained by successive convolutions i.e. a convolution by a large variance gaussian distribution, what eliminates most local minima of the function as well as of the parameter potential. After the matching, the problem is transferred to less degraded maps, etc. The potential thus acquires progressively its original sharp local minima, but the parameter usually stays in the lowest of them. The increase of sharpness of the global minimum can be measured from the increase of "lockability" of each image, computed as the determinant of its Fisher's information matrix within the family of displacements (taking the Kullback distance as a functionnal); a lockability of two different functions can also be defined to express the Hessian of their distance potential at its minimum, just as local lockabilities. Ideally, each step uses only the supplementary information.

The global minimum can be thought to be reached after a loop in the space of the degradation parameter, what we now interpret in the language of catastrophe theory.

Let us recall some definitions : given two manifolds M, N, two mappings f,g, from M to N are said to be Cp-equivalent iff there exist Cp-diffeomorphisms h,k, of M and N respectively such that f h = k g. For instance, this relation on real functions of a real variable depends on the order of the (singular) values taken at the successive critical points. Within a set of mappings, a particular mapping is stable iff it has a neighbourhood of equivalent mappings, and the unstable mappings constitute the bifurcation set. A versal unfolding of a mapping f is a k-parameter family of mappings such that any 1-parameter perturbation of f can be represented, up to equivalences, as a curve in the k-space of unfolding parameters. A universal unfolding is a versal unfolding with a minimal k. When a universal unfolding exists, the codimension k needs be non zero only for an unstable mapping; it characterises the degeneracy of f. The theory has been developped mainly for germs of mappings, it is relatively simple when restricted to functions (WASSERMAN 1973). If the evolution of a system is described by the gradient vector field of a potential function, the evolutions of the perturbed system will be described by a universal unfolding.

In the simplest case, a double welled potential of one variable has its shape controlled by two parameters. Consider

the controlled dynamical system associated with it, what includes a rule to transfer the state from one dynamical system
to another when the parameter is changed. For instance one
can bundle the different gradient vector fields into a vector
field defined on the product space states parameters, and
act on the parameter by controlling that component of the big
vector field that is parallel to the parameter axis. Starting
with a double welled potential and the state in the higher
well, it is possible to get the state in the lowest well by
slowly changing the potential, first into a single welled potential, then back to the original potential. This effect,
called hysteresis, is obtained only for loops that cross the
different parts of the bifurcation set in a certain order.
Rapidly clamping the parameter to single well shape and back
will not leave enough time for the state to leave its basin.
On a concrete example one can compute by means of the classic
methods of control theory the fastest route in the parameter
space that still brings the proper change of state.

A general potential can always be embedded in a family of
deformations of a single minimum function. The successive convolutions of multiresolution could in this regard be replaced
by interpolation with a proper single peaked image. But the Cp
theory of unfoldings does not permit to attribute definite parameter values to a given deformation (it just describes the
Cp type variations); it needs to be completed by some system
of partial coordinates in the space of functions, e.g. based
on the areas of the wells. Granted that the parameter variations are slow enough to allow the state to stay near the sin-
gularity of the moving well, still it is not sure that the
path will cross the bifurcation set the proper way : this is
the case when one uses convolution on a function with a very
narrow well. If the path is a segment in the space of parameters, the bifurcation diagram may well be distorted with the
same coordinates. Using broken lines in this space can help,
it consists in fitting and interpolating with intermediary
models that can be compared with catalysts.

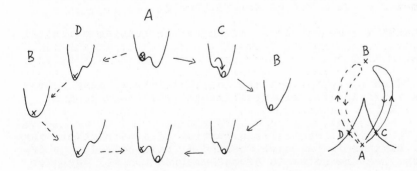

Fig.: Effect of loops in the parameter space, efficient or not.

The development of analogies with Kirkpatrick's simulated annealing (e.g. unfolding parameters and temperature) is all the
more needed that deterministic methods should remain important
in image processing due to factors like the rather low dimensionality of the problems, the existence of implemented parallel
processors, or the availability of some intermediary results.

## LOCAL MINIMA.

For a gradient system to exhibit multistability, it has to be observed long enough to reach an equilibrium, but still before a fluctuation makes it reach a global minimum. This intermediary time scale is quite relevant to biology, and the existence of local minima permits especially a system to perform memory functions, as its state depends not only on its structure, but also on its past.

As singularities, local minima have the properties of rarity important for data reduction; they can be used as signs either in a data processing view (KERGOSIEN 1983), or in a more general semiotic approach, e.g. to study such linguistic means as metonymies (KERGOSIEN 1985) or mental objects like concepts. Polysemy is an important feature of natural languages : the same word can mean different things. A particular meaning is usually choosen to be likely (besides being desirable) from two kinds of information : former meaning of the same word -a memory effect-, or coherence with a context -an interaction-. The latter effect illustrates the important function of deixis, i.e. of indicating what meaning to choose, or what detail to look at, etc., the problem of directed disambiguation, and possibly suggestion. To force a definite local minimum can be achieved by adding, possibly in a temporary way, a proper second potential, or a sequence of potentials (catalyst-likes of the previous section), so as to perform in the unfolding parameter space a loop that will drive the state where wanted. Of course concurrent suggestions on different parts of a composite concept may be inefficient since internal coherence requirements set up constraints (often of a topological nature like most problems of existence of a global object with everywhere local coherence, or continuity, or, quantitatively, low energy).

Medical imaging illustrates these considerations in the recognition of C.T. slices : the use of an adjacent recognised slice to speed up the process can use a memory effect, i.e. start from the adjacent parameters, but interactive recognition of details rather needs a set of deictic tools.

ACKNOWLEDGEMENTS : We thank  R. Azencott for reading orientation.

REFERENCES.

HORN B, BACHMAN B.(1978): Registering real images using synthetic images. In : Artificial Intelligence, Winston P.H. and Brown R.H., M.I.T. Press.

KERGOSIEN Y.L.(1983): Medical exploration of some rhythmic phenomena. In : Rhythms in Biology, Cosnard M.,Demongeot J., Le Breton editors, Lecture notes in Biomathematics vol.49, Springer.

KERGOSIEN Y.L.(1985): Sémiotique de la Nature. IVe Séminaire de l'Ecole de Biologie théorique, G. Benchetrit ed.,  C.N.R.S.

ROSENFELD A. (1984): Multiresolution image processing and analysis. Springer.

WASSERMAN G. (1974): Stability of unfoldings. Lecture notes in Mathematics vol.393, Springer.

# 5 MODELS OF BIOLOGICAL ORGANIZATION

# BOOLEAN SYSTEMS, ADAPTIVE AUTOMATA, EVOLUTION

Stuart A. Kauffman
Department of Biochemistry and Biophysics
University of Pennsylvania
School of Medicine
Philadelphia, Pennsylvania   19104

## INTRODUCTION

The past decade has seen renewed interest in non Von Neuman computation by parallel processing systems. This interest on the
part of solid state physicists and others has led to models of
pattern recognition and associative memory (1,2,3). In these
models, it is largely the dynamical attractors which are of
interest as the classes, or memories, stored in the systems.
Further, the mathematical tractability of threshold systems
with symmetric coupling, that is, in which each binary device
"fires" if a weighted sum of excitation minus inhibition exceeds some threshold, and couplings between two binary devices
are symmetrical, has focused particular attention on this subclass of automata. The marked advantage of this subclass of
automata is the existence of a potential function allowing prescription of weightings on inputs to each binary device in order to choose steady state attractors with desired properties
such as location in state space, and stability to perturbation
(1,2,3).

While symmetrically coupled threshold networks are of interest,
it is worth stressing that they comprise a small subclass of all
possible Boolean automata and that a general body of theory and
techniques for the design and construction of parallel processing
systems should cast its conceptual net as widely as possible.
The problem is to understand how to achieve parallel processing
systems with, for example, desired attractor properties, when at
present, we have no, or almost no tools to prescribe how to construct such systems. Indeed, as Wolfram has remarked concerning
cellular automata exhibiting what he calls class 4 behavior, it
is possible that no mathematical tool other than explicit simulation can be used to predict the attractors of some such systems

NATO ASI Series, Vol. F20
Disordered Systems and Biological Organization
Edited by E. Bienenstock et al.
© Springer-Verlag Berlin Heidelberg 1986

(4, and p.c.).

One approach to this problem is simulated evolution by mutation
and selection to achieve parallel processing systems with desired
properties. Organisms are, par excellance, parallel processing
systems. Consider, for example, the sophisticated behavior and
metabolism of the bacterium *E. coli* in a complex chemical environment. Via a collection of specific receptors on its surface
membrane, a complex network of structural genes coding for enzymes and other proteins, as well as regulatory genes which turn
other genes "on" and "off", and sophisticated behavioral traits,
*E. coli* can orient in a chemical environment, more toward or
away from chemical sources, switch its metabolism to cope with
or exploit the chemical resources in its vicinity, and maintain
internal homeostatic balance during the biochemical dance. Evolution has achieved not only the construction of structural genes
which code for proteins with the desired catalytic and binding
properties, but the regulatory circuitry among these genes which
coordinates their activities appropriately in any given environment. *E. coli* categorizes its rich environment, reacts to it, and
acts upon it, without benefit of potential functions to calculate
the way to design its circuitry to achieve desired attractor or
other properties. Presumably, we can learn how *E. coli* achieves
its adaptation, and learn to exploit the paradigm ourselves. The
paradigm suggests that we consider mutation and selection as the
prototype of adaptive "learning" in parallel processing systems
and assess how, and how well, such procedures work.

Evolution occurs, in a sense, within an ensemble of all genetic
systems which might be generated via the classes of mutations
which occur in a given population. For example, mutations alter
not only the amino acid composition of proteins, but can re-
arrange the genetic regulatory circuitry, changing which regulatory gene controls a given gene, or the conditions under which
that gene will be active or inactive. Adaptation is a search
procedure within this ensemble. Selection is the differential
rate of leaving offspring in the next generation as a function
of the phenotype spectrum in the population in each generation.
Selection bears a close analog to Maxwell's Demon, operating a

flap valve between two boxes, and allowing the faster molecules
in the gas to enter the right hand box. By his action, the Demon
selects for states of the gas in the total system which are displaced from thermodynamic equilibrium, hence rare in the ensemble of possible states of the system. Yet, the Demon confronts
the problem that as the number of fast molecules builds up in
the right hand box, back pressure towards the left box mounts.
If the Demon is finite, when his efficiency equals the back
pressure, the system comes to a steady state. If he is quite
weak, the steady state is relatively close to thermodynamic
equilibrium, and exhibits the typical, not rare properties of
the system. Similarly, while selection can enrich for properties
which are rare in the ensemble of genetic systems available to
it, as it does so, back pressure from mutations will drive the
selected systems ever harder toward the typical, or generic properties of the ensemble. The properties attainable, therefore, by
selection depend first upon the ensemble inside of which selection is occurring, second upon how powerful a selective force is,
or can be applied, third upon how typical or rare the desired
properties are in the ensemble explored, and how rapidly mutational processes allow exploration but also drive toward the
generic properties of the ensemble.

This general analysis of how selection occurs, suggests that it
might be valuable to consider large ensembles of Boolean dynamical systems, and determine the generic properties and adaptive
capacities of each such ensemble in turn. *A priori*, one would
expect the capacity to adapt via mutation and selection strategies, the style of adaptation, and properties which might be
achievable, would differ in the different ensembles. An analysis
of this kind might be hoped to be useful not only in the development of strategies to achieve parallel processing automata with
desired properties, but also in our understanding of evolutionary
mechanisms as well.

BOOLEAN AUTOMATA

The simplest Boolean automata consist in N binary devices, each
receiving inputs from 1 or more of the N, each realizing one of
the possible Boolean rules on its inputs, and all governed by an
internal clock such that at a clocked moment, each binary device
assesses the values of its inputs and the next value it should
adopt, and all elements adopt their new values synchronously. Any
such system with N binary devices has $2^N$ states, each a vector of
N binary values, and the network imposes a mapping from the $2^N$
states onto or into itself. Since the system is finite, and deterministic, it is a trivial property that any trajectory of states
under the mapping must eventually close on itself to form a cyclic
sequence of states. Because in general, more than one state can
have the same successor state, many trajectories can converge onto
the same state cycle. The set of states which lie on trajectories
flowing into a state cycle, plus the state cycle itself, constitute the basin of attraction of the state cycle attractor.
Each state lies in only one basin of attraction. However, the
system may have many, rather than one, state cycle attractors,
each draining its corresponding basin. A critical feature of
synchronous Boolean systems is that many binary variables can
change value at the same instant, hence successive states
along a dynamical trajectory need not be constrained to be similar to one another in a Hamming measure sense.

Among the natural questions which arise for Boolean automata
sampled from any defined ensemble of such automata are these:
1) How long are the state cycle attractors in such automata? The
minimum length is 1, the maximum number of states which might lie
on a cycle is $2^N$. 2) How many state cycle attractors does the
system have? 3) How stable is each attractor to 1 Hamming unit,
2 Hamming unit... perturbation? 4) At a given level of perturbation, how many state cycles are accessible from any state
cycle? 5) How similar to one another are the state cycles of one
automaton? 6) How much does the dynamical behavior of the automaton change if the inputs to a given binary element are altered
slightly, or if its Boolean rule is altered slightly? 7) Under
the drive of random mutations to the connections and rules of
such a system, do the definable dynamical properties of the

system change, in some sense, isotropically, or in preferred ways? 8) Given information about the typical behavior and response to mutations of members of a given ensemble of automata, can one define selection procedures to obtain automata which are members of the ensemble which have desired properties? For example, such properties might include desired attractors, desired transitions between attractors upon specified external input to the system, etc. 9) What is the balance, and how can it be characterized between the "natural" or generic properties of the ensemble, the rare properties, and the capacities for mutation and selection to achieve and maintain such properties in a population of adapting automata?


THE GRAND ENSEMBLE

Consider the ensemble of all Boolean networks with N binary variables, each receiving one input line from all N variables, and each assigned at random, one of the possible $2^{2^N}$ Boolean functions of N binary variables. Since each node receives an input from every node, there is only one possible "wiring diagram" for such a network. However, since each node has assigned to it at random one of the possible $2^{2^N}$ Boolean functions of N variables, the ensemble of all possible "completely connected" Boolean networks includes, as subensembles any other possible Boolean networks in which each variable has K (less than N) inputs or subensembles in which a constrained class of Boolean functions from among the $2^{2^N}$ possible functions of N inputs is allowed. This inclusion follows trivially from the fact that among the Boolean functions of N variables, subclasses of those functions may depend only on a specific subset K, of the N. Hence in such automata, the number of actual inputs per node is only K. Because this grand ensemble contains all possible "simpler" synchronous and autonomous Boolean automata, it is of interest to characterize the generic dynamical properties of this ensemble.

In this grand ensemble, each node is assigned at random one of the $2^{2^N}$ Boolean functions of N variables, hence equivalently,

the state into which each state maps at the next moment, is
chosen at random from among the $2^N$ states of the system. Consequently, such a network is a random mapping of the set of
$2^N$ states into itself. A number of workers have studied this
class of systems (5,6,7,8,9). The first important property of
this class is that the median state cycle length, that is, number of states on a state cycle, is about $0.5 \times 2^{(N/2)}$. Thus,
cycle lengths are on the order of the square root of the number of states. Small systems have very long cycles. With N=200,
cycles are on the order of $2^{100} = 10^{30}$.

While cycles are long, a fundamental and surprising property
of this grand ensemble, is that the number of distinct state
cycles is only linear with respect to the number of nodes, N/e.
This implies that a system with 1000 nodes and $2^{1000} = 10^{300}$
states would only have on the order of 370 dynamical attractors.
It is interesting to contrast this constrained number of attractors with the case of spin glasses, in which the number of minima
of the random energy potential increases exponentially with the
number of spins (10).

I know of no analytical or numerical studies of the stability
of the attractors in such systems, but there is every reason to
expect the following: Because the state following each state in
this class of systems is chosen at random from among the $2^N$
states, it will typically differ from its predecessor in 0.5N
of the "bits". Successor states are not Hamming neighbors, and
no topology with respect to the N dimensional Boolean hypercube
is imposed on the dynamics save that due to determinism. Consequently, 1 Hamming unit perturbation, reversing the current
value of an arbitrary one of the N "bits" in the current state
will typically, at the successive moment, lead to a successor
state differing from the successor state of the unperturbed
state by 0.5N. The probability that these two successor states
lie in the same attractor, hence flow to the same state cycle
is governed solely by the sizes of the attractors. Thus, the
stability of any state cycle attractor is identical to its size
with respect to the $2^N$ states. This is known to be a skewed

distribution (5) with a few large cycles in large attractors, and many small cycles in small attractors. Further, the length of state cycles is long with respect to the number of state cycles, and each state on such a cycle has N states differing from it by 1 Hamming distance. Therefore, in the presence of occasional, but persistent 1 Hamming unit noise, the system can typically be driven from each of its state cycle attractors to all of the other attractors, and the system will wander among its basins of attraction with transition probabilities proportional to basin size.

Because successor states are randomly chosen, states on one state cycle bear no similarity to one another, state cycles are not located preferentially in any regions of the $2^N$ Boolean hypercube, and each node changes value with probability 0.5 at each state transition.

All automata of N variables are members of the grand ensemble, specific classes of automata with N variables are therefore specialized subensembles. One limit of the specialization process lies in construction and analysis of cellular automata, in the simplest case consisting of a line or ring of nodes, each a function of some defined neighborhood of adjacent nodes, and each governed by the same Boolean rule on its inputs. The advantage of cellular automata lies in the combination of simplicity and case by case purity, allowing the dynamical consequences of each rule to be studied and revealing the existence of broad classes of rules with similar behavior. An important generalization of cellular automata consists in networks in which the "wiring diagram" of inputs between members of one automaton are chosen at random, but then fixed, while each node realizes the same Boolean rule on its inputs. That is, the rule is homogeneous throughout the automaton. A considerable body of work in this direction has been carried out (7,11,12,13) and as in the case of geometrically simpler cellular automata, reveals the existence of broad classes of Boolean rules with similar behavior. These will be alluded to in more detail below. Beyond homogeneous systems, many real physical and certainly chemical, biological, social and economic systems have the property that

the rules governing distinct "nodes" differ, thus that the systems under study are genuinely heterogeneous. A concrete class
of systems that are heterogeneous are the genetic regulatory
"circuits" underlying the highly adapted metabolic behavior of
bacteria, as well as the orderly dynamics of development from
zygote to adult in higher organisms. Clearly, analysis of such
heterogeneous systems would be useful, and the aim of the next
sections of this article is to characterize briefly some approaches to this complex set of problems.

## SUBENSEMBLES OF AUTOMATA CLASSED BY THE NUMBER OF INPUTS PER NODE

One rational approach to the generation of subensembles of the
grand ensemble described above consists in considering the ensemble of all synchronous, autonomous automata of N binary nodes,
each of which receives K (K < N) inputs from among the N. This
pattern of analysis makes connectivity the fundamental independent parameter and asks as a function of K and the size of the
automata, N, for the generic structural ("wiring diagram") and
dynamical behavior of the corresponding subensembles. The grand
ensemble corresponds to K = N.

## K = 1 AUTOMATA

Consider the case where each node has exactly 1 input, randomly
chosen from among the N. The wiring diagram of such a network is
closely analogous to the state transition diagram for K = N
nets. In both cases there is a random mapping of a set of integers into itself. The only difference is that in the state space
of K = N nets each state can have only one successor but several
states can converge on one state, while in the wiring diagram of
K = 1 nets, any node can have only 1 input, but may send inputs
to more than one node. Thus, the expected structure of the random
directed graph in the two cases are identical, except that the
direction of the arrows is reversed. The wiring diagram of K = 1
networks consist in structural feedback loops with (branching)
tails hanging from the loops. The median length of the loops is
on the order of the square root of the number of nodes, while the
number of loops in on the order of a linear function of log N.

There are four possible Boolean functions of 1 input, "tautology", "contradiction", "yes" and "no". If "tautology" and "contradiction" are included among the allowed rules, the probability is high that some nodes on the different feedback loops will be governed by these two constant functions. Each such node will fix the states of all nodes downstream from it which realize "yes" or "no", until the subsequent constant function. Thus, with high probability, each loop will fall to a dynamical fixed state, as will all descendent nodes.

If tautology and contradiction are excluded from such automata, then any feedback loop has either an even or an odd number of negating "no" Boolean functions assigned to its nodes. If the parity is even, the loop has two steady states and a variety of oscillatory state cycles corresponding to patterns of packing 1 and 0 values around the loop in arrays which repeat an integral number of times around the loop. If the parity around the loop is odd, then the loop has no steady states, and possible state cycle lengths are factors of two times the loop length (14). Because in random K = 1 networks, the lengths of different feedback loops will typically not be simple factors of one another, it is to be expected that maximum global state cycle lengths will be on the order of the product of the lengths of the feedback loops in the net. More accurately, since on the order of half the loops will have an odd number of "no" functions, maximum state cycle length will be correspondingly larger (15).

An important property of K = 1 nets using only "yes" and "no" is that the sequence of states on state cycles is entirely driven by the elements on the feedback loops. Further, since each element on the loop simply affirms or denies the state of its single input, each state of such a loop lies on a closed cycle of states and has one predecessor and one subsequence state. That is, the dynamical behavior of any such feedback loop is time reversible. Since, on such a loop each state has a single antecedent state, and two states cannot flow to the same subsequent state, there is no dynamical restoring force to return the loop, if perturbed from its current state cycle, back to that cycle. In short, since all the states of such a loop lie on state cycles, single Hamming unit perturbations either leave the loop on the same state cycle at some different "phase" or cause it to jump immediately to some other state cycle where it then remains. Thus, such nets are expected to have quite long state cycles, rather large numbers of state cycles, the cycles are expected to be unstable with respect to minimal perturbations, which should be able to drive any state cycle to a moderately large family of neighboring state cycles (15).

## K = 2 AUTOMATA

K = 2 automata, in which each node receives 2 inputs from among the N, are in some sense the simplest automata which are genuinely heterogeneous and sufficiently complex to be interesting. It happens that this class of automata exhibits very marked self organizing properties which have made it of considerable theoretical importance. I will briefly describe numerical and analytical work here, as it is documented by myself and others elsewhere (3,6,7,15-27).

The first surprise is that in such automata, the median state cycle length is on the order of the square root of the number of nodes, N. Thus a system with 10,000 nodes and $2^{10,000}$ states, typically cycles along an attractor consisting of about 100 states. Second, the number of state cycle attractors is also on the order of square root N. Third, each state cycle is stable to most 1 Hamming unit perturbations. Fourth, if, under perturbation, the system is driven from any given state cycle, it undergoes a transition to only a few "neighboring" among the possible state cycles. Fifth, most of the nodes of such an automaton (0.6N - 1.0N) settle into fixed values, 1 or 0, which are identical on all state cycles. Thus, state cycle attractors occur in a particular region of the Boolean hypercube. Sixth, typically, the nodes which fall to fixed values from a single interconnected cluster, called a forcing structure, are described below.

The response of such automata to minimal mutation, changing
a single connection in the "wiring diagram" or altering the
Boolean function at a single node, is typically to change behavior only slightly. Thus, if any single element is deleted,
typically only 0 to 10% of the elements alter their dynamical
behavior. An interesting feature of this class of automata is
that response to mutation appears to be non-isotropic. Consider
the set of binary nodes in one such automaton which happen to
be fixed in the value "0" on all state cycle attractors of the
system. Deletion from the automaton of any such node does not
alter state cycles. On the other hand, such deletions may alter
transitions induced among the state cycles by random 1 Hamming
unit fluctuations, or by specific external signals transiently
reversing the value of one or more nodes. The first point to
note, therefore, is that a class of "deletion mutations" exists
in this subensemble of automata which does not change the attractors, but can change transitions among them. If the automaton has L state cycles, the noise induced transitions among
them can be represented by a L x L matrix Markov chain, whose
elements are the transition probabilities between the i-th and
j-th state cycle. Typically, any deletion mutation of an element normally at fixed value "0", changes only a few of the
transition probabilities. The "non-isotropic" feature of this
class of deletion mutations, is that many of the single deletion mutations affect much the same subset of transition probabilities, and any one such transition probability is more
likely to be changed in one direction than the other--i.e. more
likely to be increased than to be decreased, or vice versa--by
any single deletion mutation removing an element dynamically
fixed at "0". Thus, under the drive of random mutations in this
class, the transition probabilities among attractors change in
an oriented way. It is a most interesting question whether such
non-isotropic response persists or if there are "attractors"
with respect to transition probabilities in the (open) set of
automata which can be generated by deletions (and definable
additions) of elements which do not themselves alter the state
cycle attractors of the system.

K = 2 automata were initially investigated by the author as
models of genetic regulatory systems in higher eukaryotes.
I interpret a binary node as a single gene, whose 1 and 0 values
represent active or inactive states. For a structural gene coding for a protein, active means transcriptionally active. For
a regulatory gene such as an operator sequence of the kind found
in bacteria and viruses and, almost certainly, eukaryotes, active
and inactive states refer, for example, to whether the operator
is free or bound by a repressor molecule. I then interpret a
state cycle attractor, which is one of the persistent, asymptotic patterns of activity of the model "genes" as one "cell
type" of the organism. The set of attractors become the set of
cell types, the transitions among them model pathways of differentiation during development, and metaplastic abberations in
differentiation, the similarities of state cycles become predictions about the similarities in gene expression patterns between
cell types of the organism, the existence of a core of genes
fixed active or inactive becomes a prediction of such a core,
the typical effects of mutations deleting single genes become
predictions about how many genes have activity patterns altered
by single mutants, etc. Most of the predictions are reasonably
close to biological fact, which is surprising given the extreme
simplicity of the model (16,17,18). Two particular features
might be mentioned: the number of attractors is crudely a square
root function of the number of binary elements, predicting that
the number of cell types in organisms, as a function of genomic
complexity, should increase crudely as a square root function.
While it is hard to count numbers of genes and numbers of cell
types, it does appear that numbers of cell types increases less
than linearly with numbers of genes across many phyla. A second
interesting point is that in K = 2 automata, each cycle, under
the drive of 1 Hamming unit noise, can undergo transition to only
a few neighboring state cycles, and from the latter set, further
transitions to the rather few additional neighbors of each member
of the set. If state cycles model cell types, transitions among
them model steps in differentiation, and the strong prediction
is made that any cell type can directly differentiate into only
a few other cell types, ultimately reaching many cell types

along branching pathways of differentiation. Indeed, such
branching pathways of differentiation are a universal feature
of all metazoan and metaphyton development, presumably since
the paleozoic. Thus, this universal feature of ontogeny may
not represent selection per se, but a generic property of the
ensemble of genetic regulatory systems in which evolution is
occurring, and which selection, like a finite Maxwell Demon,
is too weak to avoid. I make this point briefly here to indicate the possibility that there may prove to be true ahistor-
ical universals in biology reflecting the self organizing
properties of complex genetic systems.

Analytic insight into the behavior of K=2 automata has grown, but
remains inadequate. Two lines of approach have been taken. The
first considers constraints on the state transition matrix in such
automata while the second has focused on the particular properties
of FORCING STRUCTURES (3,15-27). The approach using forcing structures will be described briefly here. Among the Boolean functions
of K inputs, define a subclass called "canalizing", having the property that there is at least one input variable having one value,
"0" or "1", which alone suffices to guarantee that the node regulated by that Boolean function adopts one specific value, either
"0" or "1", regardless of the values of any of the K-1 other inputs.
The Boolean "OR" function exemplifies this. If the left input is
"1", then the next moment the regulated node adopts the value "1"
regardless of the current value of the right input. Canalizing
functions have the implication that cascades of nodes can be connected such that a value introduced to a node high in the cascade
is guaranteed to propagate, and generates a specific fixed value at
each descendent node. A simplest case is a set of nodes connected
by OR functions. A "1" value introduced at any node will propagate successively to all descendent nodes, switching each to the
"1" state in turn. Further, if the circuitry contains one or
more loops, any such loop has a fixed state with all elements
in the value "1". Once in that state, the loop cannot be perturbed
by alterations in values of any nodes outside the loop which
impinge upon it. Such a structure is a forcing structure and
forcing loop. The constraints needed to create a forcing connection between two nodes, from A to B, are that B realizes a

canalizing Boolean function on A, with A a canalizing input, that **A** realizes a canalizing function on its own inputs, and the value of A which can be guaranteed (the canalized value of A) is the value of A which canalizes B to its own guaranteed (canalized) value. These criteria create a transitive relation such that if A forces B and B forces C, then A indirectly forces C. Using "AND", "OR", "IF", etc. it is easy to create forcing structures in which a "1" at locus A forces a "0" at B, etc.

Among the $2^{2^K}$ Boolean functions of K inputs, the fraction which are canalizing is maximum for K=2 and decreases rapidly as K increases (7,17). For K=2, 12 of the 16 Boolean functions are canalizing. Tautology and contradiction are not counted, and Exclusive Or and If and Only If are not canalizing. For K=4, the fraction of Boolean functions which are canalizing drops to about 5%. In an automaton with K=2 inputs, the formation of connected forcing structures is a percolation problem on a random directed graph. Some fraction of the connections among the nodes will fulfill the conditions to be forcing connections. If enough of them are forcing connections, the probability is high that extended forcing structures "crystalize". Based on the results of Erdos and Renyi on the connectivity properties of random graphs (28,29), and numerical results on random directed graphs, it is clear that a kind of phase transition occurs. When the number of forcing connections is less than the number of nodes, rather small, unconnected forcing structures are present. As the number of forcing connections increases past the number of nodes, extended connected forcing structures emerge. By the time the number of forcing connections is about 3 times the number of nodes, most elements lie on a single strongly connected forcing structure (30,31).

The existence of forcing structures has been confirmed in K = 2 nets in which the input connections between the N nodes is fully random, and in the case of regular square lattice structures, where each node receives inputs from two of its 4 nearest neighbors, and sends outputs to the other two (3). The latter case is particularly attractive because the forcing structure is readily visualized as a domain  on the lattice with nodes at fixed values. It is clear that the idea will generalize to three dimensional lattices, etc. In each case, the existence of forcing structures is a percolation problem on the appropriate geometry, and depends upon the fraction of functions which are canalizing.

An important consequence of the existence of forcing structures, again easily seen on K = 2 Boolean square lattices, is that the forcing structure typically creates isolated subdomains of elements which are not part of the forcing structure, and which cannot influence one another once the forcing structure is in its fixed state, since signals cannot pass between the isolated subdomains through the fixed values of the forcing structure elements. Thus, the asymptotic behavior of such a system is given by a stable core representing the fixed values of elements of the forcing structure, and the combinatorial behavior of all the possible dynamical attractors of each isolated subdomain. Thus, if one isolated subdomain has two attractors and another three, then the pair exhibits six possible attractors.

While the analysis of forcing structures accounts for many features of K = 2 automata, it has not yet been possible to account for the number of state cycles, nor the length of state cycles in detail.

K > 2 AUTOMATA

In K = 2 automata, the median cycle length and also number of cycles is of the order of square root N. For K = N automata, state cycle length is of order $2^{(N/2)}$, while the number of cycles is of the order N/e. It would obviously be valuable to obtain scaling laws for these features as a function of K and N. In this section I suggest an approximate method to achieve a scaling law for the median cycle length of automata with K > 4.

Consider K = N automata with random choice of Boolean functions. This class consists of a fully random mapping of the set of $2^N$ states into itself. Now consider a biased random mapping . Without loss of generality, assign Boolean functions having the

property that the probability of assigning a "1" is P, (P > 0.5).
Define the "central state" to be the preferred value at each node,
hence the state "111......" The fraction of the $2^N$ states which
map into the central state as their successor is $P^N$. For P significantly greater than 0.5, this implies that very many states have
the central state as successor, hence very high convergence in
state space onto that state. States similar to the central state
have high convergence, as trivially calculated.

Utilizing the idealization that all states have convergence onto
them equal to that of the central state, and an argument similar
to Rubin and Sitgreave's (5), it is easy to obtain a minimal estimate of median cycle lengths as a function of P and N. This is

$$\text{Median cycle length } L = 0.5 \cdot B^N \text{ where } B = \frac{1}{P^{\frac{1}{2}}} \tag{1}$$

For P = 0.5, this reduces the estimate of cycle lengths for K = N
automata (19).

The next step is to link K, the number of inputs per node, with
a mean value of P for all the $2^{2^K}$ Boolean functions of K inputs,
then utilize that mean value of P to predict median cycle lengths
for automata with K inputs.

Following Walker and Gelfand (7), define the Internal Homogeneity
of a Boolean function to be the greater of the number of "1" or
"0" values which occur in the function. Thus the "OR" function of
two inputs has three "1" values and one "0" value. The I.H. is 3.
Correspondingly, the probability P of the value 1 is 0.75. By the
definition of Internal Homogeneity, the corresponding value of P
is greater or equal to 0.5.

It is simple, by inspection, to calculate the value of P for each
of the 16 Boolean functions of K = 2 inputs, hence the mean for
all 16 Boolean functions, to obtain P = 0.6875.

It can be shown that, as a function of K, the mean value of P is

$$\bar{P}_{(K)} = \frac{1}{2} + 2^{-(2^K+1)} \cdot \binom{2^K}{2^{K-1}} \tag{2}$$

As K increases, the mean internal homogeneity, hence mean value
of P, decreases toward 0.5. (2) yields: K = 2, P = 0.6875;
K = 3, P = 0.6455; K = 4, P = 0.5982; K = 5, P = 0.5699; K = 6,
P = 0.5497.

The idea lying behind the approximation utilized below is the
following: Consider K = 2 automata. Write an N x $2^N$ matrix listing
all $2^N$ states of the N variable automaton. Adjacent to this write
a corresponding N x $2^N$ matrix the i-th column of which contains
the Boolean function of the i-th node. If the i-th node realizes
the Boolean OR function, 0.75 of the $2^N$ values in that column
will be "1". Where the "1" and "0" values occur in the column
depends upon the particular two inputs to the i-th node chosen
from among the N. The distribution along the column is highly
non-random. Now consider a Boolean function of K = 10 inputs
which happen to have 0.75 "1" values. Again, the $2^N$ function
column will have 0.75 "1" values, again the positions will depend upon the particular 10 inputs, but given random choice of
the 10 inputs, the distribution will be more nearly random along
the column than for K = 2 automata. Therefore, as K increases,
it seems reasonable to approximate the distribution of the preferred value, say "1", and non-preferred value, along each function column by assigning the "1" and "0" values at random along
the column according to P. For an automaton with K inputs,
sampled at random from the ensemble of K,N automata, the value
of P for each element can be approximated by the mean value öf
P for the $2^{2^K}$ Boolean functions of K inputs. But this is equivalent to a biased random mapping of the $2^N$ states into itself
with a preferred value at each node whose probability is P. Consequently, given the mean value of P for any value of K, one has
an estimate of median cycle lengths given by (1) above.

Since, for K = 2 forcing structures are prominent, while the
fraction of Boolean functions which are canalizing drops to less
than 5% for K = 4 (7,16) and more generally the effects of specific inputs becomes less important as K increases, the expectation
is that  this approximation will be useful for K,N automata with
K > 4. This appears to be the case. Substitution of (2) into (1)

yields the following values of B: K = 2, B = 1.206; K = 3, B = 1.2446; K = 4, B = 1.293; K = 5, B = 1.324; K = 6, B = 1.349.

Note first that for K = 2 this approximation predicts that cycle lengths increase exponentially in N, i.e. $0.5 \times 1.206^N$. In fact cycle lengths increase as about the square root of N. Presumably this gross discrepancy reflects the effects of forcing structures. Walker and Gelfand have carried out numerical simulations of random automata for K = 1,2,3,4,5,6,7 [7]. For K = 4 and larger, log median cycle length is a linear function of N, as expected on these approximations, the slope increases as K increases and is reasonably closely fit by utilizing the log of the B values as a function of K given above.

This correspondence is encouraging. It appears to be the first even approximate analytic link between N,K and an important dynamical property of the corresponding ensemble of automata, except for the case K = N.

This correspondence suggests it might be possible to utilize the case K = N with biased random mappings according to P to obtain other critical properties such as the numbers of state cycles, the mean number of times a node changes value on a trajectory, the mean similarity between states on one cycle and between state cycles, etc., then apply the results to K < N automata with the corresponding value of P. In particular, J. Coste (personal communication) has obtained one estimate for the number of state cycles as a function of $P_{(K)}$:

$$N \left[ \frac{\log \left( \frac{1}{\frac{1}{2} + \alpha} \right)}{2} \right] \leqslant \text{no. cyc.} \leqslant N \left[ \frac{\log \left( \frac{1}{\frac{1}{2} - \alpha} \right)}{2} \right] \quad ; \quad \alpha = P_{(K)} - \frac{1}{2} \quad (3)$$

This predicts that, for any K, the number of cycles is a linear function of N. This is not true for K = 2. Insufficient data are available for 2 < K << N.

SELECTION AND ITS LIMITS

Study of mutation and selection strategies, what can be selected for, the limits of selection, how these vary with the ensemble

of automata in which selection is occurring, and the kinds of
mutation and selection strategies applied, is in its infancy.
On the other hand, the mathematical tools of population genetics,
developed over the past half century, are powerful, and might be
expected to be of use, either immediately, or in due course.
Appeal to population genetics at the outset, focuses attention
on the role of the population itself, and the varying fitnesses
of individuals in the population, on the selective evolution
of the population. The first point to stress, derived from the
biological domain, is that fitness applies to the phenotype of
the organism, not directly to its genotype. Selection is the
increased number of offspring derived from those organisms with
(heritably) fitter phenotypes, hence by acting on the phenotype,
selection indirectly increases the representation in the population of genotypes leading to the fitter phenotypes. Applying this
idea to automata, the wiring diagram and Boolean rules defining
the automaton correspond naturally to the genotypes, and the dynamical behavior corresponds naturally to the phenotype. Selection
will then occur by a fitness function, or surface with respect to
the dynamical properties of interest of the automaton. Hence selection induces, indirectly, a flow among the ensemble, of automata
accessible from an initial population of automata by virtue of the
classes of allowed mutations which occur.

The second fundamental feature of selection, is that it requires
variability in heritable fitness in the population in order to
work. More precisely, in the simplest cases, the rate of increase
in the mean fitness in a population is linearly proportional
to the variance in the fitness in the population. On the other
hand, the rate of increase in fitness in the simplest cases (one
locus two alleles theory) is inversely proportional to the mean
fitness in the population (32).


A third fundamental feature of selection in real or simulated
systems is that mutations are both the basic source of variability
in fitness, allowing exploration of the ensemble of accessible
systems, yet simultaneously, mutations provide a pressure which
tends to destroy accumulated positive properties, hence will tend
to drive any system under selection toward the properties which
are generic to the ensemble to be explored.

These issues can be considered more concretely with respect to a very simple model. Consider an automaton with N nodes, each receiving input wires from K specified nodes among the N. Imagine that mutation is able to reassign the origin or terminus of any such input wire, at random among the N nodes. Construct a fitness function which depends upon how closely the wiring diagram of any specific automaton in the population approximates the optimal desired wiring diagram. Let the rate of leaving offspring be linearly proportional to the fitness, supply a mutation rate, and ask whether the system can attain the optimum, or more generally, study the attractors of the system.

A simple fitness function is

$$W_{(x)} = (1-b) \cdot \left( \frac{G_{(x)}}{T} \right)^a + b \quad ; \tag{4}$$

Here, $W_{(x)}$ is the fitness of the $x^{th}$ automaton, and lies between 0 and 1, $G_{(x)}$ is the number of correct or good connections in its wiring diagram, while T is the total number of possible good connections. The ratio of $G_{(x)}/T$ measures deviation from the optimal wiring diagram. Here "a" is a real valued parameter which measures something like cooperativity. For "a" = 1, each new good connection makes an additive contribution to fitness. For "a" greater than one, successive correct connections make larger contributiuons, conversely any lapse from optimal causes a drastic loss in fitness, hence large "a" models a limit in which all connections jointly are critical. In the absence of any good connections, a basal fitness "b", a parameter between 0 and 1, might be supplied.

While the model in (4) is trivial to state, its behavior is rich, and has been reported in details elsewhere. Briefly, for "b" = 0, the selective force tending to increase the number of good connections, while mutation is a force tending to destroy good connections which is linearly proportional to the mean number of good connections. As either the mutation rate, or more importantly, the total number of possible good connections, T, increases, a bifurcation occurs such that selection cannot drive the population to the fitness peak, and the population achieves a stable compromise fitness peak part way up the slope. For "b" greater than 0, and "a" greater than 1, the model exhibits multiple stable solutions.

Perhaps the main conclusion to be drawn from this simple model is that, even with a simple fitness landscape having a single peak and a smooth surface, if the complexity of the system, T, grown large, selection may not be able to achieve the peak. Other general questions are these : For different ensembles of automata, under selection for different kinds of dynamical properties, how multi-peaked are the fitness landscapes, how continuous or discontinuous are those landscapes, how does the smoothness of landscape, the dimensionality of the system under selection, the mutational search range, etc. the influence the efficacy of a mutation selection strategy ?

Preliminary numerical results have been obtained for the following : Define ensembles of Boolean automata with K = 2, 5 or 10 inputs per node, and elementary mutations which alter one bit in the Boolean function of one node, or alter one input to one node ; define a single "target" state, or N vector of 1 and 0 values ; test whether mutation and selection can achieve a network with the target state as an attractor. A first important result is that the fitness landscape is reasonably smooth for K = 2, 5, 10, for mutations to Boolean functions, but is only smooth with respect to mutation in connections for K = 2, and clearly not for K = 10. A second important result is that the optimum is not attained under strong selection, implying many suboptimal peaks where the search becomes trapped. A third result is that partition of each automata into a subset of nodes which are "hidden" variables uncounted in the fitness measure, appears to improve the capacity of the remainder of the system to approach the optimum. These preliminary results like the analysis of Weisbuch (34) merely point in the direction of a body of theory which needs to be developed, and can be expected to have practical as well as scientific interest.

## REFERENCES

(1)     Hopfield, J.J. <u>PNAS</u> 79:2554 (1982)

(2)     Hinton, G.E. and T.J. Sejnowski. Proc. of IEEE Conf. on comp. vis. and patt. recog., pp. 448-453 (1983).

(3)     Fogelman-Soulié, F. Contribution à une Théorie du calcul sur réseaux. Thèse de l'Université Scientifique et Médicale de Grenoble (1985).

(4)     Wolfram, S. <u>Rev. of Modern Physics</u> <u>55</u>, 601-644 (1983)

(6)     Rubin, H. and R. Sitgreave. Tech. Report No.19A (Appl. Math. and Stats Lab, Stanford University (1954).

(6)     Kauffman, S.A. <u>J. Theor. Biol.</u> 22:437-467 (1969)

(7)     Gelfand, A.E. and C.C. Walker. <u>Ensemble Modeling</u>. New York : Marcel Dekker Inc. (1984).

(8)     Gelfand, A.E. and C.C. Walker. <u>Bull. Math. Biol</u>. 44:309 320 (1982).

(9)     Coste, J. Personal communication (1985)

(10)    Anderson, P.W. <u>PNAS</u> 80:3386 (1983)

(11)    Walker, C.C. Tech. Report 5, AF Grant 7-64, Electrical Engineering Research Laboratory, University of Illinois, Urbana (1965).

(12)    Walker C.C. <u>Journal of Cybernetics</u> 1:55-67 (1971).

(13)    Walker, C.C. <u>Prog. in Cybernetics and Systems Research</u> Washington, C.C. : Hemisphere, pp.43-47 (1978).

(13)    Holland, J.H. <u>Journal of the Franklin Institute</u> 270: 202-226 (1960).

(15)    Kauffman, S.A. <u>Journal of Cybernetics</u> 1:71-96 (1971).

(16)    Kauffman, S.A. <u>J. Theor. Biol.</u> 22:437-467 (1969).

(17)    Kauffman, S.A. <u>Mathematics in the Life Sciences</u> 3:63-116 (1970).

(18)    Kauffman, S.A. <u>J. Theor. Biol.</u> 44:167-190 (1974).

(19)    Kauffman, S.A. <u>Physica 10D</u> : 145-156 (1984).

(20)    Babcock, A.K. Doctoral Disseration, State University of New York, Buffalo (1976).

(21)    Atlan, H., Fogelman-Soulié, F., Salomon, J. and Weisbuch, G. <u>Journal of Cybernetics and Systems</u>, <u>12</u>, 103-131 (1981).

(22)    Cull, P. <u>Kybernetik</u> 8:31-39 (1971).

(23)    Aleksander, I. <u>Int. J. Man/Machine Studies</u> 5:115 (1973).

(24)    Sherlock, R.A. <u>Bull. Math. Biol.</u> 41:687 (1979).

(25)    SHerlock, R.A. <u>Bull. Math. Biol.</u> 41:707 (1979).

(26)    Fogelman-Soulié, F., 27-45, in Dynamical Systems and Cellular automata, edited by Demongeot J., Golès E. and Tchuente M., Academic Press, New York (1985).

(27)    Fogelman-Soulié, F. Disc. Appl.Math. <u>9</u>, 139-156, (1984).

(28)    Erdos, P. and A. Renyui. On the Random Graphs 1, Vol.6 Inst. Math. Univ. De Breceniens (1959).

(29)    Erdos, P. and A. Renyi. On the Evolution of Random Graphs publ. 5. Math. Inst. Hungarian Acad. Sci. (1960).

(30)    Berge, C. The Theory of Graphs and Its Applications. London : Methusena (1962).

(31)    Kauffman, S.A., 221-234, in Dynamical Systems and Cellular automata, edited by Demongeot J., Golès E. and Tchuente M., Academic Press, New York (1985).

(32)    Ewens, W.J. Mathematical Population Genetics. Berlin, Heidelberg, New York : Springer Verlag (1979).

(33)    Kauffman, S.A. In press, in Biology and the New Philosophy of Science. Fullerton : Univ. Calif. Press (1985).

(34)    Weisbuch, G. In C.R. Acad. Sc. <u>298</u>, 375-378, (1984).

# INVARIANT CYCLES IN THE RANDOM MAPPING
## OF N INTEGERS ONTO THEMSELVES.
## COMPARISON WITH KAUFFMAN BINARY NETWORK.

Jean COSTE

Laboratoire de physique de la matière condensée

Parc Valrose, 06034 NICE, FRANCE


Michel HENON

Observatoire de NICE

B.P. 06003 , NICE Cedex, FRANCE

According to Kauffman's idea [Kauffman 1970a,b, 1979], one considers an ensemble of P genes which may be found in two possible states $s_i$ , labelled as 0 and 1. An overall state S of the ensemble is the set $\{s_1 , s_2 ,..., s_P\}$ , which is an element of $\{0,1\}^P$ . Given a mapping of $\{0,1\}^P \longrightarrow \{0,1\}^P$ , the iteration of this mapping defines the dynamics of any initial S. In Kauffman model $s_i$ at time $(t+1)$ is determined by the states of k genes at time t, -possibly including $s_i$ itself. Therefore the dynamics is defined by the set of all gene connections and, for each gene, by the data of a Boolean function, that is by an array of $2^k$ elements whose values are either 0 or 1 (there are $2^{2^k}$ possible Boolean functions). The dynamics drives any S towards a cycle of period m $(1 \leqslant m \leqslant 2^P)$, and the problem is to find out the number and the periods of those cycles when S is varied over the various possible states. A numerical study has been performed by Kauffman for k=2 and choosing at random the set of gene connections and the P Boolean functions. It appeared that:

i)   The average number of cycles is of the order of $\sqrt{P}$

ii)  The average period of the cycles is also of the order of $\sqrt{P}$.

This remarkable result shows up some amazing simplicity in the dynamics of a large system and, in particular, helps one to understand how a so large number of interacting genes can produce only few cellular types.

NATO ASI Series, Vol. F20
Disordered Systems and Biological Organization
Edited by E. Bienenstock et al.
© Springer-Verlag Berlin Heidelberg 1986

However several problems are still open:

1° A complete demonstration of Kauffman's results is still lacking. The attempt of R.A. Sherlock [Sherlock, 1979] is semi-numerical and relies upon delicate statistical assumptions.

2° These results do not survive for $k > 2$, and a numerical study becomes obviously difficult when the connectance is increased.

3° Biological data concerning the genes connections are far from complete, and one does not even know whether the connectance is more or less constant over the genes set.

A simpler and less restrictive problem is the following. Let us remark that a Kauffman state S can be associated with a definite integer I : the digits of I, written in binary form, are the $s_i$ 's. Kauffman model is therefore a particular random mapping of $(1,..N)$ onto $(1,..N)$ with $N=2^P$ , and whose statistics concerns the set $\mathcal{C}$ of the genes connections and interactions. Instead of specifying $\mathcal{C}$ , we shall choose the images of the various integers I independently, with equal probabilities over $(1,..N)$ : in other words, we assume the statistical uniformity of the mapping. As we shall see, this model has the advantage of being analytically manageable and providing exact results ; but it cannot be related to any preimposed $\mathcal{C}$ pattern. In some sense this model is the best we could try if we knew nothing about $\mathcal{C}$. Besides it is interesting to find out how things change when passing from the restricted randomness of the Kauffman mapping to "full randomness". We have thought that the following results deserve to be known by the biologists for their general relevance in complex systems. Moreover their proofs are quite straightforward.

**Average number of period m cycles**

Let us consider the orbit of an arbitratry integer $I_o$ , that is the set of $\{I_1 , I_2 ,...., I_j...\}$ of $I_o$ successive images by the mapping. Following the orbit we must encounter some $I_k$ which coincides with some previous $I_p$ ( first coincidence ) and, for $j > k$ , the orbit will henceforth run over a cycle of period $m = k-p$ ( $1 \leqslant m \leqslant N$ ).

$$I_o \rightarrow I_1 \rightarrow \cdots \rightarrow I_p \underset{\searrow I_{k-1}}{\overset{\nearrow I_{p+1}}{}}$$

The probability $\bar{\omega}$ of the (k,p) coincidence is obtained by specifying that the

$(k-1)^{th}$ first successive images are distinct from all previous ones, and that the k iterate coincides with the $p^{th}$ one. Owing to the assumption of uniform probability of the mapping, $\bar{\omega}$ will be obtained by simply counting the number of graphs of the above type, considering all initial integers $I_o$ and the $N^N$ possible mappings. There are obviously $(N-1)(N-2) \ldots (N-k+1)$ images of the set $(I_o, I_1, \ldots I, \ldots, I_{k-2})$ such that the image of any I does not belong to the set. Moreover, since any integer not belonging to the loop can have any one of the N possible images, the total number of $M_k$ of configurations associated with the above loop is

$$M_k = (N-1)(N-2) \ldots (N-k+1) \; N^{N-k}$$

and the probability of the loop is

$$\bar{\omega}(k) = \frac{M_k}{N^N} = (1 - \frac{1}{N})(1 - \frac{2}{N}) \ldots (1 - \frac{k-1}{N}) \cdot \frac{1}{N} \tag{1}$$

Remarkably enough this probability is independent of p , which allows us to simply write it as $\bar{\omega}(k)$ .

A particular case is the one where $I_o$ belongs to the cycle, i.e. p=0 . Therefore $\bar{\omega}(m)$ is the probability that $I_o$ belongs to a period-m cycle (or "[m] cycle " ). Now the average number of points belonging to a [m] cycle is $N \times \bar{\omega}(m)$ and the average number $\mathcal{N}(m)$ of [m] cycles is $\frac{N}{m} \times \bar{\omega}(m)$, or :

$$\mathcal{N}(m) = \frac{(N-1)!}{N^{m-1}(N-m)!} \; \frac{1}{m} \tag{2}$$

In the limit of large N and (N-m)

$$\mathcal{N}(m) \approx \frac{e^{-\frac{m(m-1)}{2N}}}{m} \tag{3}$$

an approximation which fails only when m is comparable to N, that is when $\mathcal{N}(m)$ is already vanishingly small. Expression (3) shows that $\mathcal{N}(m) \approx \frac{1}{m}$ for small m values, but decreases much more rapidly when $m \gtrsim \sqrt{N}$ .


**Average number of cycles, periods, transients**

This number $\mathcal{N}$ is given by

$$\mathcal{N} = \sum_{m=1}^{N} \mathcal{N}(m) \xrightarrow[N \to \infty]{} \frac{1}{2} \log N = \left(\frac{\log 2}{2}\right) P \tag{4}$$

The argument of the preceding section shows that the probability that the orbit of any integer $I_0$ "visits" a [m] cycle after k iterations is $\bar{\omega}(k)$ (and is independent of m, provided $k \geqslant m$). Therefore, the probability $\lambda(m)$ of getting a [m] cycle is :

$$\lambda(m) = \sum_{k=m}^{N} \bar{\omega}(k) \xrightarrow[N \to \infty]{} \sqrt{\frac{\pi}{2N}} \left\{ 1 - \Phi\left(\frac{m-1}{\sqrt{2N}}\right) \right\}$$

$$\Phi(x) = \frac{2}{\sqrt{\pi}} \int_0^x e^{-t^2} dt \qquad \text{being the probability integral function.}$$

The average period of the cycle reached from a given initial integer is :

$$\langle m \rangle = \sum_{m=1}^{N} m \lambda(m) \xrightarrow[N \to \infty]{} \sqrt{\frac{\pi}{8} N} \quad , \tag{5}$$

while the r.m.s. fluctuation of this period is $[(\frac{2}{3} - \frac{\pi}{8}) N]^{1/2}$ .

One also obtains that the average number of iterations needed for reaching a m-cycle is also of the order of $\sqrt{N}$ .

## Conclusion

The above results are to be compared with those obtained by Kauffman for his binary network :

1° The number of cycles is still very small (although larger than in Kauffman model where it is of the order of $\sqrt{\text{Log } N}$ ).

2° The cycles periods are widely dipersed around $\sqrt{N}$ , and the number of iterations needed for reaching those cycles is also of the order of $\sqrt{N}$. Such a N scale would be prohibitively large in some biological applications (remember that N is of the order of $2^{10.000}$ in the genes problem).

We conclude that passing from Kauffman mapping (with connectance 2) to the "fully random" one studied here preserves the striking property that there are so few cycles, but the orbits ( cycles and transients ) become very large. A crucial question is : how does the average period of the cycles vary when the connectance is increased beyond k = 2 in the Kauffman model ?

## REFERENCES

Kauffman, S. 1970 a. "Behaviour of randomly constructed nets ". In **Towards a theoretical biology** Ed. C.H. Waddington, vol. **3**, Edimburg University Press.

Kauffman, S. 1970 b. " The organization of cellular genetic cintrol systems ". **Math. Life Sci. 3** , 63-116.

Kauffman, S. 1979 " Assessing the probable regulatory structures and dynamics of the metazoan genome. Kinetic logic ". In **Lecture notes for Biomathematics** Ed. R. Thomas, **29**, 30-61. Berlin Springer Verlag.

Sherlock, R.A. 1979 " Analysis of Kauffman binary networks " **Bull. Math. Biol. 41**, 687-724.

Note added in proof.

An evaluation of the number of the average period of the cycles in KAUFFMAN's model has been recently obtained by J. COSTE and M. HENON of Nice University, in the case where each binary variable is randomly connected to all other ones. In the following, these authors have been aware of a technical report by H. RUBIN and R. SITGREAVE on the same subject [Tech. Report n° 19A Appl. Math. and Stats. Lab., Stanford University, 1954]. Since this report is not easily obtainable (even J. COSTE and M. HENON could not have it in hands) and considering the result is important and simply derived, it appeared useful to incorporate J. COSTE & M. HENON's work in the Reports of this Conference.

# FIBROBLASTS, MORPHOGENESIS AND CELLULAR AUTOMATA

Y. Bouligand
C.N.R.S. et E.P.H.E.
67, rue Maurice Günsbourg
64200 Ivry-sur-Seine
France

## I. Introduction

Cellular automata provide models for various biological processes. Threshold automata have been invented to simulate neurons
and their electrophysiological behaviour in nervous centres
(Mac Culloch and Pitts, 1943; Hopfield, 1982). Complex behaviours such as recognition and learning can even be simulated
(Minski and Papert, 1969; Le Cun, 1986). Remarkable morpholo-
gies are also produced by cellular automata and there are numerous examples (Ulam, 1962; Conway's game "life", see Gardner,
1971-1972). The examination of pictures obtained with cellular
automata suggest new mechanisms for biological pattern formation (Wolfram, 1984).

Visual recognition and pattern production in animals are two
different processes which work quasi symmetrically relative to
time. One passes either from a pattern to its description or
the reverse. The analysis of a picture moving over the retina
leads to a series of informations which are encoded into a
neural memory by association neurons, whereas, during morphogenesis, a part of genetic information is decoded and expressed as structures and patterns. The memory of specific shapes
of the body or of the organs is encoded in certain genes and is
transmitted to the progeny, whereas neural memory and other
acquired characters are not. It is therefore clear that comparisons between genetic and neural memories seem at first view
to be quite superficial. Actually, research in cellular automata
leads to a different point-of-view, since pattern recognition

NATO ASI Series, Vol. F20
Disordered Systems and Biological Organization
Edited by E. Bienenstock et al.
© Springer-Verlag Berlin Heidelberg 1986

ahd pattern production use similar instruments and are general
problems considered in very similar terms. We intend to show
here that this conception deserves an extension to biological
systems and will be fruitful, despite dangers of comparisons.

We shall first examine the example of connective tissues and
embryonic cells from which they stem. Their morphogenetic role
is clear and some indications will be given about mechanisms.
Finally, connective cells will be compared to cellular automata, as it has been done for nerve cells.


## II. Fibroblasts and extracellular matrices

The connective tissue and its derivatives are involved at several essential steps of animal morphogenesis. The main type of
connective cell is called fibroblast and has a stellate shape,
which resembles more or less that of neurons, with long and
slender cytoplasmic projections joining between neighbouring
cells (fig. 1). These cells occupy a relatively small volume
compared to that of the intercellular space, which is filled
with several fibrous components secreted by fibroblasts : the
main one is collagen, a protein forming connective fibrils.

Organs and, indeed individuals, are more or less shaped by the
extracellular lattices of collagen fibrils, called matrices (see
Bloom and Fawcett, 1975). For instance, fig. 2 shows the beginning of intramembranous osteogenesis, or bone formation, in the
skull of a 5.5 cm cat embryo. In osteogenesis, connective fibrils secreted by cells form a denser and denser lattice which is
later mineralized by calcium phosphates (apatite). Fibroblasts
transform slightly in the course of osteogenesis and are given
new names : osteoblasts and osteocytes. Collagen fibrils adopt
precise orientations and form the organic matrix of the bone.

In compact bone, the extracellular matrix architecture is reminiscent of that of plywoods : layers of parallel collagen fibrils are superimposed and make an angle close to 90° as shown

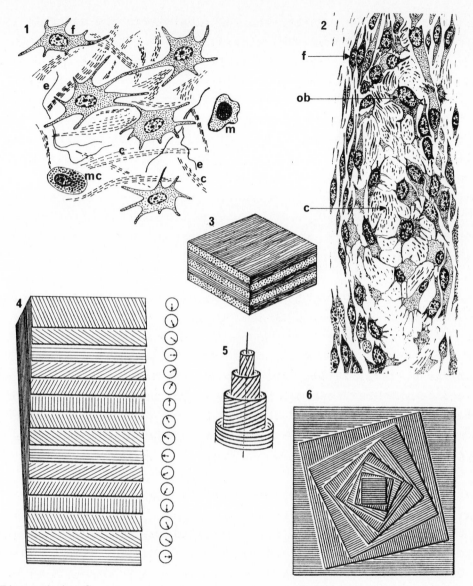

Figs. 1 to 6

Fig. 1 : Schematic representation of connective tissue. c : collagen fibril. e : elastic fibril. f : fibroblast. m : macrophage mc : mast cell. Fig. 2 : Formation of the intramembranous bone in the skull of a cat embryo. c : collagen fibrils. f : fibroblasts transforming into osteoblast. ob : osteoblast (redrawn after Bloom and Fawcett). Fig. 3 : Quasi-orthogonal packing. Fig. 4 : Regularly twisted plywood; dials show the progressive rotation. Fig. 5 : Cylindrical twist. Fig. 6 : Plywood with a $\frac{\pi}{2} + \varepsilon$ rotation angle.

in fig. 3. This is the simplest case, but there are several types of plywoods formed by collagen fibrils. For instance, in bone, there are not only right angle plywood as in fig. 3 but also various regularly twisted plywoods. Consider thin layers of horizontal parallel collagen fibrils and superimpose them, each added layer being rotated by a small angle relative to the preceeding one, as shown in fig. 4. Each orientation reappears after a turn of 180°. The layer thickness and the rotation angle from layer to layer are kept constant (Bouligand, 1965a, b, 1978). Such systems are often continuous and the orientation of fibrils rotates linearly. If n is a unit vector parallel to the fibrils at point M (x, y, z), an orthonormal frame can be found such that the components of n read :

$n_x = \cos 2\pi z/p$; $n_y = \sin 2\pi z/p$; $n_z = 0$, (p being the helical pitch).

The structure is the same as that of cholesteric liquid crystals (Bouligand, 1969, 1978). This suggests that assembly of collagen molecules into matrices may be related to the growth process of liquid crystals. This self-assembly can be studied in vitro and three different packing modes are obtained with collagen molecules (Bouligand et al., 1985).

1. Planar twist similar to that known in cholesteric liquids (fig. 3)
2. Cylindrical twist (fig. 5), similar to that considered in models of blue phases (see for instance Meiboom et al., 1981)
3. Quasi-orthogonal packing (fig. 3).

The planar twist (1) and the quasi-orthogonal packing (3) are observed in compact bones. More sophisticated plywoods also exist in other tissues. There are examples of 60° angles in certain worm skins and this leads to three main orientations of fibrils. A beautiful architecture involves angles of $\frac{\pi}{2} + \varepsilon$ ; the angle $\varepsilon$ is small (some degrees) and varies from species to species (fig. 6). This is observed in certain fish scales (Giraud et al., 1978; Meunier and Castanet, 1982) and in certain embryonic corneas (Coulombre and Coulombre, 1961). There is a wide set of such architectures observed in vivo and first attempts to reproduce them in vitro have been successful (Bouligand et al., 1985). A review of these natural plywoods appeared recently (Bouligand and Giraud-Guille, 1985).

Another important architecture corresponds to the unidirectional alignment of collagen (or of other skeletal macromolecules) in numerous organs. For instance, tendons linking extremities of muscles to skeletal structures show such unidirectional alignments of collagen molecules. Muscles themselves are made of parallel elongated cells, each containing a contractile apparatus. Muscle cells are gathered into a hierarchy of bundles separated by walls of connective tissue. In cross section, this system shows a quasi-fractal organization as can be seen from fig. 7. Other muscles, mainly smooth muscles which surround the stomach and the intestine present a plywood structure (see histology treatises : Bloom and Fawcett, 1975). The heart striated muscle has a structure which is that of a more or less twisted plywood (Brecher and Galetti, 1963; Olson, 1962). Such muscles are also partitioned by connective tissue and the orientation of muscle cells is probably specified by fibroblasts and collagen fibrils in the course of development. It appears therefore that connective tissue plays a fundamental role in morphogenesis of bones, muscles and other organs. This has been confirmed by embryologists and mainly by P. Sengel in his studies of the morphogenesis of vertebrate skin (1975). Dermis is a dense connective tissue underlying the epidermis and presenting a plywood architecture in various groups of vertebrates. The leg epidermis of a chick normally produces scales and not feathers. However, if at a certain embryonic stage, this leg epidermis is removed and grafted onto the back dermis, feathers develop from it instead of scales. Many other experiments show the effects of embryonic dermis on embryonic epidermis. It is possible, at a given embryonic stage, to recombine between chick and duck the dermis and the epidermis of the future wing or even to recombine the embryonic dermis of a mouse snout with a chick epidermis. In this latter case, the feather buds begin to

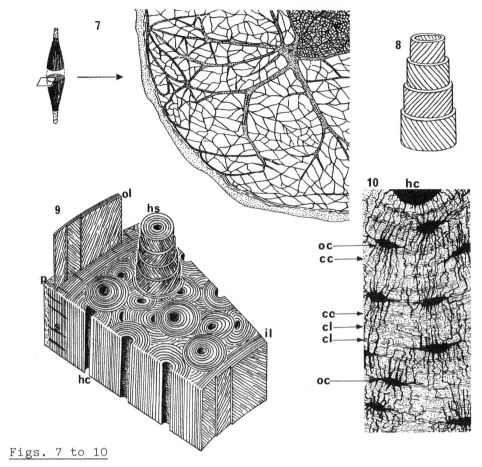

Figs. 7 to 10

Fig. 7 : Schematic drawing representing a quarter of a cross
section of a muscle. The connective tissue separates muscle
cells into a hierarchy of groups and subgroups. Fig. 8 : Left-
handed and right-handed helices in the skin of certain round
worms (the body is nearly cylindrical). Fig. 9 : Distribution
of lamellae in a sector of a long bone. hc : haversian canal.
hs : haversian system, where collagen fibrils are often arranged as in Fig. 8. Recently formed haversian systems show complete cylindrical layers, whereas ancient systems have been replaced partially by new ones. il : internal lamellae. ol : outer
lamellae. p : periosteum. s : Sharpey fibrils; these are connective fibrils penetrating perpendicularly or obliquely the outer
lamellae (drawing inspired from Bloom and Fawcett). Fig. 10 :
Section of a sector of a Haversian system, showing the osteocytes and their slender cytoplasmic projections, which show many
junctions from cell to cell. The orientation of the section is
such that collagen fibrils are seen either in cross or in longitudinal section in this sector. cc : collagen in cross section.
cl : collagen in longitudinal section. hc : haversian canal.
oc : osteocyte (inspired from Bloom and Fawcett).

differentiate, but they show an arrangement which is that of mouse whiskers.

Many other examples could illustrate the morphogenetic role of connective tissues, but most of the mechanisms are still unknown. There are however indications that these mechanisms are of three types :

. Self-assembly
. Cellular activities
. Constraints due to spatial organization

1. Self-assembly of plywood architectures.

Collagen molecules secreted by fibroblasts form triple helices and their in vitro assembly in the absence of cells has been studied extensively (Hodge and Petruska, 1963). Recently, preparations of thin sections of self-assembled gels of collagens have allowed us to look closely at the geometry of the fibril aggregates. The main architectures described above from different connective matrices are also found in these self-assembled gels : unidirectional alignment of fibrils, twisted systems, orthogonal or quasi-orthogonal superimpositions etc ... and each mode of assembly can be given a simple molecular interpretation.

2. Cellular activities.

Connective cells multiply, move and differentiate. They synthesize collagen and secrete it into the extracellular space. Most cells, and fibroblasts in particular, secrete other proteins, fibronectin for instance, which allow these cells to attach to collagen fibrils (see for instance Alberts et al., 1983). These molecular links probably help cells, not only fibroblasts, but all sorts of cells in differentiated tissues, to align along previously oriented collagen fibrils. Conversely, through these attachments, cells can modify positions and orientations of self-assembled collagen fibrils. We know for instance that hexagonal plywood, with 60° rotations are due to preferential orientations arising from the "geometry" of projections of the cell surface.

3. Constraints due to spatial organization.

Collagen of the skin of many animals form quasi-orthogonal ply-woods and the two main orientations of fibrils lie at almost 45° to the long axis of the body. This situation is well known in the dermis of lower vertebrates and their embryos (Rosin, 1946). In worms also, the outer cuticle is a protection wall made of collagen secreted by the external surface of the epi-dermis (Picken, 1962). In cylindrical worms such as nematods (= round worms, example Ascaris), the situation is very clear. Each layer of the plywood is made of inextensible fibrils for-ming either righ-handed or left-handed helices running round the body (fig. 8). The length of the body can change and the angle separating the two sets of helices varies accordingly. There is however a preferred angle (54° with the longitudinal axis) and the volume is maximal at this angle (Picken, 1961), due to the internal hydrostatic pressure which is very high (about 0.1 atm.). In this case, biomechanical constraints are involved in the orientation of collagen fibrils. A very similar geometry is observed in the plywood architecture of compact bone (fig. 9) and, here also, constraints at different organi-zation levels (haversian systems and the whole bone must be taken into account).

III. Speculations on fibroblasts considered as cellular automat

Fibroblasts form networks as do nerve cells but several diffe-rences should be pointed out. Fibroblasts present several elon-gated cytoplasmic projections which enter in contact along spe-cialized junctions but these are very different from synapses and there are no structures equivalent to axons (see Bloom and

Fawcett, 1975, p. 257). Thus, fibroblasts enter in contact only with neighbouring fibroblasts, whereas neurons join very distant cells via their axons. The concept of "closest neighbors" is problably more essential for fibroblasts than for neurons. Fibroblasts contain the classical cell organelles as nerve cells but rough endoplasmic reticulum is much less developed than in neurons. Membranes potentials are observed in fibroblasts and in all living cells. Moreover, propagating depolarizations are not an exclusive property of nerve cells. Unfortunately, electrophysiological research is mostly limited to nerve and muscle cells. However, the hypothesis of an elaborated electrophysiological behaviour of fibroblasts is quite reasonable and to compare such cells to cellular automata enters in the scheme of general models proposed by Lewis and Wolpert (1971) and Wolpert (1978), for example.

Osteocytes are derived from fibroblasts and these cells form remarkable three-dimensional networks in bones, as shown in fig. 10. This view corresponds to a section of haversian systems. Osteocytes are aligned along more or less cylindrical layers, arranged concentrically. Their cytoplasmic projections run normally to layers of the collagen plywood. These networks of fibroblasts forming Haversian Systems are involved in bone morphogenesis.

There are also examples of two-dimensional networks of connective cells. When a new skin is prepared some days before a molt in an insect, several connective cells extend onto the inner side of the epidermis (Wigglesworth, 1956). These two-dimensional networks of cells are probably involved in the morphogenesis of the new skin. Molts can also correspond to metamorphoses and, then, a radically different morphology is created.

Fibroblasts also are present along the epidermal cells at the growing edge of the shell of gastropods (and other molluscs). In this case, fibroblasts form a quasi-one-dimensional network. Shells such as those of the genus Conus present a set of dark lines forming remarkable patterns which are built up gradually from secretion of pigments by epidermal cells lining the edge of the shell. Such patterns often resemble the patterns obtained by one-dimensional cellular automata and Wolfram (1984) proposed that cells of the growing edge of the shell work similarly to such automata. The deterministic model of Wolfram differs from a model due to Waddington and Cowe (1969) involving a linear association of stochastic threshold automata. Both models provide patterns close to those observed on Conus shells.

Fibroblasts networks are possibly information processing systems. The two-dimensional network observed beneath the insect epidermis or the three-dimensional networks present in compact bone may well function as such dynamical systems. The stellate shape of fibroblasts (fig. 1) and their connections are the characteristics of a system which is well devised to specify positional information. If fibroblasts are supposed to function as cellular automata, certain asymptotic behaviours are possible and create stable patterns which can be considered as a frame of reference. According to the internal state of a fibroblast (defined probably by parameters which are often electrophysiological), certain genes are expressed and this leads for instance to secretion of collagen or associated proteins, such as fibronectin, or deposition of mineral, to cell displacement or cell division, transmitting information to other types of cells, etc... The conditions prevailing at the periphery of the network probably influence the asymptotic behaviours.

There is a resemblance between a set of stellate connective cells extending beneath the insect epidermis, at the time of renewing the skin, and the system of association neurons connecting cones and rods within a retina. In the first system, a new morphology can be produced, during metamorphosis for instance, whereas, in the second one, an optical image is analyzed. In both cases, cells forming a single layer are interconnected by stellate cells. We suggest that, in these two biological systems, similar kinds of molecules and electrophysiological processes are involved, despite the fact that information follows two opposite pathways, one centripetal, from photoreceptors to association neurons, encoding morphological information, the other centrifugal, from fibroblasts to epidermal cells, decoding a genetic information. Two complementary problems are solved therefore by these sets of cells : the first is to produce an elaborate morphology from cells which behave as cellular automata; the second one is to start from a given retinal image and to elaborate information which will be suitable for an associative memory. Certain topological aspects of these processes are similar. For instance, the letter F is recognized as such at different sizes, at different distances or at different angles of observation; the colour and contrast also can be changed. The corresponding retinal pictures are therefore different. However, what remains is the characteristic shape of an F, even when it is more or less distorted. Topological aspects prevail over metric aspects. In morphogenesis a very similar situation occurs. This has been remarkably illustrated by a large series of experiments on the presumptive limb rudiments in the embryos of amphibians. This work due to Harrison (1921) shows many topological aspects of embryonic development. These disk-shaped rudiments (future epidermis and fibroblasts) of limbs are found laterally in the body wall. At a given stage of development, these disks can be resected and limbs form however from cells at the periphery, which change their program of differentiation. The embryo can develop normally, if this experiment is performed in good aseptic conditions, but the corresponding limb is generally smaller. At a further stage, the resection of the disk leads to the absence of the limb. However, at this time, resection of only one part of the disk leads to the development of a complete but smaller limb. This means that subsets of cells in limb rudiments can form a complete limb exactly as different subsets of cones and rods in the retina receive an information which will be decrypted as belonging to a given category of patterns, the letter F for instance. Interneurons in the retina and interconnecting stellate cells beneath the future epidermis are processing systems and this processing of morphological information is mainly topological.

## Limits of these speculations

Connective tissues are involved in several aspects of morphogenesis. <u>Self-assembly</u> is an aggregation process, more or less
controlled by cells and this could be simulated by cellular
automata. There are <u>mechanical constraints</u> working at different
levels of organization and this also can be probably simulated
properly. However, there are some <u>difficulties</u> arising from
comparisons between fibroblast networks and cellular automata.
For instance, one of the most impressive steps of embryogenesis
is gastrulation and this process follows a precise program of
cell cleavages and movements. However, when an unfertilized
amphibian egg is treated by progesterone for ten minutes and is
replaced in a normal saline solution, after 36 hours, the un-
cleaved egg shows an invagination and converging movements of
cytoplasm at the surface, which strongly resemble a gastrulation. This indicates that the presence of a cellular network is
not a prerequisite at this stage for the gastrula morphogenesis. It is also well known that elaborate differentiations are
found in single cells, particularly in many protozoa. In such
cases, morphogenesis cannot be the result of the collective behaviour of several cells.

A large series of similar examples can serve as arguments
against the idea that morphogenesis can be based on cells working as cellular automata. Actually, these arguments do not
really hold, since it appears at present that morphogenesis is
never based on a unique mechanism but on several mechanisms
functioning in synergy. Physiologists are well aware that any
regulation is built on several synergetic mechanisms. Embryonic
development and growth are also regulated. This developmental
regulation is nothing else than the stability of several different mechanisms involved in a common morphogenesis.

<u>Acknowledgements</u> : Thanks are due to Dr. G. D. Mazur for very
useful discussions.

# Literature cited

Alberts B., Bray D., Lewis J., Raff M., Roberts K. and Watson
    J.D. Molecular Biology of the Cell. Garland Publishing,
    Inc. N.-Y., London (1983).
Bloom W. and Fawcett Don W. A Textbook of Histology, 10$^{th}$ edi-
    tion, W.B. Saunders (1975).
Bouligand Y. C.R. Acad. Sci. Paris, 261, 3665, 4864 (1965).
Bouligand Y. J. Physique, 30, C4, 90 (1969).
Bouligand Y. Solid State Physics, Suppl. 14, 259 (1978).
Bouligand Y., Denèfle J.-P., Lechaire J.-P. and Maillard M.
    Biology of the Cell, 54, 143 (1985).
Bouligand Y. and Giraud-Guille M.-M. In : "Biology of Inverte-
    brate and Lower Vertebrate Collagen, ed. A. Bairati and
    R. Garrone. NATO ASI Series. Series A, Life Sciences, 93,
    115, Plenum (1985).
Brecher G.A. and Galetti P.M. In : Handbook of Physiology,
    Circulation, Amer. Physiol. Soc., 2, Chapt. 23, 759 (1963).
Coulombre A. and Coulombre J. In : "The Structure of the Eye"-
    ed. G. Smalzer, Acad. Pr., 405 (1961).
Gardner M. Scientific American, 223 (4) 120, (5) 118, (6) 114;
    224 (1) 105, (2) 112, (3) 108; 226 (1) 104 (1972).
Giraud M.-M., Castanet J., Meunier J.F. and Bouligand Y. Tissue
    & Cell, 10, 671 (1978).
Harrison R.G. J. Exp. Zool., 32, 1 (1921).
Hodge A.J. and Petruska. In : "Aspects of Protein Structure",
    ed. G.N. Ramachandran, Acad. Pr., 289 (1963).
Hopfield J.J. Proc. Natl. Acad. Sci., 79, 2554 (1982).
Le Cun Y., in these proceedings (1985).
Lewis J.H. and Wolpert L., J. Theor. Biol., 25, 606 (1971).
Mac Culloch W.S. and Pitts W. Bull. Mathem. Biophys., 5, 115
    (1943).
Meiboom S., Sethna J.P., Anderson P.W. and Brinkman W.F. Physi-
    cal Rev. Lett., 46, 1216 (1981).
Meunier J.F. and Castanet J. Zoologica Scripta, 11, 141 (1982).
Minsky M. and Papert S. Perceptrons. The MIT Press (1969).
Olson R. In : "Handbook of Physiology, Circulation, 1, Chapt.
    10, 199 (1962).
Picken L. The Organization of Cells and other Organisms. Univ.
    Pr. Oxford Clarendon (1962).
Rosin S. Revue Suisse de Zoologie, 53, 133 (1946).
Sengel P. Morphogenesis of Skin, Cambridge Univ. Pr. (1975).
Ulam S. Proc. Symp. Appl. Math., 14, 215 (1962).
Waddington C.H. and Cowe R.J. J. Theor. Biol., 25, 219 (1969).
Wigglesworth V.B. Quarterly J. Microsc. Sci., 97, 89 (1956).
Wolfram S. Nature, 311, 419 (1984).
Wolpert L. Scientific American, 239 (4), 154 (1978).

# PERCOLATION AND FRUSTRATION IN NEURAL NETWORKS

A.J.Noest , Netherlands Institute for Brain Research
Meibergdreef 33, 1105 AZ Amsterdam, The Netherlands

Introduction:

There are reasons to believe that the theoretical and computational
techniques of solid state physics can be used to get an understanding of
the dynamics of the brain. Here I shall approach along similar lines a
related, but less ambitious set of problems stemming from tissue-culture
experiments with foetal rat cortex neurons [1]. Starting from a planar
array of disconnected cells, dendrites and axons grow out slowly, leading
via the formation of clusters of connected cells to a percolating network.
After a week, synapses develop at the contacts, allowing the cells to
transmit spike-signals by a discrete stochastic process. Growth of the
network is much slower than spike-propagation. Two main cell types exist:
excitatory cells that increase the spike rate of their target cells, and
inhibitory cells that decrease it.

The problem is to analyze the propagation of the spikes through the net,
and specifically to study the effects of the disordered network structure,
the balance of excitatory versus inhibitory cells and the synapse strength.

To study the dynamics of spike propagation one needs a simple model of
a neuron that still captures the essentials of the input-output relation,
i.e. the weighted ($a_{ij}$) summing and smoothing of inputs $s_j \in \{0,1\}$ and an
output firing probability $P(s_i)$ which is a thresholded, saturating function
of the inputs. The microscopic timescale of the dynamics (smoothing, spike
width and dead time) can be introduced by discretizing time. A reasonable
choice for the dynamics is of the form:

$$P(s_i(t+1)) = 1-\exp(-S_i(t)) \; ; \; S_i(t) > 0$$
$$= 0 \qquad\qquad ; \; S_i(t) < 0$$

, with $S_i(t) = \sum_j a_{ij} s_j(t)$

A simple case : $a_{ij} \geqslant 0$

First I study the simple case without inhibitory cells, which seems to
apply to the initial phase of the experiment. With $a_{ij} \geqslant 0$, the stochastic
process above is locally identical to a directed-bond percolation problem
[2] with bond-probability $1-\exp(-a_{ij})$. The geometry on which the problem
is defined here is a spacetime structure in which the sites are repeated

NATO ASI Series, Vol. F20
Disordered Systems and Biological Organization
Edited by E. Bienenstock et al.
© Springer-Verlag Berlin Heidelberg 1986

copies of the neurons, with directed bonds having a positive projection on the time direction. The bonds occur with a probability $p_t$ within time-like rows, modelling the synaptic coupling strength between pairs of cells. $p_t$ may depend on space, but is assumed independent of time ("quenched"). One may also interchange the roles of bonds and sites, or combine the two versions. Let us call this "(D+1)-percolation" (compare [3]), with D the dimension of the (spatial) network. The special cases possessing spatial translation-invariance are directed-bond percolation problems on regular (D+1)-dimensional lattices, having many applications in physics, chemistry and biology [2,3,4]. The inclusion of spatial disorder (or other types of correlated perturbations) seems worthwile not just for the particular application at hand. Presently I introduce spatial disorder by dilution, i.e. by the elimination of whole time-like rows of sites or bonds. Let the density of remaining site- or bond-rows be $p_s \in (0,1)$. Then, for $D > 1$ and short-range connections, there is a critical $p_s^*$ below which the network does not percolate spatially, hence (D+1)-percolation is impossible, except at $p_t=1$. For $p_s \geqslant p_s^*$ there should be a (D+1)-transition since the spatial network contains at least one path to infinity and it is known that there exists a directed percolation transition in (1+1)-dimensions [2]. In fact, the critical point $p_t^*$ of spatially disordered (D+1)-percolation should lie between that for the corresponding system without disorder and that of (1+1)-percolation. Thus, one expects the following phasediagram:

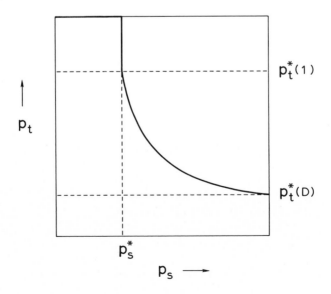

The first question that suggests itself is whether even small spatial disorder is compatible with the universality class of standard directed percolation. An argument in the style of Harris [5] shows that this is not so: Assume that there is a transition with the normal exponents. Then, the large space-time clusters sample regions $A \sim \xi_\perp^D$ of the network having fluctuations in the densities $\delta(p_t) \sim \delta(p_s) \sim \xi_\perp^{-D/2} \sim |p_t - p_t^*|^{D\nu_\perp/2}$. Selfconsistency requires $\delta(p_t) \ll |p_t - p_t^*|$ for $|p_t - p_t^*| \to 0$. Thus, one must have $D\nu_\perp > 2$. However, $\nu_\perp$ is approximately $1.1 / 0.8 / 0.6 / 0.5$ for $D = 1 / 2 / 3 / 4$ respectively, so one finds a contradiction ( $D \geqslant 4$ is marginal). For a more precise treatment of quenched disorder by RG cf.[6].

This raises the question of whether a new universality class exists. I have investigated this numerically for D=2 by MC, using up to 4096 time-steps and networks of up to 128*128 cells on square and triangular lattices with toroïdal boundary conditions. The quenched disorder consisted of cell dilutions of 0,5,10,20 and 30% . Phase transitions were found in all cases when changing $p_t$, with power-law behaviour of the singular variables near criticality. As hoped, the critical exponents were found to be the same for all non-zero diluted systems and distinctly different from those with zero dilution. This is good evidence for a new universality class characterized by the critical exponents in the left column of table 1.

Table 1.    Critical exponents of (2+1)-percolation
(and other D=2 SCA with 1 absorbing state)

| | With spatial disorder | Pure lattices |
|---|---|---|
| $\nu_\perp =$ | 1.21 +/- 0.15 | 0.84 +/- 0.05 |
| $\nu_\parallel =$ | 2.1 +/- 0.4 | 1.3 +/- 0.1 |
| $\nu_\perp/\nu_\parallel =$ | 0.58 +/- 0.03 | 0.64 +/- 0.03 |
| $\beta =$ | 1.15 +/- 0.2 | 0.63 +/- 0.03 |
| $\gamma =$ | 2.2 +/- 0.4 | 1.65 +/- 0.15 |
| $\hat{d} =$ | 1.61 +/- 0.05 | 1.81 +/- 0.04 |

,where $\hat{d} = 1 + (D\nu_\perp - \beta)/\nu_\parallel = (\beta + \gamma)/\nu_\parallel$ is the spreading dimension [7].

The general case:
One would also like to solve the case where the random network contains some fraction of inhibitory cells. With $a_{ij}$ allowed to be negative, the model can no longer be mapped onto (D+1)-percolation, but becomes instead a D-dimensional stochastic cellular automaton (SCA) [8,9] with spatial

disorder. In general, this is a very much more difficult problem because of frustration effects. Perhaps the best way to illustrate this is to map the "history" $\{s_i(t)\}$ of the SCA onto the equilibrium configurations of an Ising problem on the (D+1) spacetime structure [cf.9]. One obtains generalized spinglass Hamiltonians of the form:

$$H(\{s_i(t)\}) = K_o \cdot \sum_{it} s_i(t) + \sum_{ijt}[K_{ij} \cdot s_i(t)s_j(t+1) + L_{ij} \cdot s_i(t)s_j(t)]$$
$$+ \text{ multispin terms of degree up to } 1+(\text{cell valency}) \ .$$

A full understanding of our general problem is thus equivalent to solving a spinglass with multispin couplings in a field on a lattice with linear defects. It seems unlikely that this problem will be solved fully in the next few years, although for the (1+1)-case there are some hopes [10]. The MC-approach is of course always possible, but tends to take much time. Motivated in part by the particular application at hand, I have studied the effects of adding inhibitory (I-)cells (density=$p_I$) randomly to the network of excitatory (E-)cells (density=$p_E$), keeping $1 > p_E \gg p_s^*$ . For small values of $p_I$ , one is still in the universality class of disordered (D+1)-percolation. This can be understood by looking at an I-cell (or a small cluster) in a "sea" of E-cells which is above the (D+1)-transition. Firing of spikes by E-cells on the boundary of an I-cell will tend to be suppressed, with a delay of two timesteps. The details of this can be complicated, but the effect on the firing in the E-"sea" is larger than that of removing the I-cells alone, but smaller than that of removing the boundary E-cells also. As long as the E-"sea" minus these boundary cells percolates spatially, one expects indeed to see the same class of phase transition as with the $p_I=0$, disordered (D+1)-percolation treated before.

References:
1- F.van Huizen,H.J.Romijn and A.M.M.C.Habets,Dev.Brain Res. 19(1985)67
2- W.Kinzel,In:"Percolation Structures and Processes",
        G.Deutsch,R.Zallen and J.Adler,Eds.,Adam Hilger,Bristol,1983
3- L.S.Schulman and P.E.Seiden,In:"Percolation Structures and Processes",
        G.Deutsch,R.Zallen and J.Adler,Eds.,Adam Hilger,Bristol,1983
4- "Biological Growth and Spread",Lect.Notes in Biomathematics #38,
        W.Jaeger,H.Rost and P.Tautu,Eds.,Springer,Berlin,1979
5- A.B.Harris,J.Phys.C 7(1974)1671
6- D.Andelman and A.N.Berker,Phys.Rev.B 29(1984)2630
7- J.Vannimenus,J.P.Nadal and H.Martin,J.Phys.A 17(1984)L351-L356
8- W.Kinzel,Z.Phys.B(Condensed Matter) 58(1985)229-244
9- E.Domany and W.Kinzel,Phys.Rev.Lett. 53(1984)311
10- W.F.Wolff and J.Zittarz,In:"Heidelberg Colloqium on Spinglasses",
        Lect.Notes in Physics #192,J.L.van Hemmen and I.Morgenstein,
        Eds.,Springer,Berlin,1983

# SELF ORGANIZING MATHEMATICAL MODELS : NONLINEAR EVOLUTION EQUATIONS WITH A CONVOLUTION TERM.

Michelle SCHATZMAN
Mathématiques, Université Claude-Bernard,
69622 Villeurbanne CEDEX, FRANCE.

## 1. A SIMPLE PROBLEM.

The simplest example of a dynamical system which organizes itself through cooperation and competition has been given in this conference [8] ; I shall formalize it as follows : let A be a linear operator in the plane $\mathbf{R}^2$ ,and consider the ordinary diferential system

(1)     $\dot{x} = Ax,\ x(0) = x_0$ ,

where x is constrained to remain in the unit square

(2)     $x \quad K = [-1,1] \times [-1,1]$.

Typical trajectories are given in figures 1, 2 and 3 below. Practically, they could be approximated by the scheme

(3)     $x^{n+1} = P[x^n + \Delta t\ Ax^n]$ ,

where P is the projection on K. The projection is needed to take care of whatever happens on the boundary of K. When the eigenvectors or the eigenvalues of A vary, the number of sinks, saddles and sources varies, as can be seen on the figures.

If we assume that A is symmetric, (1) admits a Liapunov functional $\Phi(x) = -\ Ax.x$ .

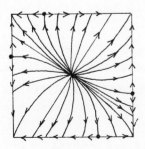

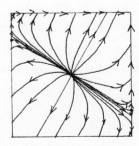

Fig. 1            Fig. 2            Fig. 3

NATO ASI Series, Vol. F20
Disordered Systems and Biological Organization
Edited by E. Bienenstock et al.
© Springer-Verlag Berlin Heidelberg 1986

There is an immediate generalization to an arbitrary number
of dimensions ; as in the two dimensional case, a stable limit
point of trajectories can only be zero, or a vertex of the hypercube $[-1,1]^N$ ; thus there are many possible limit points,
and assumptions must be made on A, as we shall see below.

## 2. PERIODIC STRUCTURES.

In 1980, N.V. Swindale [7] proposed a model of development
of ocular dominance stripes ; if u is the difference of density
at point x of fibers efferent from left eye and from right
eye, the model is

(4)        $\partial u / \partial t = (w \star u) (1 - u^2)$,

with w a symmetric continuous kernel depending only on space ;
the star denotes convolution : $(w \star u)(x) = \int w(x-x') u(x')dx'$ .
I have proposed, with E. Bienenstock, a modified model, which
gives the same kind of periodic structure as Swindale's, but
faster, and is somewhat easier to analyse mathematically. It is

(5)        $\partial u / \partial t = w \star u$      (6)    constraint : $|u| \leq 1$, a. e. in x,t.

(see [6] for a precise mathematical formulation). Around u = 0,
the problem is linear ; if one of the Fourier coefficients of w
is larger than the others, the corresponding mode grows faster
than the others, but, as soon as $|u|$ reaches 1, the modes
start interacting . If w is even, there is a Liapunov function

(7)        $\Phi(u) = - \frac{1}{2} \int(w \star u) \ u \ dx$ ,

and we can prove

Theorem. Let w be analytic in space ; a non-zero minimizer
of $\Phi$ over K = {u / $-1 \leq u \leq 1$, a.e.}   is such that $|u|$ =1 a.e.,
and u=-1 if $w \star u < 0$, u=1 if $w \star u > 0$.

Numerical simulations show square waves as the final state,
in a number of cases, but this is not necessary, as was shown
in [6, Proposition 5.9]. It has not been proved mathematically
that almost all solutions of (5),(6) converge to a local minimizer of $\Phi$ .

It turns out that minimizing $\Phi$ over K is a limiting case of
a mean field theory considered by Eisele and Ellis [5] : magnetic spins are distributed on the circle $\mathbf{T}^1$ at sites k/n ; these
spins are denoted $s_i$, $1 \leq i \leq n$, and the energy of the system

with the normalization factors $1/n$ of mean field theory, is

(8)     $H(s) = (2n)^{-1} \sum_{j,k} w((j-k)/n) \, s_j \, s_k$ ;

the system is in a state of energy $H(s)$ with probability $p(s) = Z^{-1} \exp(-H/T)$, where $T$ is the absolute temperature, and

(9)     $Z = \sum_s \exp(- H(s)/T)$.

The specific energy $\psi(T)$ is defined by

(10)     $-T^{-1} \psi(T) = \lim_n n^{-1} \log Z(n,T)$.

    Define

(11)     $i_\rho(u) = \int_0^u \tanh^{-1}(r) \, dr$ if $|u| \leq 1$, $+\infty$ otherwise.

Then, Eisele and Ellis have shown in [5] that

(12)     $\psi(T) = \inf \{- \int (w*u) \, u \, dx + T \int i_\rho(u) \, dx\}$.

    The Euler equation for the variational problem (12) is

(13)     $u = \tanh( w*u/T)$.

    If $T$ tends to zero in (13), we find

(14)     $u = \operatorname{sgn}(w*u)$,

and this is exactly the condition satisfied by local minimizers of the Liapunov function defined at (7).

    Problem (13) is interesting in its own merit, as a nonlinear bifurcation problem, and has been studied in [4].

## 3. VERY SHARP SELECTIVITY.

    Another family of dynamical systems exhibiting strong structuration properties as time goes to infinity, has been proposed by E. Bienenstock, B. Moore and myself [3]. They are of the form

(15)     $\partial u/ \partial t = w * u - f(u)$     (16) constraint $u \geq 0$.

Here $f$ is a nonlinear function of the global integral of $u$, or of some of its partial integrals, and these integrals will remain bounded thanks to $f$ ; nevertheless, a concentration phenomenon will occur, and the asymptotic state will be concentrated on a very small set.

    We have three main models, corresponding to neurophysiological problems of development.

(17)     $f(u) = (\int u \, dx)^2$ , $x \in \mathbf{T}^1$ ;

this corresponds to the development of orientation selectivity for a single neuron. The angle of the stimulus is $x$, and if $w$ is even, continuous, and has a positive strict global maximum at zero, the Liapunov function associated to this problem reaches its global minimum only if u is a single Dirac mass.

The second example, spatial distribution of preferred orientations, is a modification of (17), involving a spatial variable y, the location of the neuron :

(18)        $f(u) = (\int u(x,y) dx)^2$ , $x \in T^1$, $y \in Y$.

The global minimum of the corresponding Liapunov function is attained only for measures u carried by curves $x \to y(x)$.

Finally, if x and y are spatial locations of nervous cells, belonging to different neural sheets, one obtains a model of retinotopy, by letting

(19)      $f(u) = (\int u(x,y) dx)^2 + (\int u(x,y) dy)^2$ , $x$ & $y \in T^N$.

There is still a Liapunov function associated to this problem, and it attains its global minima on measures carried by the diagonal x-y = constant or x+y = constant.

For numerical simulations of some of the above described models, see [1].

4. FINAL REMARKS.

The results described here make use of rather heavy functional analysis ; the mathematical difficulty is due to the lack of compactness, to the large number of minima of the Liapunov functionals, and to the high symmetry of the problem.

It would be nice to think that the models described in part 3 are, somehow, the result of a passage to the limit, as in part 2, calling this limiting process mean field theory, or any other name. It is not impossible that a passage to the limit on the models presented at this conference by Bienenstock [2] or von der Malsburg [9] could be on the way.

REFERENCES.

[1] E. Bienenstock, Cooperation and competition in the central system development : a unifying approach, (1984)   in Synergetics, Springer.
[2] E.Bienenstock, this volume.
[3] E.Bienenstock, M.Schatzman, B.Moore, Nonlinear systems with a spatial convolution term in a model of development of neuronal selectivity, to appear.
[4] F.Comets, T.Eisele, M.Schatzman, On secondary bifurcations for some nonlinear convolution equations, to appear, Trans. A.M.S. Soc.
[5] T.Eisele, R.Ellis, Symmetry breaking and random waves for magnetic systems on a circle, Z. Wahrsch. Verw. Gebiete,63(1983)297-348.
[6] M.Schatzman, E.Bienenstock, Neurophysiological models, Spin models, and pattern formation, preprint, 1984.
[7] N.V.Swindale, A model for the formation of ocular dominance stripes, Proc. R. Soc. Lond. B, 208(1980)243-264.
[8] Anderson, this volume      [9] C. von der Malsburg, this volume.

# RECURRENT COLLATERAL INHIBITION SIMULATED IN A
# SIMPLE NEURONAL AUTOMATA ASSEMBLY.

H. AXELRAD[1]; C. BERNARD[1]; B. GIRAUD[2] & M.E. MARC[1]
1 - Lab. Physiologie - CHU Pitié - Paris 13 - France.
2 - Dept. Physique Théorique - CEA - Saclay - France.

## 1 - Introduction

One of the central problems that has to be resolved in view of a better
understanding of the functioning of CNS structures is that of the dynamic
properties of it's constituent neuronal nets. It clearly appears, indeed,
to experimental as well as theoretical neurobiologists that albeit the wealth
of knowledge accumulated on the properties of neurons at the molecular, membrane and cellular levels a great degree of cooperativity is present between
neuronal elements and that it is therefore the collective properties of
groups of neurons inside the structure that must be unveiled. This has at
least two important consequences. The first is that it may be difficult to
use *general* mathematical "models" to derive precise existing properties.
Indeed there does not exist such a thing as a "general" CNS structure but
on the contrary very differently organized morphological structures. The
second is that if one takes in account the ultimate scope of the CNS, which
is certainly related to an "adequate behavior" of the animal in its environment, then the temporal constraints imposed on the systems must surely
be stressed.

Since the pionneering work of CAJAL (1) the cerebellar cortex has become
the best understood part of the CNS, at least at the basic level of the
properties and organization of the elements composing it's circuitry. Compared to other central structures the cerebellar cortex has extremly specific features such as 1) a limited number of cellular types (five) with
only one efferent cell axon, that of the Purkinje cell (PC); 2) a restricted
number of inputs (two) each of which is distinctly identifiable at the morphological as well as physiological levels and 3) a topological organisation
invariant along the two axes of it's plane of extension. Another particula-
rity is the importance of inhibitory interactions between the neuronal elements, four out of five cell types, including the efferent PC, using this
modality of communication (2; 3). To gain insight on the transformation
operated by the cerebellar cortex - essentially the molecular layer - on
the incoming activity it is therefore necessary to analyze whether such

NATO ASI Series, Vol. F20
Disordered Systems and Biological Organization
Edited by E. Bienenstock et al.
© Springer-Verlag Berlin Heidelberg 1986

inhibitory interactions lead to specific modulatory operations. We have dealt indirectly with this restricted problem by a modelisation approach, whose results can be used for the interpretation of experimental data.

## 2 - Methodology

The analysis of complex systems presents a number of limits which arise from the difficulty of classifying data and variables whose dynamic interdependance are preponderant. This is exactly the case with the cerebellar cortex notwithstanding it's apparent morphological monotony. To overcome this problem we have subdivided the molecular layer neuronal net in subsystems hierarchically related. These include a) isolated PCs; b) PCs coupled by the recurrent collaterals of their axons; c) PCs coupled by interneuronal axons; d) PCs coupled by recurrent collaterals as well as by interneurons. Each of these subsystems can thus be modelized and the modulation of activity transiting through these circuits quantified. The first step, which will be detailed in this paper, was to compare activity in isolated PCs (control) and in PCs coupled by the inhibitory recurrent collaterals of their axons (test) (4; 5).

The model incorporates seven neuronal automata (NA) which can temporally evolve by sudden changes at random, through three different states namely 0, 1 and 2 for respectively "silent", "tonic" and "phasic" activity. The duration of each state can vary, also at random, between 1 and 20 ms for states 0 and 1, and 1 to 5 ms for state 2. In the control mode each of the NA has an independant functioning . In the test mode each NA is coupled to it's immediate neighbours with the following inhibitory-type rule. When a given NA is in state 2 for a time $\geq$ to 3 ms then, at the next ms, the two neighbouring NA are automatically forced into the state inferior to that in which it was (i.e. from state 2 to state 1; from state 1 to state 0) for the entire forecasted duration. If the original state was 0 then it's duration is prolonged by 20%. A number of constraints were also imposed. First and to minimize parasitic correlations we used a random number generator that was tested in the 5th dimensional hypercube (6). Second the simulation cycles were limited to 999 ms to stay in keeping with the "cerebellar time-constant" which is in the order of 200 to 300 ms. At last all statistical calculations were conducted on the five central NA so as to attenuate edge effects.

## 3 - Results

The dynamics of state shifts taking place at the level of a single NA

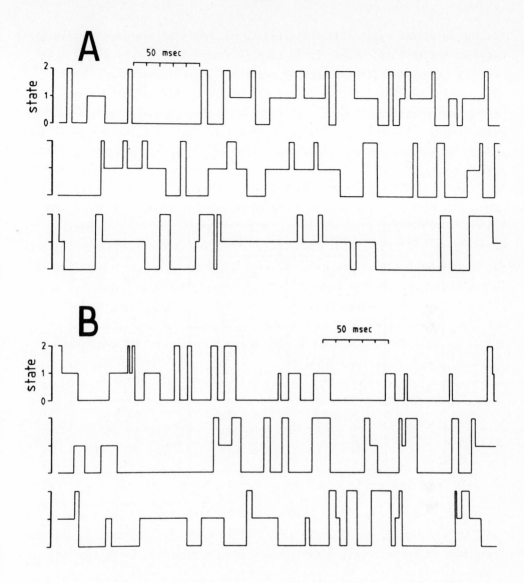

Figure 1 : Dynamic "activity" through a NA in the control, uncoupled (A) and test, coupled (B) mode. Shifts occur through three different states (0: "silent"; 1: "tonic"; 3: "phasic"). Note the important modulatory effect of inhibitory-type coupling.

during an entire cycle is illustrated in Figure 1 for the control (A) as
well as the test (B) modes. It is clearly apparent from this figure that
the inhibitory coupling between NA drastically modulates the ongoing
"activity". The quantitative aspect of this modulatory effect has been specified by a statistical analysis conducted on the cumulated data from the
five central NA for five different cycles (see table    below. N: total number of appearances of the given state; D: total duration and D$\overline{m}$: mean duration of state).

table

| state | control | test | t/c |
|:-----:|:-------:|:----:|:---:|
| 0 | N = 671<br>D = 11006<br>D$\overline{m}$ = 16.4 | N = 828<br>D = 14749<br>D$\overline{m}$ = 17.8 | N : 123.4%<br>D : 134.0%<br>D$\overline{m}$ : 108.6% |
| 1 | N = 638<br>D = 10412<br>D$\overline{m}$ = 16.3 | N = 752<br>D = 7335<br>D$\overline{m}$ = 9.75 | N : 117.9%<br>D : 70.45%<br>D$\overline{m}$ : 59.8% |
| 2 | N = 673<br>D = 3557<br>D$\overline{m}$ = 5.3 | N = 640<br>D = 2891<br>D$\overline{m}$ = 4.5 | N : 95.1%<br>D : 81.3%<br>D$\overline{m}$ : 85.5% |

In the control mode (i.e. independant NA) data for the three states fit
well with the programmed random distribution rule. By contrast, in the coupled mode three important modifications can be observed. First the total
number of state shifts is increased by more than 10%, which indicates a
greater turnover of information flow. Second this change is hierarchically
organised with a maximal effect on the "silent" state towards which the
whole activity is attracted (total number of appearances of state 0 increases
by 34%, mean duration by nearly 9%, total duration by more than 30%). On
the other hand state 1 appears as the most labile and undergoes a drastic
diminution of 40%. Third, and maybe the most noteworthy, is the emergence
of a spatiotemporal cooperativity. Indeed it appears that all of the five
NA are simultaneously in the same "silent" state for more than 5.5% of the
total simulation time. Moreover this is revealed when taking into account

## A

| shifts | control | test |
|---|---|---|
| \|0 → 1\| | 312 (15.9%) | 408 (18.6%) |
| \|0 → 2\| | 345 (17.6%) | 404 (18.4%) |
| \|1 → 0\| | 312 (15.9%) | 522 (23.8%) |
| \|1 → 2\| | 317 (16.2%) | 224 (10.2%) |
| \|2 → 0\| | 351 (17.9%) | 299 (13.6%) |
| \|2 → 1\| | 320 (16.35%) | 338 (15.4%) |
| | 1957 (100%) | 2195 (100%) |

## B

| shifts. | control | test |
|---|---|---|
| \|0→1→0\| | 141 (7.3%) | 292 (13.5%) |
| \|0→1→2\| | 167 (8.6%) | 114 (5.25%) |
| \|0→2→0\| | 183 (9.5%) | 191 (8.8%) |
| \|0→2→1\| | 161 (8.3%) | 210 (9.7%) |
| \|1→0→1\| | 145 (7.5%) | 270 (12.4%) |
| \|1→0→2\| | 161 (8.3%) | 243 (11.2%) |
| \|1→2→0\| | 161 (8.3%) | 105 (4.8%) |
| \|1→2→1\| | 155 (8.0%) | 119 (5.5%) |
| \|2→0→1\| | 163 (8.4%) | 134 (6.2%) |
| \|2→0→2\| | 180 (9.3%) | 158 (7.3%) |
| \|2→1→0\| | 168 (8.7%) | 225 (10.4%) |
| \|2→1→2\| | 147 (7.6%) | 109 (5.0%) |
| | 1932 (100%) | 2170 (100%) |

Figure 2 : Tables of observed frequencies of the six possible two letters (A) and twelve possible three letters (B) events in the two functioning modes (see text for further details).

only common periods lasting longer than 6 ms so as to reject fortuituous coincidences. At last, this temporally correlated evolution is cyclicly organised throughout the duration of a simulation run (not illustrated).

The assessment of this modulatory effect of the inhibitory recurrent collaterals and the delineation of changes it produces warrants the necessity to calculate the informational content exhibited by the system. Basically we shall consider the *shifts* from one state to the other as the information-carrier variable and the analysis will be conducted by comparison between the test and control modes, the latter being the "pure noise" reference. Thus the transition from, say, state 0 to state 2 can be accounted for by a two-letter word [02] and the transition from state 1 to state 2 and to state 0 thereafter by a three letter word [120]. Due to statistical fluctuations we will consider as physiologically significant only frequencies contrast above a 5% threshold. Figure 2 gives the observed-frequency tables for the six possible two letters (A) and the twelve possible three letters (B) events, in the two modes.

In the independantly functioning mode the six observed frequencies for the two-letter words are consistent with the raw probabilities of appearance: 1/3 (probability for the first letter) x 1/2 (probability for the second-distinct from the first-letter) = 1/6 $\simeq$ 16.67%. The same verification holds for the control three-letter words frequencies. In the test mode, on the other hand, several significative deviations are present.

First of all the instability of state 1 is confirmed. The inhibitory type rule favors indeed the 1$\rightarrow$0 shift compared to the 1$\rightarrow$2. Identically the 2$\rightarrow$1 event is enhanced compared to the 2$\rightarrow$0 jump. This explains the observed frequencies of the test column in figure 2A : 24%; 10%; 13.5% and 15.5%. A second consequence of the inscribed rule is that "a priori" frequencies of letters 0, 1 and 2 deviate from their raw value (1/3) to respectively 37%; 34% and 29%. It is therefore important to test the hypothis of independance of appearance of the second letter (j) compared to the first (i) : f(j/i), i$\neq$j, with f being the conditional frequency. These are : f(0/2) = 37/(37+34) = 52%; f(1/2) = 48%; f(2/1) = 44%; f(0/1) = 56%; f(1/0) = 54% and f(2/0) = 46%. These values must then be compared to the effectively observed conditional frequencies F which are : F(0/2) = 13.6/(13.6+15.4) = 47% which is significantly different from f(0/2) = 52% that would be obtained in the case of independant appearances. The same result holds for the other cases. It thus appears that in the recurrent inhibitory mode *the second letter of a word is strongly conditionned by the first one*. The same type of calculations can be done on the three-letter words (see figure 2B) with a conditional frequency for the third letter $\emptyset$ to be compared to F (see above). We then obtain for example $\emptyset(0/12) = 4.8/(4.8+5.5) = 47\%$ which is equivalent to $F(0/2) = 47\%$ or $\emptyset(0/01) = 13.5/(13.5+5.25) = 72\%$ which is comparable to $F(0/1) = 70\%$. We have checked that it is a general result. From the above calculations it can be concluded that in the test model the ongoing activity follows a markovian type organisation i.e. the second state is conditionned by the first, the third by the second and that one alone, etc... This may seem as partly encompassed in the programmed "inhibition" but one must note that the F and $\emptyset$ values are not trivial and can not be directly deduced from the abovementionned rule.

The last step which was conducted was to measure changes in the entropy between the two modes of functioning . The scale of information is given by the entropy of the three equiprobable letters : $\sigma_1 = -3\frac{1}{3}\log(1/3) = 1.10$. Due to the non repetition of letters the two-letter words are characterized by a reference entropy of $\sigma_2 = -6\frac{1}{6}\log(1/0) = 1.79$ which is smaller than $2\sigma_1$. Likewise the control three-letter words are characterized by $\sigma_3 = 2.48$. The real informational content transiting through the coupled model can be estimated by the classical formula $S_n = -\Sigma F_n(\lambda)\log F_n(\lambda)$ with $\lambda$ being the considered words, $F_n$ the observed frequencies and n the word size. The specific content is then $I_n = \sigma_n - S_n$. A simple calculation gives a value of 0.03 for $I_2$ and 0.06 for $I_3$. These are important values when compared to the scale (1.10) and bring convincing arguments in favor of an enhanced informational content brought about by the coupling  between NA.

4 - Conclusion

Since the first morphofunctional model of the cerebellar cortex presented by CAJAL (1) the beauty and geometrical   simplicity of the cellular arrangements of this central structure throughout the vertebrate phylum has prompted number of speculations and theories concerning the possible use of some of it's particular aspects in the global functioning of the system (7; 8; 9; 10; 11). The results presented here deal with the restricted problem of the role of inhibitory recurrent collateral coupling in the modulation of ongoing activity. Our model is based on neuronal automata, first used by Mc CULLOCH & PITTS (12). The statistical mechanics of such automata are well understood (13) and although they are a powerful tool to study self organizing processes (14; 15; 16; 17) one must keep in mind their limits when addressing biological problems.

Briefly summarized it has been shown that an inhibitory type coupling between elements in a small sized group drastically modulates the activity transiting through the system by hierarchically organizing the succession of states and allowing the emergence of collective properties which can not be deduced linearly from the rules of the elementary elements. Moreover it leads to a Markovian behavior of the system and an increase of the informational content of the "activity". These elements are in favor of a possible structural role of this type of coupling in short-term memory. It may also be hypothetized that there is a direct relationship between the transfer function of a given structure and the size of the operating neuronal aggregate.

## REFERENCES

(1) S. R. CAJAL, Histologie du système nerveux de l'Homme et des Vertébrés, t.II, 1911, Maloine.

(2) J. C. ECCLES, M. ITO & J. SZENTAGOTHAI, The cerebellum as a neuronal machine, 1967, Springer.

(3) S. L. PALAY & V. CHAN PALAY, Cerebellum Cortex, 1974, Springer.

(4) H. AXELRAD, C. BERNARD, B. GIRAUD & M.E. MARC, C. R. Acad. Sci., 1985 301 : 465-470.

(5) B. GIRAUD, C. BERNARD & H. AXELRAD, C. R. Acad. Sci., 1985, 301 : 565-570.

(6) E. YERAMIAN, Thèse de DDI, Ecole Centrale, 1985.

(7) V. BRAITENBERG & N. ONESTO, in Proc. 1st Int. Conf. Med. Cybernet, 1961 pp. 1-19, Naples : Giannini.

(8) D. MARR, J. Physiol., 202, 1969, pp. 437-470.

(9) J. S. ALBUS, Math. Biosci., 10, 1971, pp. 25-61.

(10) A. PELLIONISZ & R. LLINAS, Neuroscience, 4, 1979, pp. 323-348.

(11) M. FUJITA, Biol. Cybern., 45, 1982, pp. 195-206.

(12) W. S. Mc CULLOCH & W. H. PITTS, Bull. Math. Biophis., 1943, 5 : 115-137.

(13) S. WOLFRAM, Rev. Mod. Phys., 55, 1983, pp. 601-644.

(14) S. AMARI & M.A.ARBIB, eds, Competition and Cooperation in neural nets, 1982, Springer.

(15) J.J. HOPFIELD, Proc. Natl. Acad. Sci. USA, $\underline{79}$, 1982, pp. 2554-2558.

(16) P. PERETTO, Biol. Cybern., $\underline{50}$, 1984, pp. 51-62.

(17) G. L. SHAW, D. J. SILVERMAN & J. C. PEARSON, Proc. Natl. Acad. Sci.
USA, $\underline{82}$, 1985, pp. 2364-2368.

# CEREBELLUM MODELS : AN INTERPRETATION OF SOME FEATURES

G. CHAUVET

Laboratoire de Biologie Mathématique
U.E.R. des Sciences Médicales et Pharmaceutiques
Université d'ANGERS

## INTRODUCTION

Many authors have studied the functional capacity of cerebellar cortex and its implications on the organismic behaviour. A basic observation is the extremely regular organization of cells in the cortex with their repartition within two layers and the possible functional unity around Purkinje cell. The first theoretical model following Eccles's experimental works [1] was D. Marr's one [2]. Indeed Marr used a possible synaptic modification between Purkinje cell and parallel fibres as fundamental hypothesis, but also numerous imaginative suppositions which permit him to conceive a functional and qualitative model. The regular geometry and possible analogies with electronic devices and computation organs induce simple ideas on the functionning of a cerebellar cortex unit. After Marr, J.S. Albus [3] created a quantitative model of cerebellar cortex based on similar properties - but only three - so it was a computer approach rather than a physiological explanation of cerebellar function. That is the reason why he called his model the CMAC or Cerebellum Model Arithmetic Computer. More recent models have been considered by S. Grossberg [4] and J.C.C. Boylls [5]. For Grossberg, olivo-nuclear projections are compared with the template provided by the Purkinje cell input to cerebellar nuclear cells. Boyll's interpretation is based on temporary inactivation which immediately follows climbing fibre electrical activity.

Here is proposed an elementary model which would permit a discussion of the influence of excitatory and inhibitory synapses with *habituation rules* for an interpretation of cerebellar function.

NATO ASI Series, Vol. F20
Disordered Systems and Biological Organization
Edited by E. Bienenstock et al.
© Springer-Verlag Berlin Heidelberg 1986

## ANATOMICAL DATA

A piece of cerebellar cortex shows the organization around one Purkinje cell : Two inputs (one climbing fibre and many mossy fibres) and one unique output (Purkinje cell's axon). Mossy fibres are connected to many granule cells via the so-called glomeruli. A glomerulus contains one dendritic terminal mossy fibre and about ten dendrites of some granule cells in the granule layer. Each of them sends an axon towards the surface layer which is called parallel layer because it is made with many axons which run along the cortex. It is easy to recognize here a *bus* whose function is to carry information, like in computers, and two parallel systems including Purkinje cell, granule cells, parallel fibres, stellate (or basket) cells. Finally there exists an important divergence with a mossy fibre for 400 or 600 granule cells and according to D. Marr's calculations a major convergence from 7 000 mossy fibres to one Purkinje cell. Indeed it is considered that one Purkinje cell makes 200 000 synapses with parallel fibres and that there is one Golgi cell for 9 or 10 Purkinje cells.

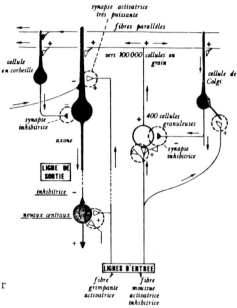

Figure 1

## AN ELEMENTARY MODEL OF THE FUNCTIONAL UNIT OF CEREBELLAR CORTEX

If only one Purkinje cell is considered as the central component of the functional unit of cerebellar cortex, it is natural to present it as on figure 1, where + symbol (resp-) signifies an activatory (resp. inhibitory) synapse. Why this localization

of synapses ? Eccles said : "The battle of excitatory and inhibitory action on Purkinje cells is fought all the time from moment to moment in every part of the cerebellum" [6]. The interpretation of such a repartition seems to be mysterious in the
absence of experimental evidence despite of Ito's results [7].

Our model is based on a theoretical study of the transmission of
potentials accross a synapse [8] using three hypotheses :
- *Synapses have to satisfy habituation rules.* There is some experimental evidence due to blocking of synapses when potential
frequency is increasing.
- With the preceding habituation rule, the stability of the system is obtained when one synapse is fixed whenever other are variable. Such hypothesis was assumed by Uttley in his informon's
theory [8].
- Without loss of generality for actual interpretation, there
is a one-to-one relation between a mossy fibre and a granule
cell. The consequences of the suppression of this hypothesis
will be analysed in another paper but the calculations are more
complex.

Notations are as in figure 2 : $X_1$, $U_1$, $X_2^i$, Z are the potential
frequencies, and $\mu_i$, $\sigma_i$ are the synaptic plasticities. The time
course of $\mu_2$ is given by (with $\mu_3(t=o) = (\mu_3)_o$) :

$$\frac{d\mu_2}{dt} = \alpha_1 \ (X_2 - \bar{X}_2)(Y - \bar{Y}) \qquad\qquad \alpha_1 < o$$

$$Y(t) = \mu_o'(t) + (\mu_2'(t) + \mu_1(t))X_1 - (\mu_3)_o X_3$$

in case of reinforcement ($X_1 = U_1$). Here :

$$\mu_o'(t) = \mu_o + n\mu_2(t)\sigma_o \qquad , \qquad \mu_2'(t) = nm\mu_2(t) \ ;$$

n and m are the number of active parallel fibres and mossy fibres respectively. Within an approximation ($\bar{Y} = \mu_o'(t)$ = long-term
mean, $\bar{X}_2 = 0$) an explicit expression of $\mu_2(t)$ can be obtained :

$$\mu_2(t) = \frac{1}{nm\sigma_1} \ [(\mu_3)_o \frac{X_3}{X_1} - \mu_1] \qquad \text{and} \quad Y \longrightarrow \mu_o'(t)$$

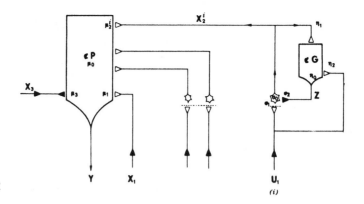

Figure 2

The interest of this quantitative study is to show that it is possible to deduce some results about Golgi cell effect : The presence of both inhibitory and excitatory synapses induces a learning of granule cells due to the required stability for the related subsystem including Golgi cell. With these hypotheses of (i) habituation rules and (ii) synaptic modification of granule cells, the following interpretation is obtained : Firstly, information arrives via spino-cerebellar tracts and acts upon Golgi cells via $U_1$. this implies a synaptic modification of $\sigma_1$ due to the influence of "external" signals like $X_2^j$, $j \neq i$, that is to say the *context* : It is a *learning phase*. Secondly, information comes with $X_1$ and $U_1$ and the system functions as a *pattern recognition device* to give $Y = \mu_0'$.

REFERENCES

1- ECCLES J.C., ITO M., SZENTAGOTHAI J., The cerebellum as a neuronal machine, Springer, New-York (1967)
2- MARR D., A theory of cerebellar cortex, J. Physiol. 202 (1969)
3- ALBUS J.S., A theory of cerebellar function, Math. Biosci., 10, 25 (1971)
4- GROSSBERG S., Cerebellar and retinal analogs of cells fixed by learnable and unlearned pattern classes, Kybernetik, 10, 49 (1972)
5- BOYLLS J.C., A theory of cerebellar function with applications to locomotion. I. the physiological role of climbing fiber inputs in anterior lobe operation, technical report 75c-6, computer and information Science dept, Univ. of Massachusetts at Amherst (1975)
6- ECCLES J.C., in cerebro-cerebellar interactions, J. Massion and K. Sasaki Eds, Elsevier, Amsterdam, p. 6 (1979)
7- ITO M., Mechanisms of motor learning, p. 418, in *Lecture Note in Biomathematics*, Vol. 44, Springer (1982)
8- CHAUVET G., Un modèle de la plasticité synaptique dans la régulation du comportement in Régulations physiologiques : Quelques modèles récents Masson Ed. (1985)
9- UTTLEY A.M., Information transmission in the nervous system, *Acad. Press* New-York (1979).

# INDEX

## A

Algebraic invariant (cf invariant)
Annealing
(cf simulated annealing)
Assembly (cf cell assembly)
Associative (memory, network)
**97, 209,233, 249, 339**
Asynchronous **155, 173**
Attraction basin
**180,187,189,201,335,342**
Attractivity **41,97,339**
Auto-association
**213,227,228,231,273**
Automata
boolean **3,33,85,339**
cellular **3,53,71,113,190,**
**327,339,367,374,383**
majority (cf majority)
monotone **49,50**
positive **101, 105**
probabilistic **4, 72**
threshold (cf threshold)
with memory **51,63**

## B

Bifurcation structure **65, 334**
Boltzmann Machine (or model)
**215, 220, 266**
Boolean (cf automata)
mapping **85**
network **53,67,85**
Bound (cf transient length)

## C

Categorization **200,211,247**
Cell assembly **248,267**
Cellular automata machine (CAM)
**4,54**
Combinatorial optimization
**255,283,295,301,321,327**
Connectivity **85**
Connection graph **36,50,67,86**
Content addressable memory
**155, 160, 249**
Core **67,68, 87, 353**

## Cortex,cortical **158,206,209,**
**381,389,390,399**
Cost function (or criterion)
**254, 284, 295**
Cycle (cf limit cycle)
length (cf period)

## D

Deconvolution **168, 306**
Degenerate,degeneracy
**161,255,259,283**
Directional entropy (cf entropy)
Discrete derivative **33**
Discrete iteration **33,39,86**
Disorder
**120,127,149,255,283,381**

## E

Energy **59,95,120,133,143,159,**
**180,215,229,254,304,344**
Entropy **14,88,113,114**
Error correcting **213, 249**
Euclidean match problem
**166, 295**

## F

Firing squad (cf synchronization)
Fixed point **35,87**
Forcing function **53,67,92,351**
Frustration
**120,149,255,283,381,384**

## G

Generalization **214,247**
Gibbs distribution **73,302**
Global transition rule
(cf transition rule)
Gradient descent
**160,317,329,330**
Graph (cf connection,iteration)

## H

Hamming distance
**208,230,292,342**
Hebb's principle (or rule)
**177,187,201,210,227,228**
**251,264,273**

# NATO ASI Series F

## Date Due